EPITAXIAL GROWTH
Part A

MATERIALS SCIENCE AND TECHNOLOGY

EDITORS

ALLEN M. ALPER

GTE Sylvania Inc.
Precision Materials Group
Chemical & Metallurgical
Division
Towanda, Pennsylvania

JOHN L. MARGRAVE

Department of Chemistry
Rice University
Houston, Texas

A. S. NOWICK

Henry Krumb School
of Mines
Columbia University
New York, New York

EPITAXIAL GROWTH
Part A

Edited by

J. W. MATTHEWS

IBM THOMAS J. WATSON RESEARCH CENTER
YORKTOWN HEIGHTS, NEW YORK

ACADEMIC PRESS New York San Francisco London 1975

A Subsidiary of Harcourt Brace Jovanovich, Publishers

ACADEMIC PRESS, INC.
111 Fifth Avenue, New York, New York 10003

United Kingdom Edition published by
ACADEMIC PRESS, INC. (LONDON) LTD.
24/28 Oval Road, London NW1

Library of Congress Cataloging in Publication Data

Matthews, John Wauchope
    Epitaxial growth.

    (Materials science and technology series)
    Includes bibliographies and index.
    1.   Epitaxy.   2.   Crystals–Growth.   I.   Title.
QD921.M37   548'.5   74-10191
ISBN 0–12–480901–4 (pt. A)

CONTENTS

3.6 Measurement of Microstains in Thin Epitaxial Films

D. L. Allinson

LIST OF CONTRIBUTORS

Numbers in parentheses indicate the pages on which the authors' contributions begin.

D. L. Allinson* (365), *Department of Physics, University of the Witwatersrand, Johannesburg, South Africa*

L. L. Chang (37), *IBM Thomas J. Watson Research Center, Yorktown Heights, New York*

D. B. Dove (331), *Department of Materials Science and Engineering, University of Florida, Gainesville, Florida*

M. H. Francombe (109), *Westinghouse Electric Corporation, Research and Development Center, Pittsburgh, Pennsylvania*

R. Ghez (183), *IBM Thomas J. Watson Research Center, Yorktown Heights, New York*

E. A. Giess (183), *IBM Thomas J. Watson Research Center, Yorktown Heights, New York*

F. Jona (309), *Department of Materials Science, State University of New York, Stony Brook, New York*

R. Ludeke (37), *IBM Thomas J. Watson Research Center, Yorktown Heights, New York*

Siegfried Mader (29), *IBM Thomas J. Watson Research Center, Yorktown Heights, New York*

D. W. Pashley (1), *Tube Investments Research Laboratories, Hinxton Hall, Hinxton, Saffron Walden, Essex, England*

Helmut Poppa (215), *Ames Research Center, NASA, Moffett Field, California*

G. H. Schwuttke (281), *IBM East Fishkill Laboratories, Hopewell Junction, New York*

Don W. Shaw (89), *Texas Instruments, Inc., Dallas, Texas*

J. A. Strozier, Jr. (309), *Department of Materials Science, State University of New York, Stony Brook, New York*

Richard W. Vook (339), *Department of Chemical Engineering and Materials Science, Syracuse University, Syracuse, New York*

J. G. Wright (73), *School of Mathematics and Physics, University of East Anglia, Norwich, England*

* Present address: Division of Inorganic and Metallic Structure, National Physical Laboratory, Teddington, England.

PREFACE

This book is the result of a suggestion made to me several years ago by Professor A. S. Nowick. It is a collection of review articles that describe various aspects of the growth of single-crystal films on single-crystal substrates. The topics discussed are the historical development of the subject, the nucleation of thin films, the structure of the interface between film and substrate, and the generation of defects during film growth. The methods used to prepare and examine thin films are described and a list of the overgrowth–substrate combinations studied so far is given.

Invaluable help at all stages of the project was provided by Academic Press. Helpful discussions of a variety of topics that included the ways in which the subject could be subdivided, and who should be asked to write on each subdivision, were held with Dr. A. E. Blakeslee, Dr. S. Mader, Dr. M. J. Stowell, and Dr. J. A. Venables.

CONTENTS OF PART B

CHAPTER 1

A HISTORICAL REVIEW
OF EPITAXY

D. W. Pashley

Tube Investments Research Laboratories
Hinxton Hall, Hinxton, Saffron Walden
Essex, England

I. Introduction

Although major interest in the study of epitaxy has existed for only 25 years, the subject had its origins as a laboratory study about 150 years ago. Interest was first aroused when it was noticed by mineralogists that two different naturally occurring crystal species sometimes grew together with some definite and unique orientation relationship, as revealed by their external forms. These observations led to attempts to reproduce the effect artificially, during crystal growth from solution, and the first recorded successful attempt was reported by Frankenheim (1836), who demonstrated the now well-known case of the parallel oriented growth of sodium nitrate on calcite.

The first systematic studies were carried out by Barker (1906, 1907, 1908) who studied, among other things, the growth of alkali halides upon each other. He found that epitaxy occurred with some combinations, and not others. The subject was restricted, at that time, in three major ways: (1) crystals had to be grown such that their orientation could be determined by optical microscopy, which was the only observational technique available; (2) growth from solution was the only readily available technique for preparing samples; (3) the understanding of crystal structure was very rudimentary, since there were no techniques available for structure analysis. Nevertheless, Barker was able to deduce that epitaxy was more likely to occur if the molecular volume of the two intergrowing alkali halides were nearly equal (i.e., low misfit).

With the discovery of X-ray diffraction in 1912, the knowledge of internal crystal structure developed very rapidly, and by the time Royer (1928) carried out his extensive and systematic studies of a wide variety of overgrowths, it was possible to examine the effect of the geometry of crystal structure on orientation. In fact it was Royer who introduced the term *epitaxy* ("arrangement on"). On the basis of his results, Royer put forward three rules of epitaxy, the most important of which is that oriented growth occurs only when it involves the parallelism of two lattice planes that have networks of identical or quasi-identical form and of closely similar spacings. The experiments indicated that epitaxy occurs only if the misfit is no more than about 15%. The misfit was defined as $100(b-a)/a$, where a and b are the corresponding network spacings in the substrate and overgrowth, respectively. This geometrical approach to the understanding of epitaxy has remained prominent to this day. A survey of the very extensive studies that have been made of epitaxy of crystals grown from solution has been given by Neuhaus (1950, 1951) and Seifert (1953).

Since the time of Royer's pioneering work, further progress in the subject has been dominated by the influence of a succession of new experimental techniques, especially those for making structural observations on crystals. Although X-ray diffraction was already available, it played little part in direct studies of epitaxy and has never been very prominent in the field. The first major advance resulted from the application of electron diffraction, which was discovered by Davisson and Germer (1927) and Thomson and Reid (1927), i.e., at about the same time as Royer was carrying out his work. Davisson and Germer used electrons of only a few tens of electron volts in energy, and thus initiated the technique of LEED. Thomson and Reid used electrons of a few thousand, or tens of thousands, of electron volts in energy, and initiated the technique of HEED; although in both cases it was many years later that these initials were used.

At the time, HEED was by far the more practical technique because it would work satisfactorily with quite a poor vacuum (10^{-3} Torr), whereas LEED required a much higher vacuum (10^{-6} Torr), which could be obtained only with great difficulty. Thus, from about 1930, HEED began to be applied to the study of epitaxy, and this involved reflection diffraction (RHEED) as well as transmission through thin films. The advantages of electron diffraction were quite clear. First, it was possible to examine extremely thin layers of overgrowths prepared by a variety of methods. Second, orientation could be determined from internal crystal structure rather than external morphology.

As a result, the years 1930–1950 saw a steady increase in the number of observed cases of epitaxy, covering a wide variety of substrates and deposits. These cases included a variety of growth techniques, especially vacuum evaporation, electrodeposition, and chemical growth (e.g., oxidation). A useful survey and compilation of data is given by van der Merwe (1949). Because of the increased scope for controlling parameters such as the conditions of growth, more effort was devoted to trying to determine the factors that control epitaxy. One of the important parameters is the substrate temperature during deposition of the overgrowth, as first demonstrated by Bruck (1936). A review of the electron diffraction studies has been given by Pashley (1956).

Attempts to interpret the experimental results continued to be dominated by the concept of a low-misfit requirement, to the extent that various ingenious models were proposed to account for observations of epitaxy associated with apparent high misfits. For example, Menzer (1938a, b, c) explained the occurrence of high apparent misfits for growth of nickel and silver on rock salt in terms of the formation of initial (221) orientations with relatively low misfits, which produced the observed (100) orientation as a result of (111) twinning. Also, Engel (1952, 1953) proposed that the high misfit could be explained by the formation of an intermediate layer of a metal salt between the metal and the rock salt. According to this mechanism, the effective misfit is quite low and there is also a means of explaining the concept of an epitaxial temperature.

An important concept was introduced by Finch and Quarrell (1933, 1934) as a result of studying the growth of zinc oxide on zinc. They concluded that the initial growth layer was strained so that its spacing matched that of the substrate (i.e., a forced zero misfit), with a compensating change in the lattice periodicity normal to the surface to maintain approximately the normal bulk density. They termed this phenomenon "basal plane pseudomorphism," now commonly known simply as "pseudomorphism." Although the evidence to support this interpretation was rather doubtful (see Section II, D), it provided a very attractive concept that stimulated both new experimental work and attempts to produce realistic theoretical models. Frank and van der Merwe (1949a, b, c) proposed a theory of epitaxy based upon the initial formation of overgrowth monolayers homogeneously deformed to fit the substrate. This theory predicted the kind of limiting misfit embodied in the rules of Royer (1928). However, there remained numerous examples of observed epitaxy for misfits much larger than could be explained by this theory. A likely reason for the lack of agreement is that many deposits do not grow as monolayers.

RHEED was able to provide structural information on layers whose average thickness was less than one monolayer (see Section II, E), and several experiments revealed that epitaxial growth often occurs without the formation of monolayers (e.g., Schulz, 1951a, 1952; Newman and Pashley, 1955). The nature of the diffraction patterns revealed that growth occurred, in many cases, in the form of isolated three-dimensional nuclei. However, detailed information on the size, shape, and distribution of the nuclei was not obtainable. Fortunately, the application of electron microscopes provided a means for further studies of such morphological factors.

With the appearance of new and improved electron microscopes in the mid 1950s, the study of epitaxy was given a completely new impetus. By means of the transmission technique (TEM), it was possible to obtain a wealth of new information on the nucleation, growth, and structure of thin films. The nucleation process deduced from RHEED was confirmed and the important morphological details were readily apparent. The way the structure changed and developed, as growth proceeded, could be examined stage by stage. The internal structure of the completed films was revealed, and, following the development of techniques pioneered by Hirsch and his collaborators (see Hirsch et al., 1965), individual lattice defects could be identified. Reviews of the electron microscope work on epitaxy have been given by Pashley (1965) and Matthews (1967).

Although the complexity of the nucleation and growth process was rapidly revealed as the number of workers studying epitaxy grew, there was little real progress toward the understanding of what really determined the occurrence of epitaxy. Theoretical interest was very much centered on the nucleation stage, especially the kinetics of nucleation (Pound et al., 1954; Bauer, 1958). There was a growing realization that surface contaminants could be playing a vital role in the nucleation process, and possibly also in controlling epitaxy.

Again, the advent of new techniques promised new hope of progress. Much improved and commercially available systems for producing ultrahigh vacua started to have an impact in the early 1960s, providing both a means of preparing specimens under very much cleaner conditions and a "new" technique for structural studies. LEED equipment, as it is known today, depended on these new UHV techniques as well as the introduction of the technique of postacceleration of the diffracted electrons to provide a visible pattern on a fluorescent screen (Scheibner et al., 1960). LEED is a very sensitive technique that allows layers of much less than one monolayer to be analyzed, and hence it has been very valuable in providing evidence on surface cleanliness, as well as on the very initial stages of growth of deposits. HEED and RHEED have also gained from the advent of UHV techniques, and they are also very sensitive; there are good reasons why LEED and RHEED should be used simultaneously (Pashley, 1970).

Further additions to the power of LEED and RHEED have resulted from the ability to incorporate facilities for providing chemical analysis. In the case of LEED this is done by measuring the energy of the Auger electrons (Harris, 1968a, b; Palmberg et al., 1967a, b) that are emitted under low-energy electron bombardment. In the case of RHEED, this is done by measuring the wavelengths of the characteristic X rays emitted under high-energy electron bombardment (Sewell and Cohen, 1967).

Today the worker on epitaxy has a wide range of powerful tools at his disposal, and he can obtain extremely detailed information on the nucleation, growth, and structure of oriented crystalline deposits. In the remainder of this chapter, the development of the subject, as outlined above, is amplified further. Particular attention is given to the work carried out between 1930 and 1965. Much of the important work carried out since 1965 is covered in the other chapters of this book.

II. The Impact of High-Energy Electron Diffraction

A. INITIAL EXPERIMENTAL ADVANTAGES

Once the technique of HEED had been demonstrated in the late 1920s, its usefulness in examining the structure of very thin films was realized. However, specimen preparation and handling techniques had not been developed for the thin single-crystal films required for transmission studies of epitaxy, and the reflection technique proved to be more appropriate. The first use of RHEED was by Nishikawa and Kikuchi (1928), who obtained patterns from calcite. The first reported applications of RHEED to the study of epitaxy were carried out in the early 1930s, and the technique was firmly established by the middle 1930s.

Clearly, the type of specimen was very different from that required for optical microscope studies, and it became possible to study a completely new range of substrate and growth materials. This was coupled with a wider range of growth techniques including electrodeposition, chemical formation of surface films (e.g., oxidation), sputtering, and vapor deposition, as discussed in detail in Chapter 2. The latter technique, especially in the form of vacuum deposition, has proved to have wide application and is the most commonly used technique today.

B. Observations on Misfit and Epitaxial Temperature

It was Lassen (1934) who first demonstrated that silver deposited on the cleavage face of a rock salt crystal could be oriented, and Lassen and Bruck (1935) who showed that heating the substrate during deposition resulted in much improved epitaxy. So firm were the ideas on the small-misfit criterion at that time, as a result of all the optical microscope evidence obtained with crystals grown from solution, that Royer (1935) challenged the interpretation of the electron diffraction patterns on the grounds that a different orientation from the observed parallel orientation would give a misfit of only $+3\%$ compared with the observed one of -27%. This work had been carried out by transmission diffraction, so Kirchner and Lassen (1935) used RHEED to confirm the observations. Electron diffraction had thus made its first real impact on the subject of epitaxy and had started to break down the apparently tidy and ordered state that had existed before.

This was followed by another important study by Bruck (1936), who showed that the orientation of a number of metals deposited on rock salt is markedly dependent upon substrate temperature. He first introduced the term "epitaxial temperature." Above this temperature, good epitaxy was obtained, whereas below it there was at least some poorly or randomly oriented deposit. The concept of epitaxial temperature has had a major influence on subsequent work, over a period of 30 years or more, and various attempts have been made to develop theoretical models of epitaxy that incorporate the concept of a sharply defined critical temperature (e.g., Engel, 1952, 1953). However, although many workers have obtained a similar kind of dependence of epitaxy upon temperature, for a variety of deposits, there is considerable disagreement between the values of epitaxial temperature that are observed. It seems clear that there is no unique and sharply defined critical temperature, and that other deposition variables can have an important influence. One that was first demonstrated by Shirai (1939) is that preheating of the substrate has a cleaning effect, which can lead to a lowering of the epitaxial temperature.

During the first 25 years of the application of HEED and RHEED, many cases of epitaxy were established, many involving misfits considerably in excess of 15%. These have been classified and tabulated by Pashley (1956). Some of the most important studies have involved a systematic examination of a series of similar deposits on a single substrate, or on a related group of substrates. The most striking evidence was obtained by Schulz (1951a, 1952) who showed that parallel orientation occurs for a wide range of alkali halides deposited, by vacuum evaporation, onto the cleavage surface of several alkali halides. The misfits cover a more or less continuous range from -39% (LiF on KBr) to $+90\%$ (CsI on LiF). Similarly, Schulz (1951a) showed that the alkali halides are also oriented by evaporation onto a mica cleavage surface, over a misfit range of zero (RbI) to -27% (KF).

C. The Influence of Surface Topography on Epitaxy

The concept of the lattice misfit between the substrate and the deposit implied knowledge of the actual crystal planes in contact. It is normally assumed that these are the planes parallel to the macroscopic surface of the substrate. Although this is probably true when cleavage surfaces are used as substrates, the extent to which cleavage surfaces can be used is very limited and other methods of surface preparation must be employed. This often raises serious experimental difficulties.

The use of metallic substrates has always been very desirable, and the specimen preparation techniques especially difficult. If flat surfaces are prepared by grinding and polishing single crystals of a metal, the worked surface layer must be removed. Chemical or electrolytic methods of removal commonly leave a matt surface that is rough on an atomic scale, and these are not very convenient as substrates. Under appropriate conditions, chemical or electrolytic polishing occurs, and such optically smooth and bright surfaces would appear to be more acceptable. However, electron diffraction evidence (Kranert et al., 1944) showed that surfaces of this kind are likely to be undulating on an atomic scale and therefore not ideal as substrates. A deliberate attempt to determine the influence of surface topography was made in the case of the growth of silver halides on silver, by direct chemical attack with halogen vapor (Pashley, 1952). RHEED was used to provide information both on the substrate surface topography (see also Pashley, 1951) and on the orientation of the silver halides. By means of additional diffraction features associated with double diffraction, it was possible to correlate specific orientations of halide with growth on specific microfacets on the substrate. It was thus shown that the orientation relationship does depend upon the atomic planes parallel to the microfacets on a substrate surface, as implied by the misfit concept. Further, electropolished surfaces were generally found to be geometrically complex, leading to complex arrangements of overgrowth orientations. Some similar results were obtained for oxide growth on zinc and cadmium (Lucas, 1952). Growth of oxide films on electropolished spherical crystals of copper (Lawless and Mitchell, 1965a, b) provided additional evidence of the influence of substrate surface orientation and surface topography on epitaxy.

Because of the difficulties of preparing flat surfaces on macroscopic crystals, single-crystal films prepared by epitaxy have been used as substrates for further epitaxial growth (e.g., Newman and Pashley, 1955; Pashley, 1959a). This type of technique is now becoming more common, due to the increased use of in situ experiments in LEED apparatus, etc. (see Section IV, B).

D. PSEUDOMORPHISM

The first proposal that pseudomorphism could occur during epitaxy was made by Finch and Quarrell (1933), in relation to the growth of zinc oxide on sputtered zinc, although epitaxy as such was not demonstrated by these experiments. It was postulated that the a-axis of the zinc oxide was contracted by about 22% from its normal value, so as to match that of the zinc substrate, with a corresponding expansion of the c-axis to maintain an approximately unchanged cell volume. Other examples included the growth of tetragonal aluminum (normally cubic) on a sputtered platinum layer (Finch and Quarrell, 1933) and electrodeposited nickel and cobalt on copper crystals (Cochrane, 1936).

However, none of this evidence was by any means conclusive, and subsequent attempts to reproduce or amplify the evidence were all negative. The most favorable condition for forming the proposed pseudomorphic zinc oxide seemed to be the oxidation of a cleavage surface of a zinc single crystal. Not only does this provide the most appropriate surface orientation, but also the zinc cleavage surface is very smooth so that RHEED allows very thin oxide layers to be detected and analyzed, much thinner than in the experiments of Finch and Quarrell (1933). Raether (1950) and Lucas (1951) showed that such layers of oxide, formed by exposure of the cleavage surface to air, have the normal lattice structure, with no detectable spacing changes.

The proposed pseudomorphic aluminum was shown (Shishakov, 1952; Pashley, 1956) to be based upon a misinterpretation of diffraction patterns, and Newman (1956) cast serious doubts on the Cochrane (1936) evidence for pseudomorphic nickel and cobalt.

Thus the concept of pseudomorphic layers, theoretically attractive though it was, had to remain a somewhat speculative concept, with no conclusive experimental support. However, the possibility of pseudomorphism occurring in a much more limited way could not be ruled out at the time (mid 1950s). Also, the theoretical model for epitaxy put forward by Frank and van der

Merwe (1949a, b, c) was based on the formation of pseudomorphic mono-layers, and their calculations appeared to define the conditions under which such monolayers would form. However, Smollett and Blackman (1951) criticized some of the conclusions on the grounds that the monolayers would not always be stable.

E. The Initial Stages of Growth

From about 1950 there was a growing realization that the key to the explanation of epitaxy is an improved understanding of the mode of growth and the structure of the very initial stages of an oriented layer. RHEED offered the best hope of obtaining more information about such thin deposits.

It had been realized for many years that RHEED was a very sensitive technique, but quantitative information on its sensitivity was lacking. It had been reported by Finch and Wilman (1937) that hydrocarbon layers as thin as 43 Å gave strong patterns and completely obliterated the pattern from the substrate. However, Schulz (1951, 1952) demonstrated for the first time that deposits of no more than 1 or 2 Å could give observable patterns. Quantitative studies by Newman and Pashley (1955), in which deposit thicknesses were measured by radioactive tracer techniques, showed that the limit of detection by RHEED was below 1 Å in favorable cases. Thus RHEED could be used to study the growth of the first atomic monolayer.

From the geometrical form of the diffraction spots, as growth of the deposit proceeds, it is possible to make qualitative, or semiquantitative, estimates of the size and shape of the crystallites that form. Schulz (1951a, b, 1952) was able to show that many alkali halide deposits on mica, or on other alkali halides, consist of three-dimensional crystallites from the earliest observable stage of growth. Only when the misfit was fairly low did the initial deposit have a form more closely approximating that of monolayers. Newman and Pashley (1955) could show quite conclusively that three-dimensional crystal-lites form well before there is even enough deposit present to form a single monolayer covering the substrate surface (e.g., for copper deposited on silver). These and many other experiments demonstrated quite clearly that the initial stages of growth consisted of three-dimensional nuclei, rather than monolayers, in very many systems. The reason why the Frank and van der Merwe (1949a, b, c) model for epitaxy did not have general application had therefore been provided, even though at that time it could have been argued that the nucleation mode was influenced by surface contamination and that monolayer growth according to the Frank and van der Merwe model would occur under clean conditions. This would have implied that epitaxy with large misfits was controlled by contamination.

Attempts were made to make accurate lattice spacing measurements on the initial deposits, to find out whether there was any tendency for pseudomorphism to occur. When the misfit is very low, say less than about 2%, accurate spacing measurements are especially difficult because the diffraction patterns from the substrate and the overgrowth almost coincide. Therefore, it was not possible to deduce with certainty whether or not pseudomorphism occurred with such small misfits. With higher misfits, certainly those above 5%, the evidence was always clear in showing that pseudomorphism did not occur. However, there were lattice strains present that were consistent with a tendency toward pseudomorphism. For example, for very thin evaporated copper deposits and chemically grown silver bromide deposits on (111) silver surfaces the deposit spacings parallel to the surface were strained by about 0.75 and 0.5%, respectively (Newman and Pashley, 1955). In both cases the spacings were modified toward the corresponding spacing in the silver substrate.

In the latter examples, the deposits were in the form of isolated nuclei of 10 to 100 Å in linear dimension, and the RHEED measurements gave average spacings throughout their volume. It was not possible to deduce how the strain varied through the thickness of the nuclei.

The use of the evaporation technique for growing deposits had the major advantage that growth could be carried out on substrates while they were under observation inside the RHEED apparatus. This in situ technique allowed the nucleation and growth process to be observed continuously, so that the detailed changes in the electron diffraction pattern could be recorded and analyzed. Evidence of this kind (e.g., Newman and Pashley, 1955) confirmed that nucleation could occur without the formation of initial monolayers, and that subsequent deposition resulted in the growth of the nuclei to form larger crystals. However, it was difficult to obtain more than a very crude picture of the geometrical form of the deposit. Fortunately, the electron microscope was about to make its impact on the subject.

III. The Use of Transmission Electron Microscopy

A. The Development of the Technique

Although the electron microscope was already available commercially by 1945, it was not applied to transmission studies of epitaxial layers until more than 10 years later. This was due partly to the inadequacy of the early instruments, partly to the lack of suitable specimen preparation and handling techniques, and partly to the absence of a real appreciation of the potentialities of electron microscopy.

The emergence of the first high-performance electron microscopes in 1954 heralded a period of rapid development of the application of transmission electron microscopy, particularly for the direct observation of crystalline specimens. Bulk samples of metals were thinned chemically, or electro-chemically, and this revealed that considerable structural detail could be obtained. At about the same time, self-supporting epitaxial layers of metals were detached from their substrates to provide specimens that were very well suited for direct examination. Also, much thinner deposits were detached on carbon-supporting films to reveal the structure of the initial deposits.

The first studies rapidly revealed the power of the technique. Not only was a wealth of geometrical information (e.g., nuclei size and shape) provided, but also there was the possibility of obtaining important crystallographic information, both as a result of carrying out electron diffraction analysis inside the electron microscope and because diffraction contrast on the microscope image revealed crystallographic features such as grain boundaries, twinned structures, and lattice defects such as dislocations and stacking faults. The technique became even more powerful when a thin-film substrate was employed (e.g., molybdenite or mica) so that the crystallographic analysis could be carried out on the substrate and deposit while they were still in contact.

Thus the period 1955–1965 saw a major breakthrough by a rapidly growing number of workers employing the transmission electron microscope to study thin epitaxial deposits.

B. Continuous Films and Lattice Imperfections

The pioneering work of Hirsch and his collaborators (1956) established that TEM studies of thin crystalline specimens could reveal the lattice imper-fections (dislocations and stacking faults) present in the films by means of diffraction contrast. Similar studies of epitaxial films of metals revealed that numerous lattice imperfections were commonly present in such specimens (Pashley, 1959b; Matthews, 1959; Bassett *et al.*, 1959; Phillips, 1960). Apart from these imperfections, the epitaxial metal layers appeared to be similar in structure to thin films prepared by thinning bulk metal.

Some of the initial interest centered on obtaining information on the geometrical properties of these imperfections, and the technique of moiré patterns from overlapping crystals (Hashimoto and Uyeda, 1957; Pashley *et al.*, 1957; Bassett *et al.*, 1958) was employed for this purpose. The moiré fringe patterns can be regarded as magnified projection images of the individual crystal lattices, and dislocations are revealed as terminating fringes (Bassett *et al.*, 1958). The density of dislocations is more readily determined from the moiré patterns than from simple diffraction contrast, and the initial evidence suggested that very high densities in the range 10^{10}–10^{11}/cm^2 were present in epitaxial metal films. This seemed likely to be a special consequence of the mode of growth of the layers.

For the commonly studied fcc metals grown on rock salt substrates, stacking faults and microtwins were found to be common, confirming the earlier evidence obtained by HEED (Bruck, 1936; Wilman, 1940) and RHEED (Kirchner and Cramer, 1938). These planar defects give rise to characteristic fringe contrast effects, which, for the case of stacking faults, had been characterized and interpreted earlier by Whelan and Hirsch (1957a, b) as a result of studies with stainless steel.

Thus the rapidly developing technique of TEM was quickly applied to great advantage to provide a wealth of new information on the structure of epitaxial metal films. While HEED and RHEED had already provided some limited evidence on imperfection structure, TEM provided, in conjunction with selected area electron diffraction and dark field imaging, precise crystallographic evidence on individual defects, such as Burgers vectors of dislocations, displacement vectors of stacking faults, and the crystallographic indices of the planes of planar defects. Analysis of this kind immediately led to attempts to understand how lattice defects were introduced into epitaxial films during their growth (e.g., Matthews, 1959; Bassett et al., 1959; Phillips, 1960).

At about the same time as TEM started to be used for studies of epitaxy, there arose a strong interest in depositing semiconductor layers on semiconductor substrates (e.g., silicon on silicon) as a means of fabricating semiconductor devices. Consequently, TEM was applied to the study of the structure of such layers, particularly their defect structure. The semiconductor deposits were normally formed either by the vacuum evaporation technique or, more commonly, by chemical vapor deposition methods. A variety of different kinds of stacking fault defects were prominent in these epitaxial layers (e.g., Theurer, 1961; Queisser et al., 1962; Booker and Stickler, 1962; Charig et al., 1962; Miller et al., 1963; Mendelson, 1964; Unvala and Booker 1964). A detailed study and analysis of these defects was made by Booker (1964). Apart from their intrinsic interest, these studies were important because of the need to reduce the numbers of imperfections in order to obtain semiconductor devices with adequate properties. An understanding of the origin of the imperfections offered the best hope of devising a means of effecting this reduction.

Thus epitaxy was showing promise of being applied successfully to a new technology. However, some qualification on the terminology of the subject is needed. Epitaxy, as originally defined by Royer (1928), referred to the oriented growth of one substance on the crystal surface of another substance. The most technologically promising semiconductor layers were those grown on similar substrates (e.g., silicon on silicon), where the misfit is zero. This would not have been classed as epitaxy by Royer, and it is in many ways misleading to do so. Unfortunately the term was even abused to the extent that some workers used epitaxy to mean growth by vapor deposition, whether or not any preferred orientation was involved. Clearly, this cannot be accepted as a valid use of the term.

However, the use of the term "epitaxy" to cover, for example, oriented growth of silicon on silicon is so well established that it has to be accepted. To clarify the position, the use of the term "autoepitaxy" is preferred for this purpose. An alternative used by some workers is to distinguish the two cases by "homoepitaxy" and "heteroepitaxy." The use of heteroepitaxy seems unnecessary and unjustified in relation to the original definition given by Royer (1928).

C. THE NUCLEATION STAGE OF GROWTH

One of the first important TEM observations was the confirmation of the earlier electron diffraction evidence (see Section II, E) that initial deposits consist of discrete three-dimensional nuclei (Bassett, 1958; Sella *et al.*, 1959). The use of rock salt, and sometimes other alkali halides, as substrates was common, as in the earlier HEED work. This had the considerable advantage of being fairly readily prepared by cleavage of synthetic crystals and of being readily water soluble so that deposits were easily detached. Also, epitaxy of a number of fcc metals is fairly readily obtained so that the study of the epitaxy of metals such as gold and silver became relatively straightforward.

TEM allowed detailed geometric information about the nuclei to be revealed directly and with comparative ease. The pictorial nature of the evidence was completely convincing, and it showed quite clearly that some new approach to the understanding of epitaxy, in addition to the Frank and van der Merwe (1949a, b, c) monolayer approach, would be necessary. One question that had been posed on several occasions, but that could not previously be answered by experiment, was the extent to which the nucleation is hetero-geneous, say on impurities or surface defects. A spectacular piece of evidence was first obtained by Bassett (1958), who revealed the now well-known preferential nucleation of gold on cleavage steps on the surface of rock salt. It was shown conclusively that the steps could be no more than monatomic in height. Apart from this step decoration effect, however, the nucleation appeared to be random, or homogeneous, in a number of systems studied.

Apart from its intrinsic interest in relation to epitaxy, this step decoration process is important because it provides a technique for revealing the fine-scale step structure on a cleavage surface, and how such structures change as a result of various kinds of surface treatment. This application of the technique was exploited very effectively by Bethge (1962a, b).

Nuclei shape and size vary from one system to another and also depend on the deposition conditions such as temperature and rate. It is probably fair to comment that there is still insufficient systematic evidence on this aspect of nucleation. There is a strong tendency for a great deal of effort to be concentrated on a relatively few systems that are especially convenient for study. However, even the early evidence showed that there is a wide variation in behavior; for example, gold deposited on silver forms thin platelike islands resembling monolayers during the very early stages of deposition (Dickson *et al.*, 1965), whereas the smallest detectable nuclei of gold on rock salt consist of near equiaxed three-dimensional nuclei of little more than 10–20 Å in diameter (e.g. Bassett, 1958).

The early TEM evidence led to considerable theoretical activity on nucleation (e.g., Walton, 1962; Walton *et al.*, 1963; Pound and Hirth, 1964; Zinsmeister, 1966), which is still continuing, and which is reviewed in Chapter 4. The main emphasis of this work has been concerned with the kinetics of nucleation from the point of view of predicting the rate of formation of nuclei and their size and distribution. However, these aspects are not necessarily too important from the point of view of epitaxy; the real need is to determine what controls the orientation of the initial nuclei, as discussed in Section V.

Apart from the influence of the more normal growth parameters such as deposition rate, surface temperature, and residual gases, some rather unexpected effects have been observed when the substrate is bombarded by electrons either just before or during deposition. Stirland (1966, 1968) showed that not only is nucleation density affected, but also epitaxy is improved by bombardment during growth of gold on rock salt. Kunz *et al.* (1966) obtained a similar effect of bombardment before deposition. Such effects as these are very conveniently studied in situ in a LEED apparatus (Palmberg *et al.*, 1967b, 1968) or a RHEED apparatus (Chambers and Prutton, 1968), where electron bombardment is readily available.

D. Postnucleation Growth

As part of a general study of the sequence of growth of epitaxial films, TEM provided important information on the growth processes that follow the initial nucleation. Examples of different kinds of growth sequences have been discussed and illustrated (Pashley, 1965), and these show that important new growth processes are involved once the initial nucleation stage is completed.

Electron microscope studies of growth sequences revealed that, in general, an initial nucleation stage was followed by a stage during which nuclei grew larger until they eventually joined together and formed a continuous deposit film. The detailed geometry of the different stages varied considerably from one substrate–deposit combination to another, and the various deposition parameters (substrate temperature, deposition rate, etc.) had a marked influence on the scale of the structures. Although processes could be postulated to explain the nucleation stage of growth, it proved difficult to understand the way that the geometrical changes occurred once nuclei became sufficiently large to intergrow with each other. Sequences of deposits of gradually increasing thickness could be examined, but it proved difficult to interpret the result because any given area of the deposit could be examined at only one stage of the growth. It therefore became highly desirable to carry out deposition inside the electron microscope, as had previously been done with RHEED (see Section II, E), in order to observe the detailed structural changes taking place on the same area of specimen.

E. In Situ Growth Studies

There were three major problems that made in situ growth experiments difficult and hazardous: (1) the extreme space limitation in the region of the specimen inside the electron microscope, together with the need to ensure no damage to the microscope and no serious reduction in image quality; (2) the relatively poor vacuum inside the electron microscope column; (3) the formation of carbonaceous contamination on the specimen during illumination by the electron beam (Ennos, 1953, 1954). The first experiments were made by McLauchlan et al. (1950) before the modern high-performance microscopes were available. The image quality was not good, and there must have been severe interference with the growth process from both the heating effect of the electron beam and the carbonaceous contamination. However, the sequences of growth found by later workers were revealed in this work.

No further experiments were carried out for nearly 10 years, when Poppa (1962) studied in situ growth of sputtered films, and Bassett (1964) and subsequently Pashley and Stowell (1962) used a simple modification to a Siemens Elmiskop I to allow in situ growth of evaporated metal films to be carried out. The removal of the specimen airlock device provided the required space, and carbonaceous contamination was avoided by heating the specimen to a minimum of about 350°C during the deposition of metals. A relatively poor residual vacuum of about 10^{-4} Torr was tolerated.

The most carefully studied systems were the growth of gold and silver on cleavage flakes of molybdenite, which resulted in well-defined (111) epitaxy of the two face-centered cubic metals. Because of the relatively poor vacuum conditions, a quantitative study of the nucleation kinetics was not justifiable with this equipment, but the initial observations revealed an unexpected phenomenon, commonly known as liquidlike coalescence. When two growing nuclei, or islands, of gold or silver touched they appeared to coalesce together just like two liquid droplets, although the deposit temperature was well below the melting point. It was subsequently confirmed, quite conclusively, that the gold and silver islands were solid (Pashley et al., 1964), and the mass transfer was explained in terms of a rapid surface diffusion process under the driving force of the surface energy minimization. In fact, the experimental observations could well have been predicted beforehand.

This kind of observation was repeated and confirmed by others (e.g., Poppa, 1964) and provided the means to understand the detailed changes that take place in the topographical structure of a film as it grows. From the point of view of understanding epitaxy, the evidence was very important because it revealed the extent to which detailed structural changes occur during growth and contribute to the structure of the complete deposited film. The two most important structural aspects were: (1) the changes in island orientation that can accompany liquidlike coalescence, and (2) the incorporation of lattice defects in the growing deposit film.

There had been evidence, since the early days of applying HEED to the study of epitaxy, that the degree of orientation of a deposit could improve as deposition proceeds (e.g., Rudiger, 1937). The in situ growth technique revealed two mechanisms whereby such orientation improvement can occur. First, as shown by Bassett (1961) and Pashley et al. (1964), individual nuclei can rotate by several degrees as they grow larger and so lead to a gradual improvement in the overall alignment of the nuclei. Second, when two nuclei of grossly different orientation coalesce, the grain boundary formed at the junction can migrate out of the composite island, thus converting the orientation of one of the islands to that of the other (Jacobs et al., 1966). In this way, randomly oriented nuclei can be converted to well-oriented nuclei as they coalesce with well-oriented nuclei. Also, one kind of oriented nuclei can be eliminated during coalescence, if an initial deposit consists of a mixture of two orientations of nuclei and coalescence favors the growth of one orientation at the expense of another. These conclusions revealed that postnucleation growth processes can be just as important as nucleation processes in determining the final orientation of a deposited film.

There had previously been much discussion and speculation on the way lattice imperfections are formed in deposited thin films. The in situ studies revealed that many imperfections arise during the coalescence stage of growth of the films, and this is discussed in full in Chapter 5.

F. INTERFACIAL DISLOCATIONS AND PSEUDOMORPHISM

As part of their theoretical model for epitaxial monolayers discussed in Section I, Frank and van der Merwe (1949a, b, c) introduced the concept of misfit dislocations at the interface between the substrate and deposit. In effect, if the natural misfit is zero, or if the actual misfit is zero because of the occurrence of pseudomorphism, there will be no interface dislocations (assuming perfect alignment between substrate and deposit). However, if there is some misfit this can be accommodated by the incorporation of an appropriate array of edge dislocations, or dislocations with significant edge components. Thus the observation and analysis of the dislocation structure at the interface provided a new means, in addition to that of electron diffraction, of obtaining information about pseudomorphism. The technique of TEM provided a powerful and unique tool for detecting and analyzing the dislocation structures.

In a series of elegant experiments, Matthews and co-workers showed that, by depositing thin-metal layers on thin-metal single-crystal films as substrates, the density of the interface dislocation structures depends on deposit thickness. For cases of low natural misfit (say less than a few percent), no misfit dislocations could be detected for low deposit thicknesses (say a few angstroms), and as the deposit thickness increased misfit dislocations were generated. This was observed for both continuous monolayer-type deposits (Jesser and Matthews, 1967) and island-type deposits (Jesser and Matthews, 1968). Thus, for the first time, there was conclusive evidence of the occurrence of pseudomorphism, and a clearer view of the conditions under which pseudomorphism can occur. The evidence is not inconsistent with the early HEED evidence summarized in Section II, D, which provided no direct proof of the existence of pseudomorphism, because pseudomorphism occurs under conditions unfavorable for the HEED and RHEED observations.

An equally important advance resulted from attempts to calculate the expected lattice strains (i.e., degree of pseudomorphism) under different conditions. Such calculations by van der Merwe (1963a, b) and Jesser and Kuhlmann-Wilsdorf (1967) predicted the limits of pseudomorphism in remarkably close agreement with the experimentally observed limits.

Although the occurrence of pseudomorphism was proved conclusively as a result of the observations on interface dislocations, there is still some uncertainty as to the extent to which some of the examples studied were influenced by the occurrence of alloying at the interface. Alloying, where it occurs, has the effect of distributing the lattice strain through a greater thickness of material on either side of the interface, and hence of making pseudomorphism more possible. In effect, the misfit at the interface is reduced. (The important subject of interface dislocations is treated fully in Chapter 6.)

IV. The Development of UHV Techniques

A. Experimental Techniques—LEED and RHEED

From the late 1920s until the early 1960s, most work on epitaxy was confined to the use of relatively poor vacuum systems, both for the vacuum evaporation of deposits and for their examination by electron optical techniques. For certain experiments (e.g., such as could be carried out in baked glassware) vacua of 10^{-7} Torr or even better could be used. This was very tedious, however, and could not be applied to the majority of work. There was always a worry as to the extent to which the results on epitaxy were influenced by contamination, and the desirability for experiments to be carried out in much cleaner vacuum systems was well recognized by many workers.

By the early 1960s, considerable advances had been made in vacuum techniques, with the use of bakeable stainless steel vacuum systems based on new types of pumps (e.g., ion pumps). These fairly rapidly became available commercially, and they allowed vacua in the range 10^{-8}–10^{-10} Torr to be obtained with fairly high reliability. As a result, there was the opportunity to prepare deposits under very much improved conditions of cleanliness.

At about the same time, a major advance was being made in the technique of LEED, when Scheibner et al. (1960) introduced the technique of post-acceleration of the electrons, after they had been diffracted, so that they could excite a fluorescent screen and so provide a visible diffraction pattern that could readily be photographed for analysis. This advance eliminated the need for the tedious and laborious technique of mapping patterns via readings from a detector that is scanned around the specimen.

As a result of the combination of these two advances, commercial LEED apparatus became available for general use in the form that is well known today. Since the technique of LEED demanded the use of UHV, it stimulated the application of UHV to surface structural studies, including those on epitaxy. Further, the high sensitivity of LEED offered a great attraction since it was claimed that it was possible to detect when a surface was contaminated even by less than one monomolecular layer of an adsorbent. Although the intrinsic high sensitivity of LEED was certainly a very major advance in technique, there are circumstances in which the sensitivity is lower than is generally the case and where RHEED can be of great sensitivity (Pashley, 1970). Unfortunately, RHEED suffers from its past reputation based on its use with poor vacuum systems; and because a clean vacuum system is not vital to the basic RHEED technique there has been far less effort devoted to developing UHV RHEED systems. Fortunately, such systems are now available, and much benefit should be gained from using both LEED and RHEED on the same specimens.

B. Film Growth in UHV

When serious attempts were first made to grow epitaxial layers in UHV, the experimenters were no doubt anticipating that they would observe improved epitaxy. This expectation appeared to be justified by the experiments of Ino et al. (1962), who showed that if a rock salt sample is cleaved inside a poor vacuum (10^{-4}–10^{-5} Torr), and immediately used as a substrate for deposition of gold, silver, or copper, the so-called epitaxial temperature is lowered. It was assumed that the improved epitaxy arose from the use of a less contaminated substrate, compared with that normally prepared by air cleavage.

However, when the same workers repeated the experiments in a vacuum of 10^{-7} to 10^{-9} Torr (Ino et al., 1964), they rather surprisingly found that the epitaxy was poorer. A similar result was obtained by Matthews and Grunbaum (1964, 1965), who also examined layers of varying thickness and found that in the very early stages of growth (average deposit thickness < 10 Å), there was little difference in the deposit orientation on a cleavage surface prepared in the UHV compared with that on a cleavage surface that had been exposed to air. The differences only developed during the subsequent stages of growth.

This careful study demonstrated what had only been partly realized previously, that the deposits of Ag, Au, or Cu on rock salt occur as a mixture of (100) and (111) orientations in the early stages. The "normal" good (100) epitaxy occurs as a result of recrystallization during subsequent growth, presumably by the grain boundary migration mechanism described in Section III, E. This recrystallization process is affected by the nucleation density, since this affects the stage at which coalescence occurs, and the nucleation density is influenced by contamination. In this particular case, contamination appears to favor good epitaxy because it increases the nucleation density.

As anticipated, nucleation on surfaces is generally influenced by contamination, and there is little doubt that meaningful experiments on the kinetics of nucleation must be carried out in very clean vacuum systems. However, it does not follow that most of the previous general work on epitaxy must be discarded once UHV systems are available. As a bare minimum, the previous work is of value because it provides a collection of data on which to base future experiments carried out under cleaner conditions. Its value, however, goes much deeper than that. It has provided a conceptual framework of interpretation, theoretical modeling, and general understanding of physical principles that should continue to make a substantial contribution to the further development of the subject.

It is, of course, important to establish the extent to which data obtained under poor vacuum conditions are valid when UHV is employed. Broadly, the same deposit orientations are found when UHV systems are employed, and the differences, where they have been identified, relate more to the influence of various deposition parameters (e.g., substrate temperature) and quantitative aspects of epitaxy (e.g., epitaxial temperature and nucleation densities).

The availability of electron optical equipment with much cleaner vacuum systems has been of special importance to in situ studies of epitaxial growth. In the case of LEED, a powerful addition to the technique has resulted from the use of Auger electrons to provide a means of carrying out chemical analysis on the specimen surfaces (Harris, 1968a, b). By using the same electron gun to generate the Auger electrons as well as the LEED pattern, diffraction analysis is carried out in conjunction with chemical analysis with a detection sensitivity significantly less than one monomolecular layer. This has proved to be a very powerful technique for assessing the extent of any contamination on a surface, and so allowing reproducibly clean surfaces to be used for systematic studies. It has also helped to identify the nature of the surface damage produced by electron bombardment of alkali halide surfaces (Palmberg *et al.*, 1967b, 1968), and its influence on epitaxy, as already discovered by Stirland (1966, 1968) as discussed in Section III, C. Thus the late 1960s saw the start of a new era in studies of the initial stages of epitaxy, with refined techniques promising to provide information at a level of detail that was previously quite unattainable. No doubt the real rewards of these new techniques have yet to be reaped.

The RHEED technique was also provided with a facility to allow chemical analysis to be carried out. In this case, the wavelengths of the characteristic X rays excited by the incident high-energy electrons are used to determine the elements present at the surface (Sewell and Cohen, 1967; Sewell *et al.*, 1969). As yet, no systematic comparison has been made of the performance of the two systems (i.e., LEED with Auger analysis compared with RHEED with X-ray analysis), but it is again likely that each technique has its own special advantages and that the greatest benefit will be derived from operating the two techniques simultaneously in the same apparatus.

Although UHV versions of RHEED and LEED apparatus are available commercially, commercial electron microscopes continue to have relatively poor vacuum systems because of their greater complexity. Also, most applications of TEM do not demand cleaner vacuum systems, so that the additional expense in manufacturing and running UHV electron microscopes would not, in general, be justified. Consequently, it has been left to individual research workers to build UHV microscopes, as required, and several have been built for the special purpose of allowing in situ thin-film growth experiments to be carried out under clean vacuum conditions (Poppa, 1965; Barna *et al.*, 1967; Moorhead and Poppa, 1969; Valdrè *et al.*, 1970). One important result produced by these instruments is that most of the effects (e.g., liquidlike coalescence and recrystallization) observed during in situ studies in poor vacuum conditions (see Section III, E) also take place in a clean vacuum. It can, therefore, be concluded with certainty that such effects are not artifacts resulting from some contamination effect.

V. Summary and Discussion

The amount of data on epitaxy that has been accumulated during this century is now very considerable, covering orientation relationships for a wide variety of substrates, deposits, and growth conditions, as well as structural information at various stages of growth. As each new technique has been applied to the problem, hopes have risen that a really meaningful understanding of the factors determining epitaxy will be obtained. Although each technique has indeed provided new and important information that has increased our knowledge, it has also served to demonstrate the enormous complexity of the problem. As a result, we can now see more clearly what is needed of a realistic theoretical model, but we can also see that the formulation of such a model is likely to be a very difficult task. It is reasonable to argue that a theoretical model must be capable of predicting the epitaxial behavior on any given substrate, if it is to be accepted as realistic. At the moment, this achievement appears to be a long way off.

Until about 1960, most theoretical approaches to the problem were based on the concept of a small misfit, although there was no very serious attempt made to develop this kind of approach further, after the limited applicability of the Frank and van der Merwe (1949a, b, c) model had been appreciated. The emphasis of the experimental approach to the problem shifted to the nucleation aspects of film growth from the mid 1950s, and this naturally led to theoretical attention being given to the determination of the orientation of very small nuclei.

However, not all deposits grow by the nucleation of isolated three-dimensional nuclei, and a growth process much more resembling the monolayer growth concept is observed in some cases (see Section III, F). The various growth modes have been classified by Bauer (1958a) as: (a) three-dimensional nucleation, which occurs when the surface energy of the deposit is greater than that of the substrate; (b) layer-by-layer growth, which occurs when the surface energy of the deposit is less than that of the substrate; and (c) a special case of (b) that arises when there is a high strain energy in the deposit film, and that leads to nucleation of three-dimensional nuclei on top of an initial monolayer. It seems likely that at least two kinds of theoretical models are required to cover these different cases, and the Frank and van der Merwe (1949a, b, c) model seems to provide the basis for explaining epitaxy during case (b), and possibly case (c). The major difficulty arises with case (a), not only because of the theoretical difficulty in dealing with three-dimensional nuclei, but also because the examples of high epitaxial misfits are largely confined to this mode of growth.

The experimental evidence from various observational techniques showed that nuclei containing well below 100 atoms can be well oriented on a substrate, but no technique, except that of field ion microscopy, appears to be capable of providing evidence of the structure and orientation of nuclei consisting of less than about 10 atoms. Consequently there is no experimental evidence on the initial nuclei consisting of just a few atoms, but theoretical models have been explored on the basis of the orientation being determined by the way clusters of no more than about 4 atoms arrange themselves on a substrate. One such model is that of Rhodin and Walton (1964), who have considered the critical size for the stability of small clusters and have estimated that the critical nucleus for a (111) stable orientation of an fcc metal on rock salt is just a pair of adsorbed atoms. For the (100) orientation it is 3 atoms. It is argued that this explains the occurrence of predominantly (111) and (110) orientations in fcc metals on rock salt.

In the past, theoretical models for epitaxy have played a major role in stimulating experiments aimed at testing out the models. These experiments have often provided a wealth of new and important information even though they have not necessarily given much support to the models. The very active theoretical interest in clusters of small numbers of atoms, and their low-energy stable configurations, has not been matched by experimental evidence because of the lack of adequate analytical techniques, and unless and until such techniques are developed there will be inadequate guidance for further realistic development of the theoretical models.

One of the major questions concerns the stability of small clusters when they grow bigger by the addition of further adatoms. It seems plausible that the configuration of lowest energy for (say) 3 atoms will not necessarily be the same as the lowest-energy configuration for (say) 6 atoms, and that a complete reorientation can occur as the size increases. It is possible, therefore, that it will be necessary to determine the most favored orientations for clusters of perhaps 10, or even more, atoms in order to predict the orientation of nuclei that are sufficiently large to be observed and analyzed (see Tick and Witt, 1971; Hoare and Pal, 1971). Clearly, this raises major theoretical problems.

A rather different approach to the question of determining the orientation of nuclei has been introduced by Distler and his co-workers. They have reported extensive experimental evidence that oriented nuclei can form on top of a continuous amorphous layer deposited on a single crystal substrate. For example, Distler et al. (1968a) obtained well-oriented nuclei of lead sulfide on rock salt cleavage surfaces with intermediate carbon layers up to 150 Å in thickness. In some cases, epitaxy occurs with intermediate amorphous layers of up to 1500 Å in thickness. The observed cases include lead sulfide on mica (Distler and Kobzareva, 1966), gold on rock salt (Gerasimov and Distler, 1968), and cadmium sulfide on rock salt (Distler et al., 1968b). These results are interpreted in terms of the effect of longe-range forces arising from some kind of active centers, but the subject rapidly aroused considerable controversy both from the validity of the experimental observations and from the interpretation of the effect. Chopra (1969) failed to observe the effect, but other workers (e.g., Barna et al., 1969) are beginning to support, at least in part, the findings reported by Distler and his co-workers. Clearly it is most important that the effect should be studied further and its general significance established.

Although considerable emphasis has rightly been placed on trying to understand how the orientation of nuclei is determined, there has been inadequate consideration of the influence of postnucleation growth processes (see Section III, D) on the orientation of a completed film. It is far from clear as to the extent that postnucleation processes do have a major influence, but certainly some cases are controlled by these recrystallization processes. Also other aspects of the internal structure, including grain size and lattice imperfection content, are very much controlled by the postnucleation processes. This appears to be an area of study that requires further attention, both from the experimental and the theoretical points of view.

During the last 20 years, considerable resources have been devoted to the study of film growth and epitaxy. It is interesting to consider whether as much as possible has been gained from these efforts, or whether some other approach would have been more fruitful. Certainly, progress has inevitably been limited by the rate at which new techniques have been developed and become available for use, and yet progress on techniques has been impressive. One striking fact that emerges from a close study of the literature is the very high proportion of publications devoted to studies of the growth of metals (predominantly fcc metals) on alkali halide cleavage surfaces (predominantly sodium chloride). This seems to have come about because of the considerable experimental convenience of these systems, coupled with the numerous special effects that have been observed and have stimulated further studies. No doubt, it is also argued that studying a few systems in great depth is more likely to provide the key to the solution of the problem of explaining epitaxy than is studying a large number of systems in a more superficial manner.

The stage now seems to have been reached at which the strategy should be changed. In fact it can be argued that a change is very much overdue. The more one studies a particular system (i.e., substrate–deposit combination), the more one sees that a number of features of that system are likely to be peculiar to that system alone, or to just a restricted class of systems. Thus the contribution of such in-depth studies of a very restricted number of systems does not make a major contribution to a general understanding of epitaxy. Therefore, a more balanced and meaningful picture seems more likely to emerge from studies of a broader range of substrate–deposit combinations. This appears to be one of the main lessons to be learned from the previous history of research on epitaxy. If the now very powerful array of experimental techniques is applied to studies of a broader range of systems, not only will the data that result represent a more balanced view of the subject, but also they will be more likely to have wider practical application for those who wish to apply epitaxy to prepare single-crystal layers, either for other kinds of experimental work or for practical applications (e.g., electrical devices).

ACKNOWLEDGMENTS

The author is indebted to Dr. M. J. Stowell for his reading of the manuscript and his constructive criticisms. This chapter is published by permission of the Chairman of Tube Investments Limited.

References

Barker, T. V. (1906). *J. Chem. Soc. Trans.* **89**, 1120.
Barker, T. V. (1907). *Mineral. Mag.* **14**, 235.
Barker, T. V. (1908). *Z. Kristallogr.* **45**, 1.
Barna, A., Barna, P. B., and Pocza, J. F. (1967). *Vacuum* **17**, 219.
Barna, A., Barna, P. B., and Pocza, J. F. (1969). *Thin Solid Films* **4**, R32.
Bassett, G. A. (1958). *Phil. Mag.* **3**, 1042.
Bassett, G. A. (1961). *Proc. Eur. Reg. Conf. Electron Micros., Delft, 1960* p. 270. De Nederlandse Vereniging Voor Electronen-Microscopie, Delft.
Bassett, G. A. (1964). *Proc. Int. Symp. Condens. Evapor. Solids, Dayton, Ohio, 1962* (E. Rutner, P. Goldfinger, and J. P. Hirth, eds.). Gordon & Breach, New York.
Bassett, G. A., Menter, J. W., and Pashley, D. W. (1958). *Proc. Roy. Soc. A* **246**, 345.
Bassett, G. A., Menter, J. W., and Pashley, D. W. (1959). *In* "Structure and Properties of Thin Films" (C. A. Neugebauer, J. B. Newkirk, and D. A. Vermilyea, eds.), p. 11. Wiley, New York.
Bauer, E. (1958a). *Z. Kristallogr.* **110**, 372.
Bauer, E. (1958b). *Z. Kristallogr.* **110**, 395.
Bethge, H. (1962a). *Phys. Status Solidi.* **2**, 3.
Bethge, H. (1962b). *Phys. Status Solidi.* **2**, 775.
Booker, G. R. (1964). *Discuss. Faraday Soc.* No. 38, 298.
Booker, G. R., and Stickler, R. (1962). *J. Appl. Phys.* **33**, 3281.
Bruck, L. (1936). *Ann. Phys. Leipzig* **26**, 233.
Chambers, A., and Prutton, M., (1968). *Thin Solid Films* **1**, 393.
Charig, J. M., Joyce, B. A., Stirland, D. J., and Bicknell, R. W. (1962). *Phil. Mag.* **7**, 1847.
Chopra, K. L. (1969). *J. Appl. Phys.* **40**, 906.
Cochrane, W. (1936). *Proc. Phys. Soc.* **48**, 723.
Davisson, C., and Germer, L. H. (1927). *Phys. Rev.* **30**, 707.
Dickson, E. W., Jacobs, M. H., and Pashley, D. W. (1965). *Phil. Mag.* **11**, 575.
Distler, G. I., and Kobzareva, S. A. (1966). *Proc. Int. Congr. Electron Microsc., 6th, Kyoto, 1966* p. 493. Maruzen, Tokyo.
Distler, G. I., Kobzareva, S. A., and Gerasimov, Y. M. (1968a). *J. Crystal Growth* **2**, 45.
Distler, G. I., Gerasimov, Y. M., Kobzareva, S. A., Moskin, V. V., and Shenyavskaya, L. A. (1968b). *Proc. Eur. Reg. Conf. Electron Microsc., 4th, Rome, 1968* p. 517. Tipographia Poliglotta Vaticana, Rome.
Engel, O. G. (1952). *J. Chem. Phys.* **20**, 1174. (1953). J.
Engel, O. G. (1953). *J. Res. Nat. Bur. St.* **50**, 249.
Ennos, A. E. (1953). *Brit. J. Appl. Phys.* **4**, 101.
Ennos, A. E. (1954). *Brit. J. Appl. Phys.* **5**, 27.
Frank, F. C., and van der Merwe, J. H. (1949a). *Proc. Roy. Soc. A* **198**, 205.

Frank, F. C., and van der Merwe, J. H. (1949b). *Proc. Roy. Soc. A* **198**, 216.
Frank, F. C., and van der Merwe, J. H. (1949c). *Proc. Roy. Soc. A* **200**, 125.
Frankenheim, M. L. (1836). *Ann. Phys.* **37**, 516.
Finch, G. I., and Quarrell, A. G. (1933). *Proc. Roy. Soc. A* **141**, 398.
Finch, G. I., and Quarrell, A. G. (1934). *Proc. Phys. Soc.* **46**, 148.
Finch, G. I., and Wilman, H. (1937). *Trans. Faraday Soc.* **33**, 337.
Gerasimov, Y. M., and Distler, G. I. (1968). *Naturwiss.* **55**, 132.
Harris, L. A. (1968a). *J. Appl. Phys.* **39**, 1419. (1968b).
Harris, L. A. (1968b). *J. Appl. Phys.* **39**, 1428.
Hashimoto, H., and Uyeda, R. (1957). *Acta. Cryst.* **10**, 143.
Hirsch, P. B., Horne, P. B., and Whelan, M. J. (1956). *Phil. Mag.* **1**, 677.
Hirsch, P. B., Howie, A., Nicholson, R. B., Pashley, D. W., and Whelan, M. J. (1965).
    "Electron Microscopy of Thin Crystals." Butterworths, London and Washington, D.C.
Hoare, M. R., and Pal, P. (1971). *Advan. Phys.* **20**, 161.
Ino, S., Watanabe, D., and Ogawa, S. (1962). *J. Phys. Soc. Japan* **17**, 1074.
Ino, S., Watanabe, D., and Ogawa, S. (1964). *J. Phys. Soc. Japan* **19**, 881.
Jacobs, M. H., Pashley, D. W., and Stowell, M. J. (1966). *Phil. Mag.* **13**, 129.
Jesser, W. A., and Kuhlmann-Wilsdorf, D. (1967). *Phys. Status Solidi.* **19**, 95.
Jesser, W. A., and Matthews, J. W. (1967). *Phil. Mag.* **15**, 1097.
Jesser, W. A., and Matthews, J. W. (1968). *Phil. Mag.* **17**, 595.
Kirchner, F., and Cramer, H. (1938). *Ann. Phys.* **33**, 138.
Kirchner, F., and Lassen, H. (1935). *Ann. Phys.* **24**, 113.
Kranert, W., Leise, K. H., and Raether, H. (1944). *Z. Phys.* **122**, 248.
Kunz, K. M., Green, A. K., and Bauer, E. (1966). *Phys. Status Solidi* **18**, 441.
Lassen, H. (1934). *Phys. Z.* **35**, 172.
Lassen, H., and Bruck, L. (1935). *Ann. Phys.* **22**, 65.
Lawless, K. R., and Mitchell, D. F. (1965a). *Mem. Sci. Rev Met.* **62** (Special number, May
    1965), 27.
Lawless, K. R., and Mitchell, D. F. (1965b). *Mem. Sci. Rev. Met.* **62** (Special number, May
    1965), 39.
Lucas, L. N. D. (1951). *Proc. Phys. Soc. A* **64**, 943.
Lucas, L. N. D. (1952). *Proc. Roy. Soc. A* **215**, 162.
Matthews, J. W. (1959). *Phil. Mag.* **4**, 1017.
Matthews, J. W. (1967). *Phys. Thin Films* **4**, 137.
Matthews, J. W. and Grunbaum, E. (1964). *Appl. Phys. Lett.* **5**, 106.
Matthews, J. W., and Grunbaum, E. (1965). *Phil. Mag.* **11**, 1233.
McLauchlan, T. A., Sennett, R. S., and Scott, G. D. (1950). *Can. J. Res.* **28**A, 530.
Mendelson, S. (1964). *J. Appl. Phys.* **35**, 1570.
Menzer, G. (1938a). *Naturwiss.*, **26**, 385.
Menzer, G. (1938b). *Z. Kristallogr.* **99**, 378.
Menzer, G. (1938c). *Z. Kristallogr.* **99**, 410.
Miller, D. P., Watelski, S. B., and Moore, C. R. (1963). *J. Appl. Phys.* **34**, 2813.
Moorhead, R. D., and Poppa, H. (1969). *Proc. Annu. Meeting EMSA, 27th, Claitors,
    Baton Rouge, Louisiana* p. 116.
Neuhas, A. (1950–51). *Fortschr. Mineral.* **29–30**, 136.
Newman, R. C. (1956). *Proc. Phys. Soc.* **B69**, 432.
Newman, R. C., and Pashley, D. W. (1955). *Phil. Mag.* **46**, 917.
Nishikawa, S., and Kikuchi, S. (1928). *Nature (London)* **121**, 1019.
Palmberg, P. W., Rhodin, T. N., and Todd, C. J. (1967a). *Appl. Phys. Lett.* **10**, 122.
Palmberg, P. W., Rhodin, T. N., and Todd, C. J. (1967b). *Appl. Phys. Lett.* **11**, 33.

Palmberg, P. W., Todd, C. J., and Rhodin, T. N. (1968). *J. Appl. Phys.* **39**, 4650.
Pashley, D. W. (1951). *Proc. Phys. Soc. A* **64**, 1113.
Pashley, D. W. (1952). *Proc. Roy. Soc. A* **210**, 355.
Pashley, D. W. (1956). *Advan. Phys.* **5**, 173.
Pashley, D. W. (1959a). *Phil. Mag.* **4**, 316.
Pashley, D. W. (1959b). *Phil. Mag.* **4**, 324.
Pashley, D. W. (1965). *Advan. Phys.* **14**, 327.
Pashley, D. W. (1970). *Recent Progr. Surface Sci.* **3**, 23.
Pashley, D. W., and Stowell, M. J. (1962). *Proc. Int. Conf. Electron Microsc., 5th, Philadelphia, 1962* (S. S. Breese, ed.), Paper GG-1. Academic Press, New York.
Pashley, D. W., Menter, J. W., and Bassett, G. A. (1957). *Nature (London)* **179**, 752.
Pashley, D. W., Stowell, M. J., Jacobs, M. H., and Law, T. J. (1964). *Phil. Mag.* **10**, 127.
Phillips, V. A. (1960). *Phil. Mag.* **5**, 571.
Poppa, H. (1962). *Phil. Mag.* **7**, 1013.
Poppa, H. (1964). *Z. Naturforsch. A* **19**, 835.
Poppa, H. (1965). *J. Vac. Sci. Technol.* **2**, 42.
Pound, G. M., and Hirth, J. P. (1964). *Int. Symp. Evaporation Condensation Solids, Dayton, Ohio, 1962* (E. Rutner, P. Goldfinger, and J. P. Hirth, eds.). Gordon & Breach, New York.
Pound, G. M., Simnad, M. T., and Yang, L. (1954). *J. Chem. Phys.* **22**, 1215.
Quiesser, H. J., Finch, R. H., and Washburn, J. (1962). *J. Appl. Phys.* **33**, 1536.
Raether, H. (1950). *J. Phys. Radium* **11**, 11.
Rhodin, T. N., and Walton, D. (1964). *In* "Single Crystal Films" (M. H. Francombe and H. Sato, eds.), p. 31. Pergamon, Oxford.
Royer, L. (1928). *Bull. Soc. Fr. Mineral. Crist.* **51**, 7.
Royer, L. (1935). *Ann. Phys.* **23**, 16.
Rudiger, O. (1937). *Ann. Phys.* **30**, 505.
Scheibner, E. J., Germer, L. H., and Hartman, C. D. (1960). *Rev. Sci. Instrum.* **31**, 112.
Schulz, L. G. (1951a). *Acta Cryst.* **4**, 483.
Schulz, L. G. (1951b). *Acta Cryst.* **4**, 487.
Schulz, L. G. (1952). *Acta Cryst.* **5**, 130.
Seifert, H. (1953). *In* "Structure and Properties of Solid Surfaces" (R. Gomer and C. R. Smith, eds.), p. 218. Chicago Univ. Press, Chicago, Illinois.
Sella, C., Conjeaud, P., and Trillat, J. J. (1959). *C. R. Acad. Sci. Paris* **249**, 1987.
Sewell, P. B., and Cohen, M. (1967). *Appl. Phys. Lett.* **11**, 298.
Sewell, P. B., Mitchell, D. F., and Cohen, M. (1969). *Develop. Appl. Spectrosc.* **7A**, 61.
Shirai, S. (1939). *Proc. Phys.-Math. Soc. Japan* **21**, 800.
Shishakov, N. A. (1952). *Zh. Eksper. Teor. Fiz.* **22**, 241.
Smollett, M., and Blackman, M. (1951). *Proc. Phys. Soc. A* **64**, 683.
Stirland, D. J. (1966). *Appl. Phys. Lett.* **8**, 326.
Stirland, D. J. (1968). *Thin Solid Films* **1**, 447.
Theurer, H. C. (1961). *J. Electrochem. Soc.* **108**, 649.
Thomson, G. P., and Reid, A. (1927). *Nature (London)* **119**, 80.
Tick, P. A., and Witt, A. F. (1971). *Surface Sci.* **26**, 165.
Unvala, B. A., and Booker, G. R. (1964). *Phil. Mag.* **9**, 691.
Valdrè, U., Robinson, E. A., Pashley, D. W., Stowell, M. J., and Law, T. J. (1970). *J. Phys. E Sci. Instrum.* **3**, 501.
van der Merwe, J. H. (1949). *Discuss. Faraday Soc.* No. 5, 201.
van der Merwe, J. H. (1963a). *J. Appl. Phys.* **34**, 117.
van der Merwe, J. H. (1963b). *J. Appl. Phys.* **34**, 123.

Walton, D. (1962). *J. Chem. Phys.* **37**, 2182.
Walton, D., Rhodin, T. N., and Rollins, R. W. (1963). *J. Chem. Phys.* **38**, 2698.
Whelan, M. J., and Hirsch, P. B. (1957a). *Phil. Mag.* **2**, 1121.
Whelan, M. J., and Hirsch, P. B. (1957b). *Phil. Mag.* **2**, 1303.
Wilman, H. (1940). *Proc. Phys. Soc. London* **52**, 323.
Zinsmeister, G. (1966). *Vacuum* **16**, 529.

CHAPTER 2

PREPARATION
OF SINGLE-CRYSTAL FILMS

2.1   Evaporation of Thin Films

Siegfried Mader

IBM Thomas J. Watson Research Center
Yorktown Heights, New York

I. Introduction

Evaporation is one of the most frequently used methods for the preparation of thin films including epitaxial films. The structural perfection and chemical purity of evaporated films is often inferior to the quality of epitaxial layers fabricated by other methods, such as chemical vapor deposition or liquid-phase epitaxy, which are described in other chapters of this book. In spite of these shortcomings, evaporation is popular because of its conceptual and experimental simplicity: The source material is transformed into its gaseous state by raising its vapor pressure through an increase in temperature. The vapor expands into the evacuated space between the evaporation source and the substrate and finally condenses on the substrate (as well as on the walls of the vacuum vessel).

This simple process can require expensive equipment if ultrahigh vacuum systems are to be used to ensure chemical purity of the deposit or if the incidence rate of the vapor at the substrate has to be precisely controlled. Deposition of multicomponent materials from several sources requires a good control of the vapor streams. An example of this is molecular beam epitaxy, described in the next section.

On the other hand evaporators can be incorporated into a variety of analytical tools such as electron diffraction instruments and electron micro-scopes. This allows in situ observations of film-growth phenomena, which are also described in detail in other chapters of this book. Most of our direct insight into thin-film growth has been gained with films prepared by evaporation.

In the present subchapter a few general aspects of evaporation are discussed. Specific results of epitaxially evaporated layers are presented in other parts of this book. Recently a comprehensive review of vacuum evaporation was published by Glang (1970). A somewhat older but very useful treatment is Holland's (1956) book. The works by Chopra (1969) and Dushman (1962) also contain important reviews of this subject.

II. Vacuum Requirements

The first requirement for evaporation is a vacuum vessel. Unavoidably it contains residual gases. It is customary to estimate the allowable residual gas pressure from the following two considerations. First, an upper limit to the vacuum pressure is set by the mean free path λ of the residual gas molecules. Note that λ should be larger than the distance between vapor source and film substrate; otherwise collisions between residual gas molecules and the evaporating material randomize the vapor beam and prevent it from reaching the substrate. In a very poor vacuum the vapor can even condense to smoke particles in the residual gas atmosphere. According to kinetic gas theory

$$\lambda = kT/2^{1/2}\pi p\sigma^2 \tag{1}$$

where k is the Boltzmann constant, T the absolute temperature, σ a collision diameter (typically of the order of a few angstroms, and p the gas pressure. For a mean free path of 50 cm the pressure has to be reduced to 10^{-4} Torr.

Second, one compares the rate of incidence r of residual gas molecules at the substrate with the incidence rate of vapor atoms. Since the gas molecules can be incorporated into the growing film, their incidence rate has to be smaller than the incidence rate of the vapor. Just how much smaller depends on the tendency of the film material to bond gas molecules and on the allowable concentration of impurities in the deposit.

The incidence rate r (in cm^{-2} sec^{-1}) is related to the gas pressure by

$$r = (2\pi mkT)^{-1/2}p \qquad (2a)$$

where m is the mass of the molecule, or

$$r = 3.513 \times 10^{22}(MT)^{-1/2}p \qquad (2b)$$

where M is the molecular weight and p the pressure in Torrs. At room temperature and at a pressure of 10^{-4} Torr (the high-pressure limit, according to the mean free path argument) the incidence rate of air is 3.8×10^{16} molecules/cm^2 sec, which is equivalent to 40 monolayers/sec. In most cases the sticking coefficient of residual gases is rather small and not all impinging molecules are trapped in the film. However it is clear that films grown in 10^{-4} Torr contain a large and uncontrolled amount of oxygen, nitrogen, and water vapor. The situation is somewhat better with films grown in 10^{-7} Torr. This seems to be the best vacuum obtained in a simple, unbaked bell-jar evaporator, evacuated with well-trapped diffusion pumps. When a film is grown with the typical growth rate of 5 monolayers/sec in such a vacuum the incident rate of residual gases is 1% of that of the vapor. For even cleaner conditions one has to employ ultrahigh vacuum (UHV) technology.

III. Vacuum Generation

Vacuum technology has brought forth a large variety of methods for producing vacuum. The field has recently been reviewed, for example, by Glang et al. (1970). One can distinguish the range of high vacuum with pressures higher than about 10^{-8} Torr and the UHV range with lower pressures.

In the UHV range one commonly uses sorption pumps, evapor–ion pumps, or sputter–ion pumps capable of an ultimate pressure of 10^{-11} Torr. The actual pressure in the vacuum chamber depends also on the desorption rates of gases from surfaces. Outgassing must be accelerated by baking the whole system for several hours at about 300°C. This places severe restrictions on the design and the materials of a UHV system. These systems are expensive and the evacuation time is prolonged. Residual gases in the UHV range mostly come from the chamber walls and consist mainly of H_2 and some CO, N_2, H_2O, and Ar.

In the high-vacuum range one uses diffusion pumps backed by rotary fore-pumps. A variety of working fluids is available with low vapor pressures leading to low ultimate pump pressures. A common handicap of all diffusion pumps is back diffusion of gases and back streaming of pump oil into the vacuum system. Therefore, the residual gases in the vacuum system tend to contain hydrocarbons from the pump fluid. Traps and baffles can reduce this contamination at the expense of pumping speed. Other contaminant sources are outgassing from elastomere gaskets and chamber walls. The great advantage of unbaked diffusion-pumped systems is the short turn-around time. An evaporator can be evacuated from atmosphere to 10^{-6} Torr in an hour or so. At this stage, water vapor is the predominant residual gas, and its partial pressure can be further reduced by cooling a large surface inside the vacuum system to liquid-nitrogen temperature (Meissner trap).

IV. Vapor Pressure and Rate of Evaporation

The next requirement for evaporation is a vapor stream, which is generated from the source material by raising its equilibrium vapor pressure through heating. (A different mechanism for transforming the source material into the gas phase is used in sputtering, namely ejection of source atoms by mechanical momentum transfer from ions bombarding the source.) The incidence rate of vapor atoms onto the substrate is again given by Eqs. (2a) and (2b) where p now denotes the vapor pressure of the evaporating material and T the temperature of the source. It is easy to see that a vapor pressure of at least 10^{-4} Torr is necessary for a reasonable film growth rate of several monolayers per second. For a majority of the elements this vapor pressure occurs at temperatures above the melting points. A well-designed evaporation source allows the evaporant material to reach the high temperature without excessive heating of the surroundings.

Vapor pressures are strongly temperature-dependent through Clausius–Clapeyron's equation

$$d \log p / dT = \Delta H / RT^2 \tag{3}$$

where ΔH is the molecular heat of evaporation. The resulting exponential temperature dependence of the vapor pressure p calls for a very constant source temperature or for a fine control of power input if the deposition rate is to be constant and reproducible.

Vapor pressure data and curves have been compiled by Honig (1962) and are also discussed by Glang (1970), Holland (1956), Chopra (1969), and Dushman (1962). Kubaschewski et al. (1967) have listed heats of evaporations and related thermochemical quantities. It is interesting that precise measurements of vapor pressures are very important for a complete thermodynamic characterization of materials. However, for evaporation of films, the vapor pressure data are only used as order-of-magnitude guides, and the actual deposition parameters are often optimized by trial and error.

Equations (2) give the incidence rate on a substrate placed in a volume uniformly filled with vapor or on a substrate facing a large area of a freely evaporating surface. In an actual evaporator the vapor sources and evaporating areas are quite small and vapor distributions are not uniform. For the limiting case of a point source, this leads to a dependence of the impingement rate on the inverse square of the distance between source and substrate and for a small evaporating area one obtains a distribution proportional to the cosine of the angle between the normal of the emitting area and the direction to the substrate. Deposit distributions on large substrates can be derived from these elements and have also been worked out for various source configurations (point source, strip source, ring source) by Glang (1970) and Holland (1956).

Finally there are kinetic effects that can reduce the deposition rate below the one expected from Eqs. (2). This is particularly the case for free evaporation from a solid surface into vacuum (sublimation). The equilibrium vapor pressure is always established in a Knudsen source, which consists of a uniformly heated enclosure of the evaporant with a small opening acting analogous to a black body radiation source. However, with free evaporation there is no enclosure that contains the vapor and allows the equilibrium pressure to build up. It is now possible that the escape rate of atoms from the surface is not only controlled by the energy of evaporation but also by the supply of atoms into suitable desorption sites, i.e., by the generation of surface steps and by surface diffusion. These effects have been treated in detail by Hirth and Pound (1963).

V. Evaporation Sources

The practical design of vapor sources is dictated by the need to concentrate the thermal energy at the source material and to avoid excessive heating of the surrounding. Power is transmitted to the evaporant by several methods. The simplest one is direct resistance heating of a small boat, strip, wire, or basket of a refractory material (W, Mo, Pt, Ta) to which a small amount of evaporant is attached. Larger charges, a few grams or more, are heated in crucibles of refractory oxides, boron nitride, or carbon by resistance heating or by rf-induction heating. It is important to choose crucible or support materials that do not react with the evaporant.

Finally, a very localized power input is achieved in an electron bombardment source where a beam of electrons is accelerated through 5–10 kV and focused onto the evaporant. The bulk of the source material rests on a water cooled hearth, and this greatly reduces chemical reaction with the support.

Most of the practical source designs are described in detail by Glang (1970).

The undesirable heat loss from the vapor source to the surrounding vacuum chamber occurs mainly through radiation and can be reduced by heat shields. This is particularly important for resistance heated crucibles. Crawford (1972) pointed out that good thermal insulation requires crowding the maximum amount of shields into the minimum amount of space. This design principle has not always been appreciated.

VI. Multicomponent Evaporation

All of the considerations above are valid for the evaporation of elements and of a few very stable compounds (e.g., halides, some oxides, and chalcogenides). Most compounds and alloys evaporate incongruently because their components have different vapor pressures. Upon heating, fractionation occurs with the more volatile component evaporating first. This produces films with compositions different from that of the source or even with graded compositions.

In principle the composition changes in alloys can be predicted from thermochemical data. The vapor pressure of component B of a binary alloy in equilibrium with the alloy is

$$P_B = f_B x_B P_B \tag{4}$$

where x_B is the molar fraction of component B, P_B its vapor pressure in equilibrium with pure B, and f_B the activity coefficient of B, an empirical parameter that characterizes the deviation of the alloy from ideal solution behavior. Activity coefficients are tabulated by Hultgren et al. (1963), while in Kubaschewski and Evans (1965) several rules are given to estimate activities from phase diagrams.

Although the composition of the vapor and the film deposit can be calculated from Eq. (4) for a given alloy source and the source composition can be adjusted to yield a desired film composition, the direct evaporation of alloys is usually not satisfactory and homogeneous films are not easily obtained. The remarkable exception is permalloy (Fe + 80% Ni), where the vapor pressures of Fe and Ni and the values of the activity coefficients happen to be just right to ensure a vapor composition equal to the source composition in the temperature range of 1500 to 1800°C. In other cases it is better to use independently heated vapor sources for each component.

A different technique is flash evaporation. Here large-scale fractionation is avoided by complete evaporation of small particles. The particles are dropped, one after the other, onto a filament that is hot enough to vaporize the least volatile component. Although fractionation occurs during the evaporation of each particle, gradients in the deposit are averaged out because there are always several particles in different states of fractionation present on the filament. A drawback of flash evaporation is the deterioration of the vacuum caused by oxide skins or gases absorbed on the surface of the powder particles. Flash evaporation was successfully used by Richards (1966) for epitaxial deposition of III–V compounds.

A variation of the concept of completely evaporating small amounts is the steady-state evaporation described by Dale (1969, 1973). A supply of liquid alloy (InSb, BiSb) is kept in an enclosed reservoir and fed with a constant flow rate through a capillary to a very hot evaporation area from where the evaporation rate is equal to the feed rate. Within the accuracy of the chemical analysis (1% or better) a 1 : 1 transfer of the source composition to the film was achieved in these cases.

VII. Substrates and Accessories

The final requirement for evaporation is a substrate on which the vapor can condense. For epitaxial films the nature and structure of the substrate is, of course, an essential part of the whole subject and numerous individual combinations of substrate crystals and overgrowth are possible. Only a few general aspects apply to all of them. One is the necessity for the substrate temperature to be adjustable, since nucleation and growth phenomena and also epitaxial relationships depend on temperature. Substrate temperatures are regulated either by direct contact with resistance heaters or by radiation (often from a quartz iodine lamp).

Most important is the preparation of the substrate surface. The methods of polishing or cleaving depend again on the particular substrate material, but in all cases it is very important that the substrate be free from contamination. A successful practice is to prepare the substrate surface in vacuum immediately prior to deposition. One can grow the substrate crystal as an epitaxial film on another crystal. For example, Ag single-crystal films grown on mica (Pashley, 1959) can be used as substrates for Au.

Some substrate crystals can be cleaved in the vacuum chamber so that the newly created surface is immediately exposed to the previously established vapor stream. This method (introduced by Walton et al., 1963; Ino et al., 1964; Sella and Trillat, 1964) has been used extensively for the deposition of metal films on alkali halides. Where cleavage in situ is not possible, a shutter should be placed in front of the substrate to protect it from contamination and uncontrolled deposition while the source is heated to the evaporation temperature.

It is often important to grow a film with a specified deposition rate or to a predetermined thickness. Several methods have been described to measure deposition rate and to control them via feedback loops to the source temperature (Glang, 1970; Chopra, 1969). They either sense the density of the vapor stream or weigh the mass that is deposited on a reference substrate. Most prominent in the former category are ionization gauges with their sensitive volume exposed to the vapor, while in the latter quartz crystal oscillators now dominate over mechanical microbalances. The resonant frequency of a quartz crystal changes when mass is attached to its surface and the frequency shift can be measured with sufficient accuracy to detect monolayer deposits.

To obtain thicknesses or growth rates on the film substrate the monitor devices have to be calibrated by an independent measurement of the thickness of a film. For thickness measurements many methods are again available, the most popular ones being based on optical interference phenomena. They have been reviewed by Pliskin and Zanin (1970).

References

Chopra, K. L. (1969). "Thin Film Phenomena." McGraw-Hill, New York.

Crawford, C. K. (1972). *J. Vac. Sci. Technol.* **9**, 23.

Dale, E. B. (1969). *J. Vac. Sci. Technol.* **6**, 568.

Dale, E. B. (1973). *J. Appl. Phys.* **44**, 1101.

Dushman, S. (1962). "Scientific Foundations of Vacuum Technique," 2nd ed. Wiley, New York.

Glang, R. (1970). *In* "Handbook of Thin Film Technology" (L. I. Maissel and R. Glang, eds.), Chapter 1. McGraw-Hill, New York.

Glang, R., Holmwood, R. A., and Kurtz, J. A. (1970). *In* "Handbook of Thin Film Technology" (L. I. Maissel and R. Glang, eds.), Chapter 2. McGraw-Hill, New York.

Hirth, J. P., and Pound G. M. (1963). "Condensation and Evaporation, Nucleation and Growth Kinetics." Macmillan, New York.

Holland, L. (1956). "Vacuum Deposition of Thin Films." Wiley, New York.

Honig, R. E. (1962). *RCA Rev.* **23**, 567.

Hultgren, R., Orr, R. L., Anderson, P. D., and Kelley, K. K. (1963). "Selected Values of Thermodynamic Properties of Metals and Alloys." Wiley, New York.

Ino, S., Watanable, D., and Ogawa, S. (1964). *J. Phys. Soc. Japan* **19**, 881.

Kubaschewski, O., Evans, E. L., and Alcock, C. B. (1967). "Metallurgical Thermochemistry." Pergamon, Oxford.

Pashley, D. W. (1959). *Phil. Mag.* **4**, 324.

Pliskin, W. A., and Zanin, S. J. (1970). *In* "Handbook of Thin Film Technology" (L. I. Maissel and R. Glang, eds.), Chapter 11. McGraw-Hill, New York.

Richards, J. L. (1966). *In* "The Use of Thin Films in Physical Investigations" (J. C. Anderson, ed.), p. 419. Academic Press, New York.

Sella, C., and Trillat, J. J. (1964). *In* "Single Crystal Films" (M. H. Francombe and H. Sato, eds.), p. 201. Pergamon, Oxford.

Walton, D., Rhodin, T. N., and Rollins, R. W. (1963). *J. Chem. Phys.* **38**, 2698.

2.2   Molecular-Beam Epitaxy[†]

L. L. Chang and R. Ludeke

IBM Thomas J. Watson Research Center
Yorktown Heights, New York

I. Introduction

The term "molecular-beam evaporation" has been used loosely in the literature. There are works using this name inappropriately; yet others, using in essence this technique, are called otherwise. The confusion arises because of the lack of, and the difficulty in arriving at, a unique definition. In one extreme, molecular-beam evaporation is thought of as the deposition in a vacuum system resulting from the condensation of molecular vapors however created. Almost any technique of evaporation could then be classified as that of molecular beam. In the other extreme, the term is thought to be applicable only when the solid source undergoes molecular vaporization without dissociation. Few compound materials would strictly satisfy such vaporization behavior.

A molecular beam is defined as a directed ray of neutral molecules or atoms

† Research sponsored in part by the Army Research Office, Durham, North Carolina.

in a vacuum system. The beam density is low and the vacuum high so that no appreciable collisions occur among the beam molecules and between the beam and the background vapor. The beam is usually produced by heating a solid substance contained in an effusion cell. The orifice dimension of the cell is small compared to the mean free path of the vapor in the cell so that flow of the molecules into the vacuum chamber is by effusion (Kennard, 1938). Quasi-equilibrium exists in the cell so that both the vapor composition and the effusion rates of the beam are constant and are predictable from thermodynamics, in contrast to the case of free evaporation (Somorjai and Lester, 1967). The beam is guided by the orifice and possibly by other slits and shutters onto a substrate where the situation is usually far from equilibrium. Under proper conditions, governed mainly by kinetics, the beam would condense resulting in nucleation and growth. This process of film deposition is here defined as molecular-beam evaporization. Needless to say, molecular-beam epitaxy (MBE) requires that the film be epitaxic; and as the title implies, only such films are of interest in this review. To take full advantage of the MBE technique, it is essential to have an ultrahigh vacuum system to provide a clean environment, and a spectral mass analyzer to monitor the environment in addition to the component vapor species of interest. It is further desirable to have in situ evaluation methods, such as electron diffraction, scanning electron microscopy, and Auger spectroscopy, the selection depending on the objective of the investigation. It should be understood that the above definition of MBE together with the various desirable features was given to serve mainly as future guidelines. In considering practical experiments or reviewing work in the literature some of the requirements would have to be relaxed.

The development and utilization of the molecular-beam method in the study of molecules and atoms were pioneered by Dunoyer (1911) and Stern (1920). Discussion of this method in general and its use in uncovering solid–vapor scattering processes in particular has been given by Ramsey (1956) and Stickney (1967), respectively. The origin of its use in the epitaxial growth of thin films is difficult to assess partly because the MBE technique had not been previously defined and partly because the experimental conditions in early works were usually not clearly specified. The materials under consideration in this work are the II–VI, III–V, and IV–VI compound semiconductors. Schoolar and Zemel (1964) were perhaps the first to grow epitaxial PbS on NaCl by use of effusion cells, although Elleman and Wilman (1948) had succeeded earlier in preparing the same material by simple vacuum sublimation. Using multiple molecular beams, Miller and Bachman (1958) grew CdS and Gunther (1958, 1966) described in general the growths of both II–VI and III–V materials. However, it was not until later that epitaxy was achieved for CdS by Zuleeg and Senkovits (1963) and for GaAs by Davey and Pankey (1964, 1968) by use of variously modified-beam techniques. Arthur

(1968a, b) first reported the kinetic behavior of Ga and As on GaAs that led to the understanding of the growth mechanism. Subsequently, epitaxial films of GaAs and related compounds of superior quality and extreme smoothness have been achieved by the molecular-beam technique by Arthur and Lepore (1969), Cho (1971a), Cho et al. (1970), Esaki et al. (1972), and Chang et al. (1973). In fact, it is this series of successes that has aroused the current interest in what is now known as MBE.

There are a number of features of MBE that are generally considered advantageous in growing semiconducting films. The growth temperature is relatively low, which minimizes any undesirable thermally activated processes such as diffusion. The growth rate is relatively slow, which makes possible precise thickness control. The introduction of various vapor species to modify the alloy composition and to control the dopant concentration can be conveniently achieved by adding different beam cells with proper shutters. These features become particularly important in making structures involving junctions, which are of primary technological interest in semiconductor materials. It is obviously due to such interest that semiconductors become the subject of this review. We choose to discuss compound semiconductors not only because of their wide range of properties that attract an ever-growing attention, but also because of the complication in achieving stoichiometry, which requires special consideration. The growth of films of elemental semiconductors such as Ge and Si is generally simple in comparison, and may be readily achieved by conventional evaporation techniques. Of special interest, it should be mentioned, is the growth of epitaxial Si from a molecular beam of silane, first reported by Joyce and Bradley (1966).

In what follows, a typical MBE system is first described. Since such a system is basically a vacuum evaporation system, the subject of another chapter in this book, only features pertinent to the molecular-beam technique will be described. The vaporization behavior of the compound semiconductors, which determines the composition and rates of the effusing species, is discussed in some detail. GaAs, CdTe, and PbTe are taken, respectively, as representatives of the III–V, II–VI, and IV–VI compounds. The deposition processes are described next in terms of the condensation characteristics for stoichiometric growth of the compounds of interest. Epitaxy and nucleation, treated in a later chapter of this book, are more or less taken for granted. Finally, experimental results of growing these compound films are reviewed, and their properties briefly described.

II. Molecular-Beam Epitaxy System

A molecular-beam epitaxy system is basically a vacuum evaporation apparatus. The many fundamental aspects of constructing and operating an evaporator have been discussed in various articles, for example, the classic text by Holland (1958). Since the purity and thus the quality of films to be deposited in an evaporator are directly related to the frequency of their reaction with the background gaseous species, systems with ever improving vacuum are always in demand. Both the knowledge and the technology of high vacuum have advanced greatly in recent years. The developments of new materials and components and of clean and efficient pumps have made available ultrahigh vacuum systems with pressure $<10^{-9}$ Torr. Review articles devoted to these subjects have been written by Caswell (1963) and by Glang *et al.* (1970).

What may be considered a standard MBE system is shown schematically in Fig. 1. This is based essentially on the system used in the authors' laboratory for the growth of semiconducting gallium arsenide films (Chang *et al.*, 1973). The system is pumped by a combination of ion pumps and a titanium sublimation pump, the latter being enclosed in a liquid-nitrogen shroud. The source ovens are resistively heated effusion cells made usually of materials such as graphite or boron nitride. Shown in the figure are two ovens, although as many as six have been incorporated. In the growth of compound semiconductors, as will be seen later, the materials contained in the ovens could be the compounds themselves, their components, or different elements to be used as doping impurities. Individual thermocouples are embedded in the ovens. Shutters are provided for each source, and the entire assembly is surrounded

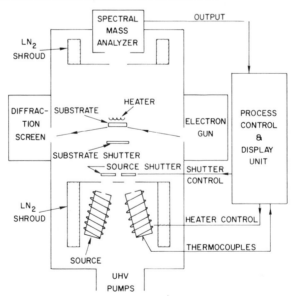

FIG. 1. A schematic diagram of a molecular-beam epitaxy system.

by another liquid nitrogen shroud as shown. The substrate holder is located along the center line of the system, a few inches away from the openings of the ovens. The substrate is usually a monocrystalline material that has been cleaned, polished, and etched. It may or may not be the same material as that to be deposited, depending on whether homoepitaxy is of interest. The substrate, during deposition, is kept at elevated temperatures, which are usually necessary for epitaxial growth. It can also be heated before deposition primarily for cleaning and afterwards for various heat treatments.

In addition to sources and substrate, which are necessary parts in any evaporation system, monitoring, analyzing, and controlling equipment should be considered essential in a modern MBE system. As shown in Fig. 1, a spectral mass analyzer is located on axis at the top of the chamber, behind a liquid nitrogen shroud. It is capable of detecting the vaporizing as well as residual atoms or molecules and thus monitors both the thickness of the film and the environment under which the film is grown. The control unit could consist of simple analog controllers that correct the temperatures of the source cells based on the thermocouple readings. It could also be a sophisticated, digital computing system when a high degree of precision in both composition and thickness is required. Based on the actual rates sensed by the mass analyzer, the computing system not only controls the cell temperatures but also commands the shutter operation. The details of such a control system are in the original article describing the growth of a semiconductor superlattice (Chang *et al.*, 1973a; Esaki and Tsu, 1970). For in situ examination of the growth, the apparatus further incorporates a scanning high-energy electron diffraction system, employing the small glancing angle, reflection mode of operation. The diffraction patterns can be viewed on a screen for qualitative information (HEED) or their intensities recorded by magnetically scanning the diffracted electrons that pass through an aperture in the screen (SHEED). These patterns, as discussed in many texts and articles (Holloway, 1966), contain a great deal of information about the crystalline structure, the smoothness, and, in some cases, the reconstructed surface structures under various experimental conditions.

The entire vacuum system is bakeable to 250°C. After an overnight bakeout the system with all the components installed as described above reaches 2×10^{-10} Torr, using only the main ion pump. The spectrum of residual gas species is shown in Fig. 2 for masses up to 50, beyond which no additional peaks are observable on the sensitivity scales shown. The overall transmission efficiency of the quadruple mass analyzer used is quite uniform within this mass range, so that the spectrum displays directly the relative abundance of the various residuals. The major species are CO at mass 28, H_2O at 18, and H_2 at 2. Of secondary importance are CH_4 at 16, CO_2 at 44, and peaks attributed to cracking at mass numbers 12 and 17. With liquid-nitrogen cooling, both

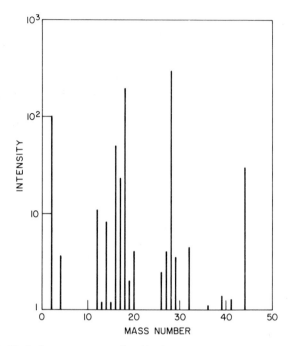

FIG. 2. Typical mass spectrum of residual gaseous species in the MBE system.

CO_2 and H_2O drop drastically; however, carbon monoxide remains the dominant impurity and is difficult to remove from the system.

During operation of the MBE system, the quantities of most interest are the arrival rates at the substrate of the various vapor species. The arrival rates together with the substrate temperature determine the surface conditions under which the film is deposited. These quantities, in conjunction with the sticking coefficients, control both the growth rate and the doping concentration. The effusion rates can be determined from the partial pressures using the Knudsen–Langmuir equation (Knudsen, 1909; Langmuir, 1913) modified by Clausing (1932) and Motzfeldt (1955). The arrival rates can then be estimated simply from the cosine law of distribution if secondary effects such as adsorption and reevaporation can be neglected (Ruth and Hirth, 1964).

As an example, let us consider a semiconductor with a mass of 50 and a density of 5 gm/cm^3 in a system having a source–substrate distance of 5 cm and a source orifice of 5×10^{-2} cm^2. An equilibrium vapor pressure of 10^{-5} atm at 1000°K would produce an arrival rate of the order of 10^{15} cm^{-2} sec^{-1}, which corresponds to a growth rate of ~ 1 Å sec^{-1} if a unit sticking coefficient is assumed. Under similar conditions, a dopant source maintained at 10^{-10} atm would result in a concentration of $\sim 5 \times 10^{17}$ cm^{-3}. The temperatures necessary to achieve these rates largely determine the design of the MBE system and thus the feasibility of the MBE technique in preparing a certain material.

III. Vaporization of Binary Semiconductors

The phase relations for a binary system, M and X, can be completely specified by a three-dimensional p (pressure)–T (temperature)–x (composition) diagram, where the composition refers, to be specific, to the atom fraction of the nonmetallic element X. The information is usually presented in two-dimensional, T–x and p–T projections. It is such relations that determine the conditions for melt and solution growth, and serve as guidelines for vapor growth of the material. In addition to their importance in crystal preparation, phase equilibria considerations are of extreme value for heat-treatment experiments. The compositional defect state of the material is specified by such considerations and thus governs its properties either directly or indirectly through interactions with other impurities.

For film deposition by the MBE technique, phase information applies to the source where thermal equilibrium is maintained. The T–x diagram is of interest mainly because it continuously specifies the state of the source. However it is the p–T diagram that predicts the vapor pressures and thus the effusion rates of the various species. For the compound semiconductors under consideration, vaporization upon heating of the materials proceeds in general according to the reaction

$$MX(s) \rightarrow (1-\alpha)\left[M(g) + \frac{(1-\beta)}{2}X_2(g) + \frac{\beta}{4}X_4(g) \right] + \alpha MX(g)$$

where α is the fraction of molecular vaporization, and β the fraction of dissociation into tetramer species. The constant for each reaction at a given temperature can be obtained, and thus the vapor pressure of each species determined. The different kinds of compound semiconductors have quite different behavior of vaporization and are discussed separately in the following sections. The variety of behavior obviously makes the discussion applicable to a great many other nonsemiconducting compounds. At the substrate where the situation is far from equilibrium, as mentioned above, the phase information cannot be applied directly. It serves, however, as a useful guideline for the deposition conditions when specific kinetic information is lacking, as in most cases.

A. GALLIUM ARSENIDE AND OTHER III–V COMPOUNDS

Gallium arsenide, because of its many desirable electrical and optical properties, is the most studied and understood material of all the compound semiconductors. The T–x and p–T diagrams of the Ga–As system are plotted in Fig. 3. Figure 3a shows the $T-x$ or condensed phase diagram with a single compound formation corresponding to GaAs. The liquidus curve gives the temperature at which a liquid of a given composition is in equilibrium with a solid and a vapor, which is not explicitly shown in this type of diagram. The dotted curve near $x = 0.5$ represents the solidus, which bounds the region of nonstoichiometry. This region has been greatly exaggerated to show its existence, although it is much narrower and is not precisely known. The region is, in general, not symmetrical with respect to the $x = 0.5$ line, depending on the statistics of the formation of vacancies. In fact, it may lie completely off the stoichiometric composition in some compounds. Figure 3a has been constructed by Thurmond (1965) by thermodynamically analyzing the solubility data obtained by Hall (1963) and Koster and Thoma (1955). Some data on the solidus curve from measurements of the lattice constant of annealed GaAs have been reported by Straumanis and Kim (1965) and by Potts and Pearson (1966).

The $p-T$ diagram of the Ga–As system is shown in Fig. 3b where the partial pressures are usually plotted semilogarithmically versus reciprocal temperature. These loops are generally referred to as "three phase lines," since solid, liquid, and vapor coexist along the loops. Taking the species As_4 as an example, the upper and lower legs of the loop give As_4 pressures for GaAs in equilibrium with an As-rich liquid, and with a Ga-rich liquid, respectively. As the temperature is decreased, the As_4 pressure represented by the upper leg approaches that in equilibrium with pure As; and that represented by the lower leg approaches the hypothetical pressure of As if it were in equilibrium with pure Ga. The vapor pressure curves for the Ga–As system were first constructed by Thurmond (1965) using the data of Drowart and Goldfinger (1958), Gutbier (1961), and Richman (1963). Later, Arthur (1967) obtained new spectrometric data of the vapor pressures, reporting a significant increase of the As_2 pressure with respect to that of As_4. Additional vapor pressure measurements have been reported by De Maria et al. (1970) and Lou and Somorjai (1971). The plot shown in Fig. 3b follows that of Arthur (1967).

It is seen from the figure that species of Ga and of both As_2 and As_4 coexist and that the pressure of As (As_2 and As_4) lies above the pressure of Ga over most of the temperature range shown. Upon heating GaAs in a Knudsen cell in this range, only the pressures labeled Ga-rich are of interest, since the volatile As always vaporizes first leaving a Ga-rich solution in the cell. The reaction of vaporization, neglecting molecular vaporization by setting $\alpha = 0$ in the general equation, becomes

$$GaAs(s) \rightarrow Ga(g) + \frac{1-\beta}{2} As_2(g) + \frac{\beta}{4} As_4(g)$$

A Ga-rich liquid phase is always formed, but both the pressures of As and Ga

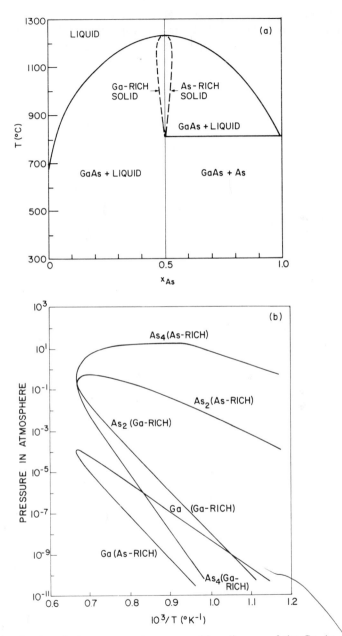

FIG. 3. (a) Condensed phase temperature-composition diagram of the Ga–As system (after Thurmond, 1965). (b) Equilibrium partial pressures of the Ga–As system (after Arthur, 1967).

remain constant at a given temperature as long as a small percentage of As is maintained in the liquid, as determined from Fig. 3a. The contribution to the total As pressure from As_4 is small compared to that from As_2 except at very high temperatures, and will be neglected later for simplicity of presentation. At very low temperatures, i.e., below 637°C in Fig. 3b where the pressure of Ga equals that of $2As_2 + 4As_4$, congruent vaporization or sublimation occurs. This will be considered in more detail in the II–VI compounds, because the low vapor pressures involved here make it of little practical interest.

Other III–V compound materials are expected to behave similarly, although precise $T-x$ and $p-T$ curves are generally lacking. We refer to the review article by Weiser (1962) and the references therein. Recent works on the vaporization of these compounds include GaP by Thurmond (1965), AlAs by Hoch and Hinge (1961), AlP by De Maria *et al.* (1968), and InP by Panish and Arthur (1970). Hildenbrand and Hall (1964) have studied the nitride of boron and of aluminum, and Thurmond and Logan (1972) that of gallium. For the nitride compounds, the nitrogen vapor consists of only the dimer species.

B. Cadmium Telluride and Other II–VI Compounds

The compound CdTe is taken as representative of the II–VI materials. The $T-x$ and $p-T$ diagrams of the Cd–Te system are shown in Fig. 4a and 4b,

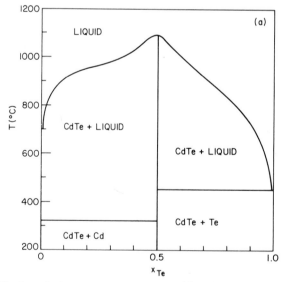

Fig. 4. (a) Condensed phase temperature–composition diagram of the Cd–Te system (after Jordan, 1970). (b) Equilibrium partial pressures of the Cd–Te system. Dotted lines indicate the pressures of Cd and Te_2 under conditions of congruent vaporization (after Jordan and Zupp, 1969b).

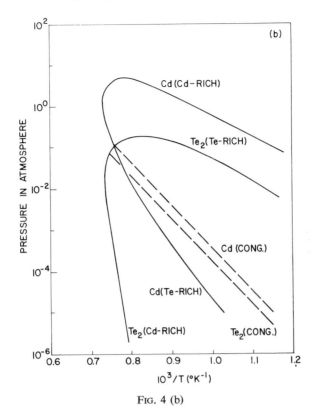

FIG. 4 (b)

respectively. These curves have been calculated by Jordan (1970) and Jordan and Zupp (1969b) by assuming a regular associated solution and using primarily the experimental data of Kulwicki (1963). Other data by De Nobel (1959), Lorenz (1962), and, most recently, Brebrick (1971) are in general agreement with the calculated results. A narrow solidus region, not shown in the figure, again exists at $x = 0.5$ as in the case of GaAs. The vapor species, unlike GaAs, contain the metallic atoms and only the diatomic molecules of the nonmetallic element.

It is noticed from Fig. 4b that over most of the temperature range the loops of the partial pressures of the two species overlap. On heating the compound, the solid may shift its composition such as to equal its vapor composition without a liquid phase being formed. The total pressure of the system becomes a minimum in this case. This fact, together with the equation of vaporization

$$CdTe(s) \rightarrow Cd(g) + \tfrac{1}{2}Te_2(g)$$

determines the partial vapor pressures. The situation is known as *congruent vaporization* or *sublimation*. The pressures are shown as dotted lines in Fig. 4b, with the pressure of Cd being twice that of Te_2 for a narrow solidus region. In the case where CdTe is used as a source in a Knudsen cell, the result is slightly modified. The steady state is reached for the effusion rates rather than for the pressures of the two species. From the Knudsen–Langmuir equation taking into account the difference in their masses, the ratio of the pressures in the cell becomes 1.33 instead of 2.

For the liquidus curves and partial pressures of the other II–VI compounds, we refer in general to the classic work by Goldfinger and Jeunehomme (1963) and a review article by Lorenz (1967). ZnTe has been studied in detail by Jordan (1970), Jordan and Zupp (1969a), Brebrick (1971), and Lee and Munir (1967), and CdSe, most recently, by Sigai and Wiedemeier (1972). Floegel (1969) has reported data on selenium compounds and Boev *et al.* (1969) on Zn chalcogenides. A series of articles has been published recently by Munir (1970) and co-workers (Munir and Mitchell, 1969; Mitchell and Munir, 1970; Seacrist and Munir, 1971) on ZnS, CdS, HgS, and CdSe. HgSe was studied by Brebrick (1966) and HgTe by Brebrick and Strauss (1965).

C. Lead Telluride and Other IV–VI Compounds

The IV–VI compounds, for which PbTe has been the most widely studied, represent another important group of semiconductors. The characteristics of

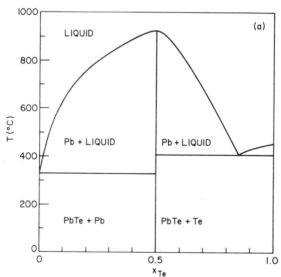

Fig. 5. (a) Condensed phase temperature–composition diagram of the Pb–Te system (after Hultgren *et al.*, 1963). (b) Equilibrium partial pressures of the Pb–Te system. Dotted lines indicate the pressures of Pb and Te_2 under conditions of congruent vaporization (after Brebrick and Strauss, 1964a).

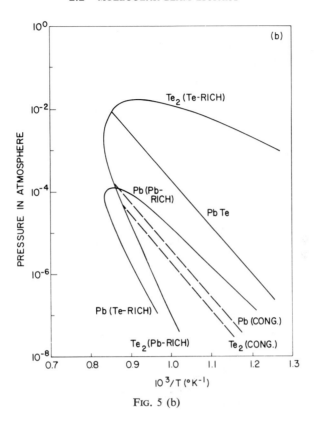

FIG. 5 (b)

vaporization show yet another different behavior. Figure 5 shows the T–x and p–T diagrams of the Pb–Te system. The liquids curve is taken from that constructed by Hultgren *et al.* (1963). Although the solidus region is still too small on the scale and can only be drawn as a line at $x = 0.5$, it is generally larger than that of the other compounds discussed earlier. Furthermore, since electrical conduction in this kind of material is usually dominated by the degree of nonstoichiometry rather than by the concentration of foreign impurities, details of the solidus region and its associated vapor pressures can thus be determined from electrical measurements (Brebrick and Gubner, 1962; Fujimoto and Sato, 1966). SnTe and GeTa are extreme examples with a remarkably wide solidus region (Brebrick and Strauss 1964a, b), the former exhibiting in addition the unusual property of having this region completely in the Te-rich side.

A glance at the vapor pressure curves shown in Fig. 5b for the Pb–Te system indicates that they are somewhat similar to those in Fig. 4b. Both Pb and Te$_2$ vapors exist, and their pressure loops overlap over most of the temperature range. There is an important difference, however. The molecular species, PbTe, is also present, having a significant vapor pressure relatively independent of composition. Hence the compound upon heating undergoes congruent sublimation with pressures of Pb and Te$_2$, shown as dotted lines in Fig. 5b, smaller than the pressure of PbTe. The equation of vaporization with $\alpha = 1$, becomes simply

$$PbTe(s) \rightarrow PbTe(g)$$

The compound continues to vaporize molecularly as long as congruent sublimation can be maintained. Such behavior, similar to that of an elemental material, explains the wide use and the early success of the evaporation technique in preparing stoichiometric PbTe films (Elleman and Wilman, 1948; Schoolar and Zemel, 1964).

The vapor pressure curves shown in Fig. 5b are based on those obtained by Brebrick and Strauss (1964a) using partly the data by Pashinkin and Novoselova (1959). The curves were completed by using the congruent vapor pressures and the fact that they represent the minimum of the total pressures. Other vapor pressure data on PbTe have also been reported by Fujimoto and Sato (1966) and most recently by Hansen and Munir (1970), who worked on single-crystal materials. In other compounds such as SnTe, however, the pressure loop of Te$_2$ is above that of Sn so that congruent sublimation is not possible (Brebrick and Strauss, 1964b). The source upon heating would progressively become Sn-rich similar to the behavior of GaAs. However, unlike GaAs, the molecular vapor of SnTe remains existent and dominant. For the various compounds in this group of materials, one is referred to the review article by Strauss and Brebrick (1968) that summarized all the information regarding the phase diagrams and the vapor pressures. Works since then include those by Blair and Munir (1970a, b) on PbS and SnSe, Hansen *et al.* (1969) on PbSe, and Pashinkin *et al.* (1969) on GeSe.

IV. Deposition Process

While the arrival rates of various species can be predicted from the geometry and vapor pressures of the sources, film growth is governed by a series of events in a deposition process that is best described by a kinetic formulation. Briefly, these events are: adsorption or capture of the impinging molecules, surface diffusion in which the absorbed molecules spend a characteristic residential time on the surface, and finally association of the molecules resulting in nucleation and growth. The theory of nucleation has been treated in various works and is presented in Chapter 4 of this volume. It is sufficient here to give a brief account. The adsorbed molecules first combine to reach a critical size and become stable nuclei on the substrate. The critical size depends on a variety of factors, including the substrate temperature, the arrival rates, and the affinity with the substrate. Subsequent molecules may condense directly onto these nuclei, thereby decreasing the rate of formation of new nuclei and increasing the size of existing nuclei to form islands that may coalesce upon contact. After a continuous film has been established by filling in voids between the net of contacted islands, the process of film growth is believed to be by direct molecular condensation onto low-energy sites. Impurities, kinks on an atomic ledge, dislocations, and grain boundaries may provide such suitable sites. The growth process at the initial stage, therefore, must be distinguished between the heteroepitaxial and homoepitaxial cases. Nucleation in the latter case is equivalent to continuous film growth, provided the substrate is atomically clean.

The situation is schematically illustrated in Fig. 6, where deposition rate is plotted against arrival rate at a given substrate temperature. Figure 6a

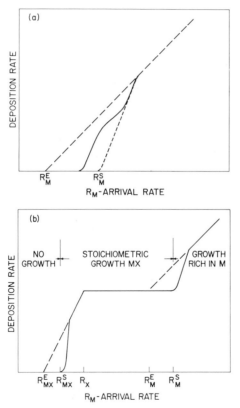

FIG. 6. Schematic plot of deposition rate versus arrival rate of component M for (a) one-component system and (b) two-component system with a fixed arrival rate of component X. Superscripts E and S indicate reevaporation and supersaturation, respectively.

represents the simple case of a single component M. The dashed line indicates the asymptotic deposition rate, which is applicable to homoepitaxial growth at all stages and to heteroepitaxial growth after completion of nucleation. The intercept R_M^E is the reevaporation rate at the given substrate temperature. Supersaturation, represented by an arrival rate $R_M^S > R_M^E$, is usually necessary to initiate nucleation for heteroepitaxial growth on an ideal crystalline substrate, as shown by the dotted line in the figure. The intermediate solid curve indicates the case where other low-energy sites, such as surface impurities, are available for nucleation but have a density less than the nuclei density for an ideal substrate. The situation for a system having two components to form compound MX is shown similarly in Fig. 6b. A fixed arrival rate of R_X is used that is smaller than R_X^E, the reevaporation rate of X itself, but larger than R_{MX}^E, the reevaporation rate of the compound. Nucleation starts when R_M exceeds the supersaturation value R_{MX}^S. The rate of deposition is proportional to R_M upon completion of nucleation and becomes a constant when it is limited by the arrival rate of R_X. Only when R_M reaches the critical value of R_M^S, greater than its own reevaporation rate R_M^E, does the deposition rate again increase, resulting in M-rich growth. It is clear that, under proper conditions, stoichiometry can be achieved as indicated in Fig. 6b.

This discussion would serve as a general guideline for growing a stoichiometric binary compound. However, in actual situations, the deposition is ultimately determined by the kinetic reaction at the substrate surface. The overall process of adsorption, diffusion, and growth can be described by a parameter known as the *sticking coefficient*. It is defined for each species as the fraction of the total impinging molecules that sticks to the surface and is incorporated into the film. *Condensation coefficient* is sometimes also used for this quantity, although, strictly speaking, it should refer to the fraction that is immediately captured by, or not specularly reflected from, the surface. There are various theoretical treatments of this quantity, for example, that given by Eyring *et al.* (1964). It is, however, usually determined experimentally under specified conditions for either homoepitaxial or prenucleated heteroepitaxial deposition.

Within the realm of the present work, we are mainly concerned with the sticking coefficients of the various vapors and their interrelationship in growing stoichiometric compound semiconductors of the different kinds, as discussed in the previous section. The simplest case is that of molecular vaporization, such as PbTe mentioned earlier. To grow PbTe on a NaCl substrate, which is usually used, the sticking coefficient is not accurately known but is believed to be of the order of unity. Its precise value, in this case, would only modify somewhat the growth rate. Stoichiometry, however, is generally assured if the compound is used as the source. There are secondary effects due to the component Pb and Te_2 vapors as can be seen from Fig. 5b.

Depending on their precise sticking coefficients, these vapors, together with those from additional sources containing the component elements, can be used for fine control of the film composition.

The general principle described earlier for a binary system in Fig. 6b is directly applicable to the deposition of a typical II–VI compound. This group of materials, as has been seen, is usually characterized by congruent evaporation, having comparable partial pressures of the two components. CdSe, for which Gunther (1966) and Junge and Gunther (1963) have made detailed studies, is used for illustration. The sticking coefficients on prenucleated substrates of both Cd and Se are determined as a function of their respective arrival rates and the substrate temperature. The results are shown in Fig. 7a. For an arrival rate of $5 \times 10^{16}/cm^2$ sec, for example, no sticking would occur for Cd above 200°C and for Se above 170°C. At the higher temperature both vapors, impinging simultaneously, would stick to some extent to form the compound. A plot similar to Fig. 6b in this case is shown in Fig. 7b for a fixed rate of Cd of $2 \times 10^{16}/cm^2$ sec and two substrate temperatures. At 160°C, the film is rich in Cd at low Se rates and rich in Se at high Se rates. Only in a very narrow region in the vicinity of a Se rate of $5 \times 10^{16}/cm^2$ sec, is the film stoichiometric. The situation at 200°C is quite different, however. There is Se-rich growth above a rate of $10^{17}/cm^2$ sec as before, but no growth at all below a rate of $10^{16}/cm^2$ sec. A wide region exists in between, which is the region of stoichiometry. The maximum deposition rate in this region corresponds to a sticking coefficient for CdSe of ~ 0.25.

One can repeat such experiments at each temperature by varying the Cd rate and plot regions of stoichiometry using the rate of the components as the coordinates, as has been done by Gunther (1966). A difficulty in practice is, of course, a decrease in the sticking coefficients as the substrate temperatures are increased. However, a high temperature is desirable not only for stoichiometry, but also for crystalline perfection. To achieve a fast growth rate, as in the case of CdSe in Fig. 7, high arrival rates are used with the undesirable effect of beam scattering, which is responsible for the drop in the deposition rate. On the other hand, a low substrate temperature, although it results in improved sticking and faster growth, usually produces nonstoichiometric and polycrystalline films. Subsequent heating is sometimes performed to improve stoichiometry and crystallinity.

For noncongruently evaporating compounds such as the III–V semiconductors, the situation is expected to be complicated since one component is generally substantially more volatile than the other. Although congruent evaporation may be possible under equilibrium conditions over a narrow temperature region, as shown for GaAs in Fig. 3b, the situation for a freely evaporating surface, such as that at the growing film, is different. Consequently, the more volatile component must be supplied in an amount greater

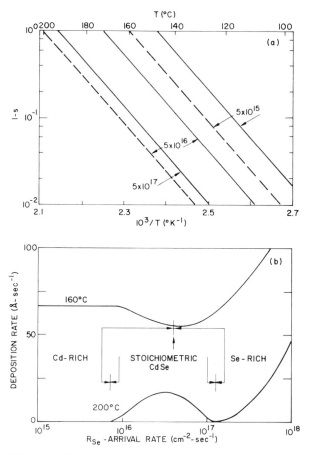

FIG. 7. (a) Sticking coefficient s of Cd (dashed lines) and Se (solid lines) on prenucleated substrates as a function of substrate temperature. Arrival rates of Cd and Se are used as the parameter (after Gunther, 1966). (b) Deposition rate versus the arrival rate of Se to illustrate the regions of stoichiometric growth for a fixed arrival rate of Cd but two different substrate temperatures, as indicated (after Gunther, 1966).

than the stoichiometric ratio during film growth. The guidelines for achieving stoichiometric compounds described earlier are still useful and have been applied by Gunther (1962, 1966) to GaAs and the indium compounds. However, a complete understanding was obtained of the growth processes for stoichiometric GaAs from the kinetic studies by Arthur (1968a, b). These studies entailed the reflection of pulsed As_2 and Ga molecular beams from GaAs surfaces as a function of substrate temperature and Ga coverage. The results are briefly summarized in Fig. 8a. The Ga beam is not reflected below

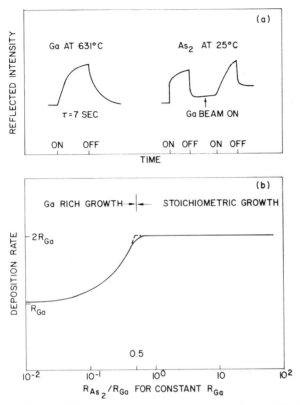

FIG. 8. (a) Sticking coefficients measured through reflection of pulsed Ga and As$_2$ beams from GaAs surface. The As$_2$ beam is totally reflected unless a prior Ga coverage exists (after Arthur, 1968a). (b) Deposition rate versus the arrival rate of As$_2$ to illustrate the threshold in terms of the arrival rates ratio for stoichiometric growth.

477°C, indicating a sticking coefficient of unity. At higher substrate temperatures, a finite surface life τ of the Ga atoms is measured as shown for 631°C. The As$_2$ beam, on the other hand, is totally reflected unless there is a prior deposit of Ga. By analyzing the data it is concluded that the sticking coefficient of As$_2$ is proportional to the Ga coverage, being essentially unity for a Ga covered surface, but decreasing rapidly to zero otherwise.

Stoichiometric GaAs is formed as long as the rate of collision between adsorbed Ga and impinging As$_2$ is equal to the arrival rate of Ga. For arrival rates R_{Ga} and R_{As_2}, the growth rate of GaAs is given by (Arthur, 1968a): $(R_{Ga} + \Gamma)/[1 + (1/\tau k R_{As_2})]$, where Γ is the rate of dissociation of GaAs, and k the constant for the reaction between adsorbed Ga and As$_2$, estimated to be $\sim 10^{-15}$ cm^2. Neglecting Γ, and assuming $R_{As_2} \sim 10^{15}$/cm^2 sec and a substrate temperature $<600°C$ for which $\tau > 10$ sec, the growth rate is then essentially equal to the Ga arrival rate. Figure 8b shows schematically the deposition rate as a function of the ratio of As_2 to Ga arrival rates, for a fixed Ga arrival rate assumed to be much greater than the reevaporation rate of Ga from the surface. Below $R_{As_2}/R_{Ga} = 0.5$, the growth is rich in Ga with a rate proportional to R_{As_2}. Stoichiometric growth occurs at and beyond this ratio with a constant deposition rate until, in principle, As begins to condense. In practice, stoichiometric GaAs film has been obtained for a ratio of 0.4 to 50 (Arthur, 1968a; Cho, 1971a; Ludeke, 1973), and the upper limit has not really been established. Therefore, although the high volatility of As makes it seem difficult to deposit stoichiometric films, the relationship between the sticking coefficients makes this system easy, convenient, and flexible. The growth rate of GaAs is essentially determined by the Ga arrival rate, and its stoichiometry is ensured as long as sufficient As molecules are supplied. One can thus use for source material either the compound itself, its component elements, or their combination, and is thereby able to control the growth rate and the surface environment in terms of the relative arrival rates. This result is not unique to GaAs, but is also valid for GaP (Arthur, 1968a,b) and, perhaps, for some other III–V compounds.

V. Experimental Results

In this section experimental results for compound semiconductors grown by the molecular-beam technique are reviewed. Only stoichiometric, epitaxial films as deposited are of interest; films requiring extensive postdeposited heat treatments being excluded. Because of the space limitation, only a brief account can be made on the growth method, the film morphology, its structure and orientation, and the electrical and optical properties. While the results reported in the literature are voluminous and diverse, an attempt is made to emphasize the work judged to be representative and the features considered to be characteristic of each group of the compound materials.

A. Gallium Arsenide and Related Compounds

The growth of stoichiometric GaAs was pioneered by Gunther (1962, 1966), as mentioned in the previous section. Davey and Pankey (1964, 1968), using a modified method, succeeded in achieving epitaxy on both GaAs and Ge substrates. It was reported that the GaAs film was monocrystalline with little or no twinning when deposited between 425 and 450°C on low-order surfaces of Ge and nonpolar surfaces of GaAs.

Homoepitaxial growths of high-quality GaAs have been made by the MBE technique by a number of workers including Arthur and Lepore (1969), Cho

(1971a), Cho *et al.* (1970), Esaki *et al.* (1972), and Chang *et al.* (1973). The basic setup, similar among the various workers, consists of two Knudsen ovens containing GaAs, primarily as a source for As_2 vapor, and Ga, which largely determines the growth rate. Alternatively, the As vapor can also be supplied in the form of As_4 from a pure As source. Additional ovens containing material suitable for n- and p-type doping are used to control the electrical properties of the film. The substrate is monocrystalline GaAs, polished, etched, and preheated to above 600°C in the vacuum system for cleaning. The deposition temperature varies from 520 to 600°C. The growth rate is usually in the range of 1 to 10 Å/sec.

Figure 9 shows electron micrographs of Pt–C replicas of the surfaces of a (100) GaAs substrate prior to deposition and of a GaAs growth of 1000 Å. While the substrate reveals the expected roughness on this scale, the film is microscopically featureless. Corresponding diffraction patterns along the [1$\bar{1}$0] azimuth are also shown in the figure. It is noticed that the pattern of the film is streaked, which indicates that diffraction occurs predominantly in the surface, as opposed to bulklike diffraction effects seen on the substrate, and suggests smoothness on an atomic scale. Analysis from the intensity data by

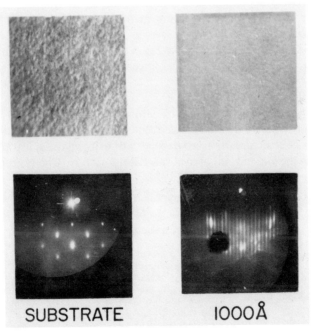

SUBSTRATE   1000Å

FIG. 9. Comparison of (100) GaAs surfaces between the substrate prior to deposition and a deposit of 1000 Å. Upper photographs are Pt–C replicas with a magnification of 38,000, and lower photographs are reflected HEED patterns at [1$\bar{1}$0] azimuth.

SHEED indicates that the extent of coherently scattering regions is about 0.4 μm (Dove *et al.*, 1973). The observation of the surface-smoothing process strongly suggests that the mechanism of growth is via a two-dimensional step propagation. Indeed, evidence that growth on ($\bar{1}\bar{1}\bar{1}$) GaAs proceeds along ⟨112⟩ steps has been reported (Cho, 1970b). Since the lowest energy site is a kink site at a surface step for such a growth mechanism, and since the step density is highest at the surface irregularities, initial growth will be largest at these locations with the tendency to reduce the surface roughness. The thickness required for such surface smoothing, clearly dependent on the substrate preparation, is usually in the range of 300 to 1000 Å, beyond which the HEED pattern undergoes no further changes.

As the surface begins to smooth out, additional diffraction streaks may appear at fractional intervals between the elongated bulk spots, as can be seen in Fig. 9. They represent diffraction from a rearrangement of surface atoms into an ordered array with a lattice spacing greater than that of the bulk. The rearrangement is the consequence of the lowering of the surface energy from a high-energy truncated arrangement to a lower, energetically more favorable, arrangement. Chemically one may look at it as a hybridization process of the unsatisfied dangling bonds, with an accompanying rearrangement to accommodate the new bond angle. The pattern in Fig. 9 is that of the $\frac{1}{4}$-order along the [1$\bar{1}$0] azimuth. The pattern observed in the orthogonal [110] azimuth exhibits a $\frac{1}{2}$-order and is shown in Fig. 10. These combine to indicate a *c* (2 × 8) structure on the (100) surface, since it takes two azimuthal sightings to determine the symmetry of a surface structure. Figure 10 also illustrates [110] azimuth patterns of 1000-Å overgiowths on (110) and ($\bar{1}\bar{1}\bar{1}$) GaAs substrates. The pattern shown on the ($\bar{1}\bar{1}\bar{1}$) surface also exhibits a $\frac{1}{2}$-order, while there is no apparent structure due to surface reconstruction on the (110) surface.

The surface structure observed depends critically on the substrate temperature and on the Ga and As_2 arrival rates at the surface. Cho (1970b) and Cho

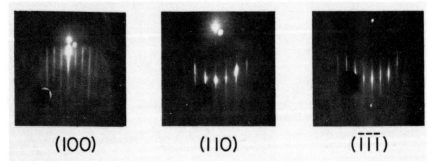

(100)   (110)   ($\bar{1}\bar{1}\bar{1}$)

FIG. 10. HEED patterns at [110] azimuth for a 1000-Å film deposited on variously oriented GaAs surfaces to illustrate reconstructed surface structures.

and Hayashi (1971a) have studied ($\bar{1}\bar{1}\bar{1}$) and (111) surfaces. They have observed a ($\bar{1}\bar{1}\bar{1}$)–(19)$^{1/2}$ structure at relatively high substrate temperatures and low ratios of R_{As_2}/R_{Ga}, and a ($\bar{1}\bar{1}\bar{1}$)–2 structure under opposite conditions. Only the (111)–2 structure has been observed on the (111) face. For (100) surfaces, Cho (1971b), Ludeke (1973), and Chang $et\ al.$ (1973) have observed the c (2 × 8) and c (8 × 2) structures at relatively high and low R_{As_2}/R_{Ga} ratios, respectively, as well as a transitional (3 × 1) structure at intermediate ratios of arrival rates. A detailed study using separate Ga and GaAs sources has indicated that the existence of the various structures depends not only on the substrate temperature and the ratio of R_{As_2}/R_{Ga} but also on the absolute rates (Ludeke, 1973; Chang $et\ al.$, 1973b). This is shown in Fig. 11, where the region of the (3 × 1) structure is seen to widen when the substrate temperature is increased and the Ga-rate is decreased; the c (8 × 2) structure occurs whenever the ratio is below 0.44. Other structures have also been observed, including the (1 × 6) structure at low evaporation rates (Cho, 1971b; Ludeke, 1973) and a (4 × 4) structure along the ⟨100⟩ azimuths at very high As_2 to Ga ratios (Ludeke, 1973; Chang $et\ al.$, 1973).

The carrier mobility of the epitaxial GaAs films has been measured on films grown on semi-insulating GaAs substrates, for which Hall measurements can be conveniently performed. This quantity, because of its sensitivity to the

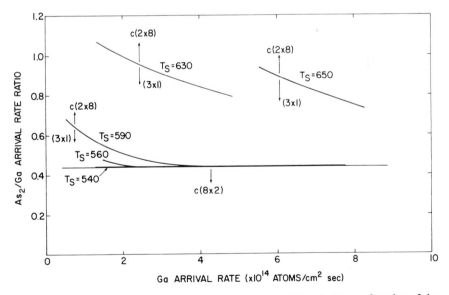

FIG. 11. Regions of reconstructed surface structures on (100) GaAs as a function of the arrival rate of Ga and the ratio of arrival rates of As_2 to Ga, using substrate temperature as a parameter.

presence of impurities and defects, has commonly been used to evaluate film quality. Figure 12 is a plot of mobility versus carrier concentration for both n- and p-type films. Below a concentration of 10^{17} cm^{-3} the films were unintentionally doped, the doping resulting, most likely, from residual impurities in the system. Above 10^{17} cm^{-3}, the films were doped with Sn and Ge for n-type and with Ge and Mg for p-type conductivity. Germanium, being an amphoteric impurity in GaAs, can be used as either a donor or an acceptor depending on the ratio of As$_2$ to Ga during growth (Cho and Hayashi, 1971b; Chang *et al.*, 1973). Conventional acceptors such as Zn and Cd, because of their low sticking coefficients at the growth temperature, are difficult to incorporate into GaAs by MBE techniques (Cho and Hayashi, 1971b). The data shown in Fig. 12 are those of Chang *et al.* (1973b) with the addition of recent results from Mg-doped films. They are in the same range as those reported earlier (Cho, 1971a). Also shown for comparison are data obtained from high-quality, liquid-phase epitaxial GaAs by Vilms and Garrett (1971).

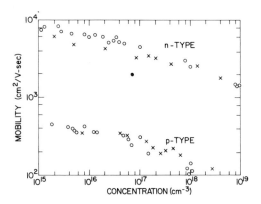

FIG. 12. Hall mobility versus carrier concentration of GaAs at room temperature for MBE data ($\times$). Results of liquid phase epitaxy ($\bigcirc$) from Vilms and Garrett (1971) are also shown for comparison.

Extensive photoluminescence measurements have been reported by Cho (1971a) and Cho and Hayashi (1971a) on GaAs films doped with various impurities. It can be said in general that, in addition to the near-gap emission, a peak with an energy about 0.1 eV below the energy gap has usually been observed at 80°K. Its intensity is greatly reduced in films grown at low ratios of As$_2$ to Ga rates and with a doping concentration beyond 10^{17} cm^{-3}. Chang *et al.* (1973b) have extended cathodoluminescence measurements to 5°K and observed a narrow-gap emission peak with a half-width of 8 meV. Also observed was a peak at an energy of 0.02 eV below the energy gap together with its phonon replica, as well as peaks at still lower energies associated with lattice vacancies (Hwang, 1968; Chang et al., 1971). Using the sharpness and intensity of emissions in the near-gap region as a gauge for material quality, as emissions at low energies represent defects, the films grown by MBE are much superior to the substrates used and are comparable to the liquid-phase materials.

By using the same technique but adding new sources or replacing the Ga and GaAs sources, other III–V compounds have been grown similarly, including GaP, GaAsP, and GaAlAs alloys. The vaporization and condensation kinetics of GaP are similar to those of GaAs (Arthur, 1968; Thurmond, 1965). Arthur and Lepore (1969) have grown both GaAsP and GaP epitaxial films on GaAs substrates kept as low as 430°C and on GaP substrates at 600 to 630°C. Cho et al. (1970) have also succeeded in growing homoepitaxial GaP on (111) and ($\bar{1}\bar{1}\bar{1}$) surfaces and reported surface structures similar to those observed in GaAs. Heteroepitaxial growth of GaP on CaF_2 substrates has been made by Cho (1970a) and Cho and Chen (1970). It was observed that the epitaxial temperature increased with the Ga arrival rate, consistent with the theory of heteroepitaxial nucleation (Lewis, 1971).

Both Cho et al. (1970) and Chang et al. (1973) have made epitaxial GaAlAs films on (100) GaAs substrates by use of an additional Al oven in the usual GaAs setup. The growth conditions are similar to those of growing GaAs. The surface structures, although a systematic study is lacking, seem to be different, however. In growing $Ga_{0.1}Al_{0.9}As$ alloy, for example, with a ratio of arrival rates of 2 for As_2 to Al and a substrate temperature of 650°C, the (3×1) surface structure is the dominant one while no other structures are clearly observable by varying the deposition conditions (Ludeke and Chang, 1972). Physical measurements on GaAlAs that have been reported include optical reflectivity (Cho and Stokowsk, 1971), Raman spectroscopy (Tsu et al., 1972), and cathodoluminescence with near-gap emission for Al composition up to 50% (Chang et al., 1973a).

There is great current interest in the materials of GaAs and GaAlAs because of their close lattice matching yet different energy gaps and dielectric constants. Examples of their use are the heterojunction laser (Hayashi et al., 1971), the dielectric waveguide for integrated optical circuitry (Cho and Reinhart, 1972; Tracy et al., 1973), and the periodic, multilayered structure known as a superlattice (Esaki et al., 1972; Lebwohl and Tsu, 1970). By use of the MBE technique, such a periodic structure has been made by Chang et al. (1973) and Mayer et al. (1973). Interesting transport properties have been observed in structures having a period of less than 100 Å thick (Esaki et al., 1972; Chang et al., 1973). Other work of growing epitaxial films involving the III–V compounds has been reported. Yan and Young (1971) have used a close-packed evaporation geometry to grow GaAs on sapphire. Kosicki and Kahng (1969) have grown GaN on both GaAs and Al_2O_3 substrates, using a modified MBE technique by employing a N_2 gaseous beam through a discharge. Recently Ludeke *et al.* (1973) have succeeded in making alternating epitaxial films of GaAs and a metal, Al, on GaAs substrates.

B. CADMIUM AND ZINC CHALCOGENIDES

With the conflicting requirement on the substrate temperature to achieve both growth and good crystallinity, it is generally difficult to grow thick films of this group of materials because of the low sticking coefficients of the component elements (Ritter and Hoffman, 1963; Gunther, 1966). When evaporation is used in preparing the material, therefore, special techniques are usually employed, including flash evaporation (Richard, 1966), evaporation by localized electron-beam heating (Holt and Wilcox, 1972), and what is known as the "hot-wall" technique (Coller and Coghill, 1969). The first two techniques are clearly outside the realm of the MBE. The hot-wall technique together with its many variations (Muravyeva *et al.*, 1970b) must be considered separately, however. Basically, the hot-wall technique employs a long heated tube, sealed at one end where the source material is located, and open at the other end. In one variation, the sample is placed close to the opening of the tube and is heated to nearly the same temperature so that the deposition is carried out under essentially isothermal conditions (Genthe and Aldrich, 1971; Muravyeva *et al.*, 1972). This technique, then, cannot be considered as MBE but is rather akin to that of vapor transport. In other variations, the sample is placed farther away from the opening and kept at a lower temperature, the mouth of the tube is either open (Zuleeg and Senkovits, 1963; Escoffery, 1964a) or obstructed by quartz wool (Deasley *et al.*, 1970; Calow *et al.*, 1972), or by one or more perforated lids (Ueda and Inuzuka, 1969; Muravyeva *et al.*, 1970a). In the former open-ended case the evaporation is only somewhat restricted as compared to free evaporation. In the latter cases, consideration of a particular technique as being MBE depends on whether the definition, given earlier in terms of the orifice dimensions and vapor pressures, is satisfied or at least nearly satisfied in practical situations.

The material CdS is perhaps the most studied in this group of compound semiconductors, polycrystalline films having been prepared by use of molecular beams by Miller and Bachman (1958). Zuleeg and Senkovits (1963) have employed the open-end, hot-wall technique to grow epitaxial CdS using the compound as a single source. The substrate was lead–glass, chemically polished, degassed, and kept between 25 to 300°C during deposition. The degree of orientation increases with substrate temperature as expected. The film is hexagonal with the c axis normal to the substrate so that the densely packed planes are parallel to the surface. The monocrystalline region extends over 100 μm in average diameter. Resistivities of the films vary from 10^{-1} to

10^8 Ω cm, decreasing with increasing source temperature. Values of the mobilities from 10 to 100 cm^2/V sec have been obtained.

Although the film reported by these authors displays the usual hexagonal structure, Escoffery (1964a,b), using a similar technique, has observed cubic structure of CdS film grown on mica. Ueda and Inuzuka (1969) have grown CdS on NaCl substrates from the multiple-slit type of molecular beam source. Only the hexagonal phase is observed below a substrate temperature of 100°C, and the two phases coexist up to 400°C. The epitaxial relationship is expected to be the same as that observed by Chopra and Khan (1967) using a simple vacuum evaporation technique. For the hexagonal phase, the relationships are predominately (0001) CdS on (001) NaCl with [10$\bar{1}$0] CdS parallel to [110] NaCl and (0001) CdS on (0001) mica with [10$\bar{1}$0] CdS parallel to [10$\bar{1}$0] mica. The cubic phase of CdS grows in parallel orientation on (100) NaCl substrates, but exhibits (111) growth on (0001) mica, with [11$\bar{2}$] parallel to [10$\bar{1}$0] mica. The energy difference between the hexagonal and cubic structures of the II–VI compounds is generally small, so that both growth habits often coexist (polymorphism). Selective, polymorphic growths, by proper choice of the experimental conditions during the epitaxial deposition process, have been reviewed by Khan (1970). Muravyeva et al. (1970b) have investigated the dependence of such phase composition of all the Cd and Zn chalcogenide films on both the source and substrate temperatures in variously designed molecular-beam systems.

A detailed study of ZnSe epitaxy on Ge substrates has been reported by Calow et al. (1972). The setup is similar to that used in CdS (Zuleeg and Senkovits, 1973) but quartz-wool plugs were used inside as well as at the open end of the heated tube. The degree of ordering, growth rate, and surface morphology of the film have been investigated as a function of the orientation and temperature of the substrate, of the source temperature, and of the vacuum in which the growth occurs. It was found that the epitaxial temperature decreases with improvement of vacuum, from 420°C at 10^{-5} to 10^{-6} Torr to below 300° at 10^{-7} to 10^{-8} Torr, under otherwise identical conditions. This finding, although it is not new, points to the need for clean, ultrahigh vacuum systems in the growth of epitaxial semiconductor films. It was also found that correlation exists between the degree of ordering of the ZnSe film and the known presence of well-defined surface structures of the Ge substrate. The observation suggests that an ordered arrangement of surface atoms at the time of nucleation plays an important role for growth of monocrystalline films.

Muravyeva et al. (1970) have studied the growth and reported the electrophysical properties of epitaxial films of Cd and Zn chalcogenides from Knudsen or modified Knudsen cells. The substrates used were mainly mica, but also included Ge, GaAs, and CdS. In addition to the observation of polymorphic growth mentioned earlier, a general relation between the temperature of evaporation and that of epitaxy was found. The evaporation temperature increases first with epitaxy temperature, reaches a maximum, and then decreases as the epitaxy temperature is further increased. The existence of the maximum seems to be due to recrystallization of mica and to the desorption from its surface of the adsorbed gases. The epitaxial temperature associated with the maximum is in the vicinity of 300 to 320°C where the most perfect films, in terms of resistivity and mobility, can be grown. Corresponding evaporation temperatures, however, depend on the compound, varying, for example, from ~ 700°C or a growth rate of 1 Å/sec for ZnTe to ~ 1200°C and 200 Å/sec for CdS. A correlation of the temperature condition with the molecular weight of the chalcogenides has also been observed.

Other work of interest includes that of ZnS on Ge by Deasley et al. (1970) and those using an isothermal hot-wall technique. This technique does not, strictly speaking, qualify as molecular-beam evaporation. However, it is an important and widely used technique, because the geometry and the high substrate temperature involved allow growth of thick films with good physical properties. Genthe and Aldrich (1971) have grown ZnSe on GaAs; Behrndt and Moreno (1971), ZnS on GaAs; and Muravyeva et al. (1972), Cd chalcogenides on mica substrates.

C. Lead Chalcogenides and Alloys

The generally predominant molecular sublimation of the IV–VI compound, as discussed above, makes the film deposition by evaporation technique a relatively easy task. As long as its dominance can be maintained, it makes little difference in practice whether or not the vapor is generated in a truly molecular-beam oven. Elleman and Wilman (1948) have succeeded early in depositing PbS on heated NaCl substrates in a vacuum of only 10^{-3} Torr. Schoolar and Zemel (1964) have made high-quality epitaxial PbS films evaporated from molecular-beam ovens made of quartz containing finely pulverized galena. The substrates were cleaved, synthetic single crystals of NaCl heated to temperatures of up to 300°C. The formation of nuclei occurs at preferred surface sites during the early stages of growth. The mechanism of surface diffusion is believed to be assisted by the existence of water vapor (Matthews and Grunbaum, 1965). Hydrated ions are formed, which enhance the diffusion process and, because of their strong binding, reduce the possible interaction of the water with the film material. A systematic study has been extended from PbS to other lead chalcogenides and SnTe (Zemel et al., 1965). In the meantime, successful epitaxial growths of the lead salts have been reported on mica (Makino, 1964; Egerton and Juhasz, 1967) and other varieties of substrates (Semiletov and Voronina, 1964).

The growth and morphology of PbTe have been studied in detail by Lewis and Stirland (1968). The deposition was performed at a rate of 1 Å/sec in a vacuum of about 10^{-6} Torr using PbTe evaporated from a silica crucible. The substrate was cleaved (001) NaCl, held between 20 and 250°C during deposition. It was found that below 100°C islands of the deposit develop a dendritic structure for a thickness of about 50 Å. At 200°C, however, square islands were observed with edges lying along $\langle 100 \rangle$ directions. The islands are particularly dense and have great area coverage and thickness at cleavage steps. Figure 13 demonstrates the growth sequences up to a thickness of 300 Å for which diffraction patterns are also shown. Single orientation is distinctly indicated, (001) PbTe on (001) NaCl with [110] PbTe parallel to [110] NaCl.

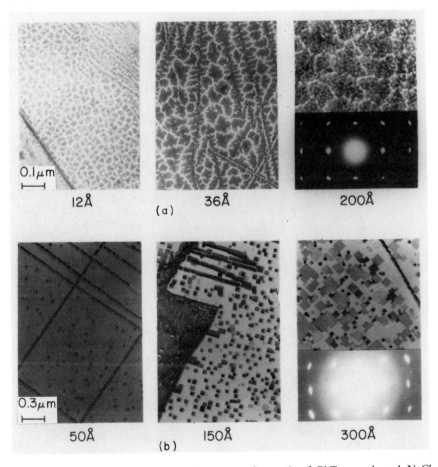

FIG. 13. Micrographs illustrating the stages of growth of PbTe on cleaved NaCl substrated at (a) 20°C and (b) 200°C. Diffraction patterns taken at the indicated thickness are also shown (after Lewis and Stirland, 1968).

The growth behavior is ascribed to a combined effect of the growth rate on ⟨100⟩ edges and the interchange of peripheral molecules between grown islands. Such behavior is expected in other systems like PbTe on NaCl, where the structural similarities between substrate and film suggests an approximate homoepitaxial system.

On mica substrates, the nucleation and growth of PbSe have been studied by Poh and Anderson (1969). The growth rate was varied from 0.5 to 5 Å/sec and the substrate temperature from 300 to 410°C. Figure 14 shows the stages of growth at a substrate temperature of 410°C. It is seen that, initially, nuclei are of tetrahedral shape with (111) bases and (100) sloping faces; the relationship to the substrate being (111) PbSe on (001) mica with [1$\bar{1}$0] PbSe parallel to [100] mica. Subsequently, as the thickness of deposition increases, (100) square nuclei appear with their [110] direction parallel to the [110] direction of the tetrahedral nuclei. The effect of an increase in the substrate temperature is to enhance the appearance of the square nuclei and to facilitate their coalescence. Also shown in Fig. 14 are the electron diffraction patterns corresponding to the two types of nuclei.

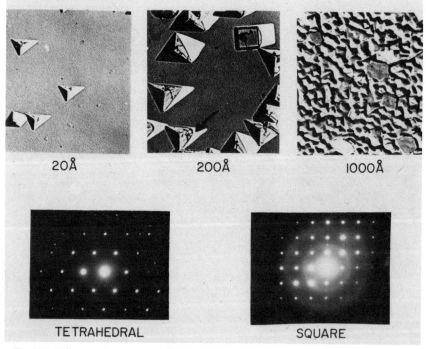

FIG. 14. Micrographs at a magnification of 14,000 illustrating the stages of growth of PbSe on cleaved mica substrate at 410°C. Both tetrahedral and square nuclei exist as seen from the micrographs and the diffraction patterns (after Poh and Anderson, 1969).

Other work in the growth of lead salts includes that by Spinulescu-Carnaru *et al.* (1969) on the study of crystalline structure of PbTe on polished glass substrate, and that by Sumner and Reynolds (1969) on the observation of defect structures of PbTe on NaCl during initial growth and subsequent stages of coalescence and closure. The effects of oxygen on epitaxial PbTe, PbSe, and PbS films have been reported by Egerton and Juhasz (1969). The isothermal hot-wall technique described previously has also been used in growing this group of compound semiconductors.

Electrical properties of the lead salts, including resistivity, Hall coefficient, and mobility, have been reported by Zemel *et al.* (1965) and by Mitchell *et al.* (1964). The films were usually grown at a rate about 100 Å/min onto freshly cleaved NaCl or KCl substrates heated in the range of 200 to 250°C. Figure 15 shows the best results of mobilities given for these materials and those for SnTe as a function of temperature. While both n- and p-type films were obtained for PbTe and PbSe, only n-type behavior was observed for PbS, the carrier concentration being of the order of 10^{18} cm^{-3}. For SnTe, because of its wide solidus region being completely in the Te-rich side, only p-type films are usually obtained, with carrier concentrations of the order of 10^{20} cm^{-3}. Their mobilities approach the values of the bulk materials, indicating high quality of the epitaxial films. For precise control of the concentration, the film can be annealed at various partial pressures of the group IV or VI elements, as mentioned earlier in connection with the deviations from stoichiometry. Summer and Reynolds (1969), for example, have found that their as-deposited lead salts were usually n-type and that reproducible p-type materials were obtained by postdeposition diffusions with Te. Alternatively, the concentrations can be varied during deposition by a fine control of the arrival rates using

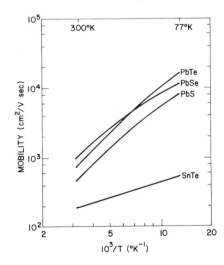

FIG. 15. Temperature dependence of the Hall mobility of Pb chalcogenides and SnTe (after Zemel *et al.*, 1965).

additional ovens containing the elemental components. Recently, Duh *et al.*
(1973) have succeeded, by using a secondary sulfur source, in growing p-type
PbS with a carrier concentration at 77°K ranging from 3×10^{16} to 8×10^{18}
cm^{-3} and a mobility of 6000 cm^2/V sec.

The influence on the mobility of a number of experimental variables has
been studied. The substrate temperature seems to be the most important
parameter. In growing PbS on both NaCl, KCl, and LiF substrates, it has been
found that the mobility peaks at a value of 500 cm^2/V sec in the temperature
region around 220 to 230°C (Zemel, 1969). This behavior can be explained by
a model that employs defects generated by the differences in thermal expansion
coefficients between the film and the substrate or by the introduction of
additional defects due to strain (Zemel, 1969). On the other hand, Palatnik
and co-workers (1965, 1966) have observed a range of temperatures where a
minimum in mobilities occurs for PbTe and PbSe deposited on NaCl and KCl
substrates having relatively large numbers of imperfections. This is attributed
to the transition of growth from (111) to (100) nuclei on (100) NaCl substrates.

The effect of atomic hydrogen in producing large shifts in concentrations
and mobilities in Pb chalcogenides has also been reported (Brodsky and Zemel,
1967, McLane and Zemel, 1971). Other electrical properties, including
magnetoresistance and surface effects, have been studied by Zemel *et al.* (1965)
and Brodsky and Zemel (1967). Zemel (1969) has summarized these results and
also the results on reflectance, transmittance, and refractive indices of the lead
chalcogenide films. Photoconductivity, an important technological property,
has been reported by Sumner and Reynolds (1959) on PbS, by Schoolar (1970)
on PbTe, and recently by Schoolar and Lowney (1971) on PbSe.

Because of the ease with which epitaxial films of the lead salts can be grown,
and because of the wide properties that a mixture of the salts may provide,
particularly in the application of infrared detection, there is growing interest
in the growth of alloys of the IV–VI compounds. Both a single oven containing
a premixed alloy source or a double oven containing separately the two
compounds to be mixed can be employed. Bis and Zemel (1966) and Bis *et al.*
(1966) have used the double-oven setup in growing PbTeSe and obtained high-
quality alloys with the p-type material even superior to bulk crystals. Using
both single- and double-oven techniques, Bylander (1966) has obtained
PbSnTe. The composition of the source was found to be essentially preserved
during the film deposition in the single-oven technique. The interest in this
alloy, because of the decrease in the band gap as the Sn composition increases,
has led to further work by Farinre and Zemel (1970) using the single-oven and
by Bis (1970) using the double-oven technique. Similar systems that have been
studied include PbGeTe by Holloway *et al.* (1971), and PbSnSe by Strauss
(1967) and by Tao and Wang (1972) on CaF_2 and BaF_2 substrates.

ACKNOWLEDGMENTS

The authors wish to thank many of their colleagues at the IBM Research Center who have contributed to this work, particularly L. Esaki, W. E. Howard, G. Schul, and R. Tsu. Thanks are also due B. Lewis and J. C. Anderson for supplying the original photographs of Figs. 13 and 14.

References

Arthur, J. R. (1967). *J. Phys. Chem. Solids* **28**, 2257.

Arthur, J. R. (1968a). *J. Appl. Phys.* **39**, 4032.

Arthur, J. R. (1968b). *Proc. Int. Mater. Symp. Struct. Chem. Solid Surfaces, Berkeley* p. 46–1.

Arthur, J. R., and Lepore, J. J. (1969). *J. Vac. Sci. Technol.* **6**, 545.

Behrndt, M. E., and Moreno, S. C. (1971). *J. Vac. Sci. Technol.* **8**, 494.

Bis, R. F. (1970). *J. Vac. Sci. Technol.* **7**, 126.

Bis, R. F., and Zemel, J. N. (1966). *J. Appl. Phys.* **37**, 228.

Bis, R. F., Rodolakis, A. S., and Zemel, J. N. (1966). *Rev. Sci. Instrum.* **36**, 1626.

Blair, R. C., and Munir, Z. A. (1970a). *High Temp. Sci.* **2**, 169.

Blair, R. C., and Munir, Z. A. (1970b). *J. Amer. Ceram. Soc.* **53**, 301.

Boev, E. I., Benderskii, L. A., and Mil'kov, G. A. (1969). *Zh. Fiz. Khim.* **43**, 1393.

Brebrick, R. F. (1966). *J. Phys. Chem. Solids* **27**, 1495.

Brebrick, R. F. (1971). *J. Electrochem. Soc.* **118**, 2014.

Brebrick, R. F., and Gubner, E. (1962). *J. Phys. Chem. Solids* **36**, 1283.

Brebrick, R. F., and Strauss, A. J. (1964a). *J. Chem. Phys.* **40**, 3230.

Brebrick, R. F., and Strauss, A. J. (1964b). *J. Chem. Phys.* **41**, 197.

Brebrick, R. F., and Strauss, A. J. (1965). *J. Phys. Chem. Solids* **26**, 989.

Brodsky, M. H., and Zemel, J. N. (1967). *Phys. Rev.* **155**, 780.

Bylander, E. G. (1966). *Mater. Sci. Eng.* **1**, 190.

Calow, J. T., Kirk, D. L., and Owen, S. J. T. (1972). *Thin Solid Films* **9**, 409.

Caswell, H. L. (1963). *In* "Physics of Thin Films" (G. Hass, ed.), Vol. 1, p. 1. Academic Press, New York.

Chang, L. L., Esaki, L., and Tsu, R. (1971). *Appl. Phys. Lett.* **19**, 143.

Chang, L. L., Esaki, L., Howard, W. E., and Ludeke, R. (1973a). *J. Vac. Sci. Technol.* **10**, 11.

Chang, L. L., Esaki, L., Howard, W. E., Ludeke, R., and Schul, G. (1973b). *J. Vac. Sci. Technol.* **10**, 655.

Cho, A. Y. (1970a). *J. Appl. Phys.* **41**, 782.

Cho, A. Y. (1970b). *J. Appl. Phys.* **41**, 2780.

Cho, A. Y. (1971a). *J. Vac. Sci. Technol.* **8**, 531.

Cho, A. Y. (1971b). *J. Appl. Phys.* **42**, 2074.

Cho, A. Y., and Chen, Y. S. (1970). *Solid State Commun.* **8**, 377.

Cho, A. Y., and Hayashi, I. (1971a). *Solid State Electron.* **14**, 125.

Cho, A. Y., and Hayashi, I. (1971b). *J. Appl. Phys.* **42**, 4422.

Cho, A. Y., and Reinhart, F. K. (1972). *Appl. Phys. Lett.* **21**, 355.

Cho, A. Y., and Stokowsk, S. E. (1971). *Solid State Commun.* **9**, 565.

Cho, A. Y., Panish, M. B., and Hayashi, I. (1970). *Proc. Symp. GaAs Related Compounds, Aachen* p. 18.

Chopra, K. L., and Khan, I. H. (1967). *Surface Sci.* **6**, 33.

Clausing, P. (1932). *Ann. Phys.* **12**, 961.

Coller, L. R., and Coghill, H. D. (1969). *J. Electrochem. Soc.* **107**, 973.

Davey, J. E., and Pankey, T. (1964). *J. Appl. Phys.* **35**, 2203.

Davey, J. E., and Pankey, T. (1968). *J. Appl. Phys.* **39**, 1941.

Deasley, P. J., Owen, S. J. T., and Webb, P. W. (1970). *J. Mater. Sci.* **5**, 1054.

De Maria, G., Gingerich, K. A., and Piacente, V. (1968). *J. Chem. Phys.* **49**, 4705.

De Maria, G., Malaspina, L., and Piacente, V. (1970). *J. Chem. Phys.* **52**, 1019.

De Nobel, D. (1959). *Philips Res. Rep.* **14**, 361.

Dove, D. B., Ludeke, R., and Chang, L. L. (1973). *J. Appl. Phys.* **44**, 1897.

Drowart, J., and Goldfinger, P. (1958). *J. Chem. Phys.* **55**, 721.

Duh, K., Lopez-Otero, A., and Zemel, J. N. (1973). *Bull. Amer. Phys. Soc.* **18**, 325.

Dunoyer, L. (1911). *C. R. Acad. Sci. Paris* **152**, 549.

Egerton, R. F., and Juhasz, C. (1967). *Brit. J. Appl. Phys.* **18**, 1009.

Egerton, R. F., and Juhasz, C. (1969). *Thin Solid Films* **4**, 239.

Elleman, A. J., and Wilman, H. (1948). *Proc. Phys. Soc.* **61**, 164.

Esaki, L., and Tsu, R. (1970). *IBM J. Res. Develop.* **14**, 61.

Esaki, L., Chang, L. L., Howard, W. E., and Rideout, V. L. (1972). *Proc. Int. Conf. Phys. Semicond. 11th, Warsaw* p. 431.

Escoffery, C. A. (1964a). *Rev. Sci. Instrum.* **35**, 913.

Escoffery, C. A. (1964b). *J. Appl. Phys.* **35**, 2273.

Eyring, H., Wanlass, F. M., and Eyring, E. M. (1964). *In* "Condensation and Evaporation of Solids" (E. Rutner, P. Goldfinger, and J. P. Hirth, eds.), p. 3. Gordon and Beach, New York.

Farinre, T. O., and Zemel, J. N. (1970). *J. Vac. Sci. Technol.* **7**, 121.

Floegel, P. (1969). *Z. Anorg. Allg. Chem.* **370**, 16.

Fujimoto, M., and Sato, Y. (1966). *Jap. J. Appl. Phys.* **5**, 128.

Genthe, J. E., and Aldrich, R. E. (1971). *Thin Solid Films* **8**, 149.

Glang, R., Homlwood, R. A., and Kurtz, J. A. (1970). *In* "Handbook of Thin Film Technology" (L. I. Maissel and R. Glang, eds.), Chapter 2. McGraw-Hill, New York.

Goldfinger, P., and Jeunehomme, M. (1963). *Trans. Faraday Soc.* **59**, 2851.

Gunther, K. G. (1958). *Z. Naturforsch.* **13a**, 1081.

Gunther, K. G. (1962). *In* "Compoind Semiconductors" (R. K. Willardson and H. L. Goering, eds.), p. 313. Van Nostrand-Reinhold, Princeton, New Jersey.

Gunther, K. G. (1966). *In* "The Use of Thin Films in Physical Investigations" (J. C. Anderson, ed.), p. 213. Academic Press, New York.

Gutbier, H. (1961). *Z. Naturforsch.* **169**, 268.

Hall, R. N. (1963). *J. Electrochem. Soc.* **110**, 385.

Hansen, E. E., and Munir, Z. A. (1970). *J. Electrochem. Soc.* **117**, 121.

Hansen, E. E., Munir, Z. A., and Mitchell, M. J. (1969). *J. Amer. Ceram. Soc.* **52**, 610.

Hayashi, I., Panish, M. B., and Reinhart, F. K. (1971). *J. Appl. Phys.* **42**, 1929.

Hildenbrand, D. L., and Hall, W. F. (1964). *In* "Condensation and Evaporation of Solids" (E. Rutner, P. Goldfinger, and J. P. Hirth, eds.), p. 399. Gordon and Breach, New York.

Hoch, M., and Hinge, K. S. (1961). *J. Chem. Phys.* **35**, 451.

Holland, L. (1958). "Vacuum Deposition of Thin Films." Wiley, New York.

Holloway, H. (1966). *In* "The Use of Thin Films in Physical Investigations" (J. C. Anderson, ed.), p. 111. Academic Press, New York.

Holloway, H., Hohnke, D. K., and Logothetis, J. (1971). *J. Vac. Sci. Technol.* **8**, 146.

Holt, D. B., and Wilcox, D. M., (1972). *Thin Solid Films,* **10**, 141.

Hultgren, R., Orr, R. L., Anderson, P. D., and Kelly, K. K. (1963). "Selected Values of Thermodynamic Properties of Metals and Alloys." Wiley, New York.

Hwang, C. J. (1968). *J. Appl. Phys.* **39**, 5347.

Jordan, A. S. (1970). *Met. Trans.* **1**, 239.

Jordan, A. S., and Zupp, R. R. (1969a). *J. Electrochem. Soc.* **116**, 1264.

Jordan, A. S., and Zupp, R. R. (1969b). *J. Electrochem. Soc.* **116**, 1285.

Joyce, B. A., and Bradley, R. R. (1966). *Phil. Mag.* **14**, 289.

Junge, H., and Gunther, K. G. (1963). "Physik und Technik von Sorptions—und Desorptions Vorgagen bei Niederen Drucken." p. 123. Rudolf A. Lang-Verlag.

Kennard, E. H. (1938). "Kinetic Theory of Gases." McGraw-Hill, New York.

Khan, I. H. (1970). *In* "Handbook of Thin Film Technology" (L. I. Maissel and R. Glang, eds.), Chapter 10. McGraw-Hill, New York.

Knudsen, M. (1909). *Ann. Phys.* **28**, 999.

Kosicki, B. B., and Kahng, D. (1969). *J. Vac. Sci. Technol.* **6**, 593.

Koster, W., and Thoma, B. (1955). *Z. Metall.* **46**, 291.

Kulwicki, B. M. (1963). Ph.D. Thesis, Univ. of Michigan, Ann Arbor, Michigan.

Langmuir, I. (1913). *Phys. Rev.* **2**, 329.

Lebwohl, P. A., and Tsu, R. (1970). *J. Appl. Phys.* **41**, 2664.

Lee, W. T., and Munir, Z. A. (1967). *J. Electrochem. Soc.* **114**, 1236.

Lewis, B. (1971). *Thin Solid Films* **7**, 179.

Lewis, B., and Stirland, D. J. (1968). *J. Crystal Growth* **3–4**, 200.

Lorenz, M. R. (1962). *J. Phys. Chem. Solids* **23**, 939.

Lorenz, M. R. (1967). *In* "Physics and Chemistry of II-VI Compounds" (M. Aven and J. S. Prener, eds.), p. 73. Wiley (Interscience), New York.

Lou, C. Y., and Somorjai, G. A. (1971). *J. Chem. Phys.* **55**, 4554.

Ludeke, R. (1973). *Symp. Appl. Vac. Sci. Technol., Tampa, Florida.*

Ludeke, R., and Chang, L. L. (1972). Unpublished results.

Ludeke, R., Chang, L. L., and Esaki, L. (1973). *Appl. Phys. Lett.* **23**, 201.

McLane, M., and Zemel, J. N. (1971). *Thin Solid Films* **7**, 229.

Makino, Y. (1964). *J. Phys. Soc. Japan* **19**, 580.

Matthews, J. W., and Grunbaum, E. (1965). *Phil. Mag.* **11**, 1233. 2322.

Mayer, J. W., Ziegler, J. F., Chang, L. L., Tsu, R., and Esaki, L. (1973). *J. Appl. Phys.* **44**, 2322.

Miller, R. J., and Bachman, C. H. (1958). *J. Appl. Phys.* **29**, 1277.

Mitchell, D. L., Palik, E. D., and Zemel, J. N. (1964). *Proc. Int. Conf. Phys. Semicond. Paris* p. 325.

Mitchell, M. J., and Munir, Z. A. (1970). *High Temp. Sci.* **2**, 265.

Motzfeldt, K. (1955). *J. Phys. Chem.* **59**, 139.

Munir, Z. A. (1970). *High Temp. Sci.* **2**, 58.

Munir, Z. A., and Mitchell, M. J. (1969). *High Temp. Sci.* **1**, 381.

Muravyeva, K. K., Kalinkin, I. P., Aleskovsky, V. B., and Bogomolov, N. B. (1970a). *Thin Solid Films* **5**, 7.

Muravyeva, K. K., Kalinkin, I. P., Sergeeva, L. A., Aleskovsky, V. B., and Bogomolov, N. B. (1970b). *Inorg. Mater.* **6**, 381.

Muravyeva, K. K., Kalinkin, I. P., Aleskovsky, V. B., and Anikin, J. N. (1972). *Thin Solid Films* **10**, 355.

Paltnik, L. S., and Sorokin, V. K. (1966). *Sov. Phys.-Solid State* **8**, 869.

Palatnik, L. S., Sorokin, V. K., and Lebedeva, M. V. (1965). *Sov. Phys.-Solid State* **7**, 1374.

Panish, M. B., and Arthur, J. R. (1970). *J. Chem. Thermodynam.* **2**, 299.

Pashinkin, A. S., and Novoselova, A. V. (1959). *Russ. I. Inorg. Chem.* **4**, 1229.

Pashinkin, A. S., Ukhlinov, C. A., and Novoselova, A. V. (1969). *Vestnik. Mosk. Univ. Khim.* **24**, 107.

Poh, K. J., and Anderson, J. C. (1969). *Thin Solid Films* **3**, 139.

Potts, H. R., and Pearson, G. L. (1966). *J. Appl. Phys.* **37**, 2908.

Ramsey, N. F. (1956). "Molecular Beams." Oxford Univ. Press (Clarendon), London and New York.

Richard, J. L. (1966). *In* "The Use of Thin Films in Physical Investigations" (J. C. Anderson, ed.), p. 71. Academic Press, New York.

Richman, D. (1963). *J. Phys. Chem. Solids* **24**, 1131.

Ritter, E., and Hoffman, R. (1963). *J. Vac. Sci. Technol.* **6**, 773.

Ruth, V., and Hirth, J. P. (1964). *In* "Condensation and Evaporation of Solids" (E. Rutner, P. Goldfinger and J. P. Hirth, eds.), p. 87. Gordon and Breach, New York.

Schoolar, R. B. (1970). *Appl. Phys. Lett.* **16**, 446.

Schoolar, R. B., and Lowney, J. R. (1971). *J. Vac. Sci. Technol.* **8**, 224.

Schoolar, R. B., and Zemel, J. N. (1964). *J. Appl. Phys.* **35**, 1848.

Seacrist, L., and Munir, Z. A. (1971). *High Temp. Sci.* **3**, 340.

Semiletov, S. A., and Voronina, I. P. (1964). *Sov. Phys.-Dokl.* **8**, 960.

Sigai, A. G., and Wiedemeier, H. (1972). *J. Electrochem. Soc.* **119**, 911.

Somorjai, G. A., and Lester, J. E. (1967). *Progr. Solid State Chem.* **4**, 1.

Spinulescu-Carnaru, I., Draghici, I., and Petrescu, M. (1969). *Thin Solid Films* **3**, 119.

Stern, O. (1920). *Z. Phys.* **2**, 49.

Stickney, R. E. (1967). *Advan. At. Mol. Phys.* **3**, 143.

Straumanis, M. E., and Kim, C. D. (1965). *Acta Cryst.* **19**, 256.

Strauss, A. J. (1967). *Phys. Rev.* **157**, 608.

Strauss, A. J., and Brebrick, R. F. (1968). *J. Phys. Coll. Suppl. 11–12* **29**, C4.

Sumner, G. G., and Reynolds, L. L. (1969). *J. Vac. Sci. Technol.* **6**, 493.

Tao, T. F., and Wang, C. C. (1972). *J. Appl. Phys.* **43**, 1313.

Thurmond, C. D. (1965). *J. Phys. Chem. Solids* **26**, 785.

Thurmond, C. D., and Logan, R. A. (1972). *J. Electrochem. Soc.* **119**, 622.

Tracy, J. C., Wiegman, W., Logan, R. A., and Reinhart, F. K. (1973). *Appl. Phys. Lett.* **22**, 511.

Tsu, R., Kawamura, H., and Esaki, L. (1972). *Proc. Int. Conf. Phys. Semicond. Warsaw* p. 1135.

Ueda, R., and Inuzuka, T. (1969). *In* "Growth of Crystals" (N. N. Sheftal, ed.), Vol. 8, p. 171. Consultants Bureau, New York.

Vilms, J., and Garrett, J. P. (1971). *Solid State Electron.* **15**, 443.

Weiser, K. (1962). *In* "Compound Semiconductors" (R. K. Willardson and H. L. Goenig, eds.), Vol. 1, p. 471. Van Nostrand-Reinhold, Princeton, New Jersey.

Yan, G., and Young, L. (1971). *Solid State Electron.* **14**, 1003.

Zemel, J. N. (1969). *In* "Solid State Surface Science" (M. Green, ed.), Vol. 1, p. 291. Dekker, New York.

Zemel, J. N., Jensen, J. D., and Schooler, R. B. (1965). *Phys. Rev.* **140**, A330.

Zuleeg, R., and Senkovits, E. J. (1963). Extended Abstracts, 123rd Meeting of the Electrochem. Society, Pittsburgh, Pennsylvania, No. 95.

2.3 Electrodeposition

J. G. Wright

School of Mathematics and Physics
University of East Anglia, Norwich, England

I. Introduction

Electrodeposition was undoubtedly one of the earliest forms of thin-film preparation. While the precise starting point for this technique is in some dispute, it must have occurred after Volta's discovery of electricity in 1799, but before 1838, since by that date several papers on electroforming had been published. The art of electrodeposition quickly assumed a position of great importance, and as early as 1842 the French Academy of Sciences offered three prizes for advances in the subject.

Initially work was performed on a purely empirical basis and was confined to the deposition of single elements. Indeed even during the second half of the 19th century approximately 90% of the published papers on electrodeposition concerned elements. Since 1900 electrodeposition has been studied in a more scientific manner, but the processes involved are still not fully understood. At the present time some 30 elements and over 100 binary and tertiary alloys are known to be capable of electrodeposition (Brenner, 1963). The complex nature of the problem is easily recognized by reference to Brenner's classic work on electrodeposition, which covers 1400 pages, and took 15 years to prepare.

Even today, significant advances are being made in the deposition of both elements and alloys (Benninghoff, 1970).

The earliest structural investigations of thin films (Wood, 1931, 1935; Cochrane, 1936; Finch and Sun, 1936), using both X-ray and electron diffraction techniques, were concerned with electrodeposited layers. By 1939 many deposits, some of which were epitaxial, had been investigated. As the quality of vacuum systems improved, the emphasis moved slowly from electrodeposition to vacuum deposition. This movement was probably due to the relative simplicity of vacuum deposition and the generally held view that electrodeposition is less clean than vacuum deposition (Pashley, 1965). In the last 20 years the use of vacuum deposition for investigating epitaxial growth has increased considerably, while electrodeposited epitaxial layers have attracted few workers. This situation is reflected in books concerned with thin-film deposition and reviews on film growth where electrodeposition techniques are often ignored, stated to be well understood and therefore unnecessary of detailed discussion, or regarded as unimportant because of having application only to metals growing on metals. In fact, electrodeposition is not well understood, particularly in the area of epitaxy, and may have considerable use. In regard to the latter point, two specific areas spring to mind: (a) the case of metal-on-metal epitaxial growth problems (see Section V), and (b) the investigation of epitaxial growth on planes other than cube planes or hexagonal basal planes, which are the planes most easily available for vacuum deposition.

II. Basic Principles of Electrodeposition

Many workers have prepared electrodeposited layers by following recipes with little consideration to the mechanisms involved. Indeed, some early workers considered the deposition to be determined only by the conductivity of the electrolyte and the potential drop (pd) across the cell. This led to some erroneous results (Cochrane, 1936) and was due, not to knowledge being unavailable (see, for example, Falkenhagen, 1934), but more probably to a lack of communication between thin-film depositers and industrial platers and electrochemists. As far as epitaxial work is concerned, Newman (1956) was one of the first to consider the deposition mechanisms.

In order to understand the deposition process, some discussion is required of the potentials that are set up in electrolytic cells at the electrode–electrolyte interface. In the same way that a pd exists between two dissimilar metals when placed in contact with each other, a pd exists between a metal and the ions in an electrolytic solution with which it is in contact. If there are several species of ions in the solution, each species will have a different potential with respect to the metal. It is beyond the scope of this chapter to deal other than briefly with the problem, and the reader is referred to an extremely lucid account given by Raub and Müller (1967).

The electrode potential is in part determined by the concentration of ions in the solution and, for a solution of ions of valence z and activity a, is given by the Nernst equation

$$E = E_0 + (RT/zF) \ln a$$

where T is the temperature, R the gas constant, and F the Faraday constant. The activity a is defined by $fc = a$, where c is the concentration in moles and f the dissociation factor. If $a = 1$, $E = E_0$, and therefore E_0 is the electrode potential for a normal solution, known as the *standard potential.*

This potential is essentially a static potential measured at zero current flow. During deposition the electrode potential will differ from the static potential due to polarization effects. Ideally the current-carrying ions pass across the liquid–solid interface without performing work, constituting a reversible electrode process hence giving a constant electrode potential independent of current. However, there will usually be some kinetic inhibition, which causes the process to be irreversible and gives a variation in electrode potential with current. The difference between the dynamic potential and the static potential given by the Nernst equation is called an *overvoltage* when speaking in terms of the ions, or a *polarization* when referring to the electrode. The polarization can be considered to arise from two separate situations: (i) the penetration polarization, i.e., the inhibition of the flow of ions through the interface, and (ii) concentration polarization brought about by the inhibition of diffusion processes of ions moving toward the interface and/or by the inhibition of chemical reactions that are necessary in the electrode vicinity. In general, several kinds of polarization effects will exist at a single electrode for each of the separate ion species in the electrolyte.

The standard electrode potentials are not quantities that may be measured absolutely but are measured against a second electrode. For the purposes of definition the reference electrode is taken to be a hydrogen electrode, the potential of which is defined to be zero. Electrode potentials that are positive with respect to this are regarded as more noble (since the noble metals have the highest positive potential), while those electrode potentials that are negative are regarded as less noble. Some values of the more important standard electrode potentials are given in Table I together with some electrode potentials measured at the lowest voltage for deposition to commence.

The measurement of current density in a plating cell as the cathode potential is varied shows that the initial current is very small until the cathode potential reaches a critical value at which point the current increases sharply as shown in Fig. 1a. The point V_A is the point at which deposition commences. As the applied emf is increased further the current will continue to increase until a

TABLE IA

STANDARD ELECTRODE POTENTIALS[a]

Ion/electrode	Potential	Ion/electrode	Potential
Tc^{2+}/Ti	−1.63	Ag^+/Ag	+0.799
Zn^{2+}/Zn	−0.763	Al^{3+}/Al	−1.66
Fe^{2+}/Fe	−0.44	Cr^{3+}/Cr	−0.74
Cd^{2+}/Cd	−0.402	Fe^{3+}/Fe	−0.036
Co^{2+}/Co	−0.28	Au^3/Au	+1.50
Ni^{2+}/Ni	−0.25	In^+/In	−0.25
Sn^{2+}/Sn	−0.136		
Pb^{2+}/Pb	−0.126	Cr^{3+}/Pt	+1.33
Cu^{2+}/Cu	+0.337	Fe^{3+}/Pt	0.77
Cu^+/Cu	+0.521	Sn^{2+}/Pt	0.154
Hg^{2+}/Hg	+0.854	Cu^{2+}/Pt	0.153

[a] Values taken at 25°C and 1 atm (Raub and Müller, 1967).

TABLE IB

EQUILIBRIUM ELECTRODE POTENTIALS DEMONSTRATING THE MAGNITUDE OF POLARIZATION EFFECTS (ΔV)[a]

Half-cell reaction	Electrode potential	ΔV
$Zn(CN)_4 + 2e^- = Zn + 4CN^-$	−1.26	0.50
$ZnO_2^{2-} + 2H_2O + 2e^- = Zn + 4OH^-$	−1.216	0.45
$Cu(NH_3)_2 + e^- = Cu + 2NH_3$	0.12	0.22
$AgCN + e^- = Ag + CN$	0.017	0.78
$Ag(NH_3)_2 + e^- = Ag + 2NH_3$	0.373	0.43
$Ag(CN)_2 + e^- = Ag + 2CN^-$	−0.31	1.11

[a] From Brenner (1963).

limiting value I_A is reached. This current is limited by the maximum possible arrival rate of ions at the cathode due to diffusion mechanisms. No further increase in current will take place unless a second ion species, for example, hydrogen, commences deposition at a higher potential V_B. Thus the potential current curve for a cathode immersed in a solution containing several ions will be a series of steps. The onset of deposition of a given metal ion will vary from one type of bath to another, as will the deposition of different metals. This is demonstrated in Fig. 1b.

It is particularly important at this point to consider the question of hydrogen. Since practically all electrodeposition takes place from aqueous solution, it

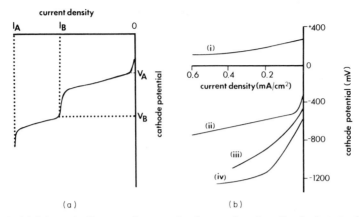

FIG. 1. (a) Schematic diagram of current density as a function of cathode potential for a solution depositing two species. (b) Cathode potential–current density curves of some common electrolytes: (i) acid copper electrolyte, (ii) Watts nickel electrolyte, (iii) silver cyanide electrolyte, (iv) copper cyanide electrolyte (after Raub and Müller, 1967).

is highly likely that hydrogen will be generated at the cathode either as the direct liberation of hydrogen ions from solution by

$$2H^+ + 2e^- \rightarrow H_2$$

or by the reduction of water molecules by

$$2H_2O + 2e^- \rightarrow 2OH^- + H_2$$

However, the discharge of hydrogen will depend on the hydrogen electrode potentials for the reaction under consideration. Fortunately, there are usually significant hydrogen overvoltages that prevent hydrogen discharge until similar or higher voltages are reached than those necessary for metal deposition. If it were not for these overvoltages, certain metal depositions would not take place, an example being the deposition of zinc from a sulfate bath. In general, the magnitude of the hydrogen overvoltage decreases with increasing temperature and depends on the cathode material and solution pH. According to Smith (1948), the magnitude of the hydrogen overvoltage against different cathode materials increases in the following order: Pd, Pt, Fe, Au, Ag, Ni, Cu, Ti, Sn, Pb, and Hg.

While considering the electrode processes, one important advantage of electrodeposition should be mentioned, which concerns substrate cleaning processes that may take place at the cathode. If a metal is placed in aqueous solution, it will in all probability have or obtain an oxide coating. It is sometimes possible for this to be removed at the onset of deposition. An example is the case of copper, where the oxide reduction

$$Cu_2O + 2H^+ + 2e^- \rightarrow 2Cu + H_2O$$

takes place once the deposition potential is applied. This may be one of the reasons why extremely good epitaxial growth of some metals is observed on copper.

III. Bath Composition, Purity, and Throwing Power

To date a substantial number of metal–metal epitaxial systems have been grown by electrodeposition. Table II lists some of the more common metals and baths from which successful deposition has been achieved. Unfortunately, much of the early work failed to note all the important plating parameters, the most notable omission being solution pH.

This last factor is of the utmost importance, since it has serious effects on the deposit purity. In much of the earlier work the current efficiency may well have been less than 100%. Thus production of hydrogen at the growing surface may have taken place, which would either be evolved as free gas or absorbed in the deposit.

In terms of impurity content the amount of hydrogen absorbed may be quite low but, if sufficient hydrogen is available, may rise to the normal bulk solubility limit as is demonstrated in Fig. 2. This, in fact, demonstrates the validity of the argument (Brenner, 1963) that the observed structures and solubilities observed in electrodeposits are the same as those observed in bulk. Although the assertion is in general true, at least one instance of a deposited metastable alloy has been observed (Dutta and Clarke, 1968). We shall also see in Section V that in epitaxial work a number of metastable structures have been observed in very thin layers due either to strong substrate influences or to impurity stabilization.

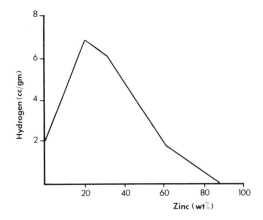

FIG. 2. Hydrogen content of nickel–zinc deposits (after Raub and Sautter, 1959).

TABLE II

A Selection of Plating Baths That Have Been Used to Deposit Single Elements Epitaxially

Element	Bath composition (gm/liter)	Current density (mA/cm²)	Temp (°C)	pH	Reference
Ag	$36AgCN$; $52KCN$; $38K_2CO_3$	10	20	—	Finch and Sun (1936)
As	$50As_2O_3$; $20Na_4P_2O_7 \cdot H_2O$	70	50	—	Finch and Sun (1936)
Au	$2AuCl_3NaCl$; $15KCN$	1	50	—	Finch and Sun (1936)
Bi	$47Bi_2O_3CO_3H_2O$; $19H_2ClO_4$	5	50	—	Finch and Sun (1936)
Cd	$32CdO$; $75NaCN$	10–150	20	—	Finch and Sun (1936)
Co	$300CoSO_4 \cdot 7H_2O$; $3NaCl$; $6H_3BO_3$	0.5–30	20–60	1.8–6.0	Cochrane (1936), Newman (1956), Goddard and Wright (1964)
Cr	$250CrO_3$; $2.5H_2SO_4$	100–250	50	—	Cleghorn et al. (1968)
Cu	$200CuSO_4 5H_2O$; $30H_2SO_4$				Cochrane (1936)
Fe[a]	$350Fe(NH_4)(SO_4)_2 \cdot 12H_2O$	1–10	20	<2.7 or 4.0–5.5	Whyte (1965), Wright (1971)
Ni	$300NiSO_4 \cdot 7H_2O$; $3NaCl$; $6H_3BO_3$	1–30	20	1.8–6.5	Cochrane (1936), Reddy (1958), Wright (1972)
Pt	$13.3H_3PtCl \cdot 6H_2O$; $45(NH_2)_3HPO_4$	4	70	—	Finch and Sun (1936)
Sn	$11.5SnCl_2 \cdot 2H_2O$; $62KOH$; $13KCN$	5	50	—	Finch and Sun (1936)
Zn	$96ZnSO_4 \cdot 7H_2O$; $4NaCl$; $6H_3BO_3$	—	—	—	Cochrane (1936)

[a] This solution must be freshly prepared each day.

It is obvious that all ions in solution will have a tendency to move toward one of the electrodes under the influence of any applied electric field. It is therefore extremely important that unwanted cation species should, as far as possible, be precluded from an electrolytic solution used for epitaxial work. If this is not possible, attempts must be made to adjust the electrode potentials by chemical complexing so that deposition of the unwanted species does not take place. In terms of metallic impurities it may be seen by reference to Section VI that very small impurities in the solution could manifest themselves, under adverse conditions, as high-level impurities in the plating. An example of this situation is the deposition of cobalt impurities in zinc (Maja and Spinelli, 1971).

Under good plating conditions (Brenner, 1963), it is quite possible to achieve electroplates that have impurity concentrations as low as 100 or even 10 ppm. Such values are lower than the normal content of nonmetallic impurities in high-purity metal samples. This level of purity, which includes trapped gases, in all probability exceeds most vacuum deposits, which may only be reduced to such levels in the best UHV and with considerable care. Thus the argument that electrodeposition is not sufficiently clean for epitaxial work may be seen to be difficult to substantiate. However these extremes of purity are only obtained using solutions that have been thoroughly purified.

The codeposition of nonmetallic impurities is only accomplished by introducing the component in the plating bath in the form of certain compounds (Raub and Müller, 1967) such as the deposition of sulfur from solutions containing thiosulfate but not from solutions containing only sulfate. Organic substances in general result more readily in codeposition than inorganic substances, and small traces may have marked effects as is demonstrated by the addition of 10^{-5} moles/liter of n-decylamine in copper baths (Damjanovic et al., 1965). While the addition of such materials has great value in industrial plating as brighteners and hardeners, etc., they should be omitted from epitaxial work concerned with pure materials.

It has been shown that the pH of the solutions is extremely important. For this reason plating solutions have a buffer added whose purpose is to regulate small changes in solution pH (see for example, Davies, 1967). This may be particularly important in the cathode region where composition changes occur at the interface. One of the most common buffers is boric acid, particularly in sulfate baths. However, it has been shown (Raub and Müller, 1967) that boric acid may be incorporated in the plating of silver with a consequent change in properties. Thus for work on epitaxy it may be wise to omit buffers altogether.

One further aspect indirectly concerned with composition is the problem of stirring. The cathode layers that are established during deposition may be quite thick (up to 100 μm), and in this region the composition may be different from that in the bulk solution. If the bath is vigorously stirred, this layer may be significantly disturbed and cause changes in the cathode overvoltages. The

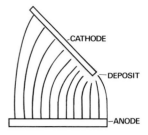

FIG. 3. Variation in electric field and consequent deposit thickness due to bath geometry.

desirability of such changes depends upon the system under investigation, but stirring is sometimes a useful variable.

The last property of the bath to receive some discussion here is the throwing power. In general, the field strength, and hence the current density, at a cathode will be a function of position and determined by the bath geometry, e.g., Fig. 3. However, a change in local current density may result in a local change of polarization, and thus the deposition rate may not be so adversely affected as initial considerations would suggest. The degree of self-compensation of a bath is essentially described by the throwing power of a bath. Solutions of high throwing power would produce only a small change in thickness along the wedge shown in Fig. 3 (typically alkaline or cyanide baths), whereas solutions of poor throwing power could produce large thickness variations (sulfate and acid baths, particularly chromium baths). Mathematical expressions for throwing power have been derived by several authors, e.g., Der Tau Chin (1971).

IV. Basic Plating Parameters

Once a particular bath has been chosen and carefully prepared, the three parameters that affect the growth of the deposit are current density, pH, and temperature. These parameters, in conjunction with the crystallographic plane of the substrate, control whether or not epitaxial growth takes place, the degree of twinning, and the surface morphology (Lawless, 1967).

In general, it is found that as the current density is changed, so does the type of deposit. For good epitaxial growth the current density should normally be small, of about a few milliamperes per square centimeter or less, but varies from one material to another. If the current density is increased, twinning may occur, or even random polycrystalline growth as shown in Fig. 4.

Apart from the crystallographic structure, the current density may have some effect on the surface morphology. It has been well established by many workers (Lawless, 1967) that many epitaxial layers exhibit ridges, pyramids, or dendrites on the surface. The precise nature of these protrusions depends not only on the material of the deposit but also on the crystallographic plane of the substrate. Although these surface features become more obvious as the

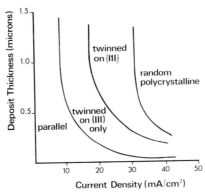

FIG. 4. Orientation of silver deposits on electropolished (111) plane of silver (after Setty and Wilman, 1955).

film becomes thicker, Garmon (1960) and Lawless (1965) have both shown that roughness may be present even in films only 100 Å thick. The precise reason for the existence of the features is not known, but could possibly be due to different electrode potentials existing for different crystallographic faces, hence leading to preferential growth on some faces inclined to the plane of the substrate. Only in recent years has the nucleation rate been studied in detail, particularly with reference to different crystallographic planes (Tanabe and Kamastri, 1971).

It has already been noted that the solution pH has in the past been one of the most seriously overlooked parameters. Essentially it is a measure of the hydrogen-ion concentration in the solution, and therefore one might expect that as the pH is altered, so will the hydrogen electrode potential and consequently the relative rates of deposition of metal and hydrogen. Baths operating at 100% current efficiency for metal deposition should be used for epitaxial work where possible.

In general, the third parameter, temperature, is much less important than either the current density or solution pH. Deposits are sometimes improved with increase in temperature. However, the fact that hydrogen overvoltages decrease with temperature can result in more hydrogen being liberated at the cathode. It is customary for most epitaxial plating to be performed at room temperature.

V. Epitaxial Growth of the Elements

The mechanisms of electrodeposition have been discussed together with the parameters that control the deposition. To the uninitiated it may seem that electrodeposition is a more formidable technique than vacuum deposition. It

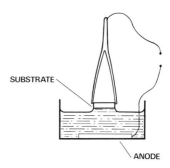

SUBSTRATE

ANODE

Fig. 5. Experimental arrangement for preparation of epitaxial layers.

is true that a full understanding is difficult; however, the actual process of deposition is relatively straightforward (see Fig. 5). Normally deposition takes place onto metal single crystals. These are electropolished (Jacquet, 1956) and removed from the polishing bath without switching off the current, hence preventing backdeposition (Takahashi, 1953). Immediately after removal from the bath the sample is thoroughly washed. Any excess liquid is drained, and the sample is placed face down in the plating solution directly over the center of the cathode, taking care to preclude air bubbles from the sample face. Deposition at a preset current density, pH, and temperature may now take place. With practice, the time taken for the removal from the polishing bath, washing, and placing in the plating bath should be only tens of seconds. Thus the single crystal will (even if it is copper) have only very thin oxide layers on the surface, which may be removed by cathodic reduction during the initial part of the plating process. It is possible to obtain very uniform deposits of accurately known thickness at a constant rate of deposition.

A list of some of the observed epitaxial growth systems observed to date is given in Table III. Unfortunately much of the earlier work was performed without reference to the pH, current density, and current efficiency of the solution, and reproducible results may not be easy to obtain. The deposition of cobalt from sulfate solutions (Cochrane, 1936; Newman, 1956; Fukada, 1958; Goddard and Wright, 1964) demonstrates the need for these parameters to be carefully controlled. Figure 6 shows the current efficiency versus pH for such deposition, demonstrating that those parameters are interdependent. Furthermore cubic cobalt is deposited only in a region where the current efficiency is less than 100% and consequently appears to depend on the codeposition of hydrogen for its stabilization in thick layers.

It is important to note that the growth of metastable structures by epitaxy is not solely confined to vacuum deposition but has also been observed in electrodeposits (Chopra, 1969; Wright, 1971).

For small misfits, extremely thick single-crystal deposits can be obtained, for example, several microns of nickel on copper (Lawless, 1967). Such films

TABLE III

Epitaxial Growth Systems[a]

| | | Orientation | | |
| | | Parallel plane | Parallel direction | Reference |
Deposit[b]	Substrate			
α-Co	Co	$\|10\bar{1}0\|$ Co $/\!/$ $\|110\|$ Cu	$\langle 0001 \rangle$ Co $/\!/$ $\langle 001 \rangle$ Cu	Goddard and
		$\|0001\|$ Co $/\!/$ $\|111\|$ Cu	$\langle 1000 \rangle$ Co $/\!/$ $\langle 1\bar{1}1 \rangle$ Cu	Wright (1964)
β-Co	Cu	$\|110\|$ $/\!/$ $\|110\|$ Cu	$\langle 001 \rangle$ Co $/\!/$ $\langle 001 \rangle$ Cu	Newman (1956),
		$\|211\|$ Co $/\!/$ $\|211\|$ Cu	$\langle 1\bar{2}0 \rangle$ Cu $/\!/$ $\langle 1\bar{2}0 \rangle$ Cu	Fuhuda (1958),
		$\|100\|$ Co $/\!/$ $\|100\|$ Cu	$\langle 001 \rangle$ Cu $/\!/$ $\langle 001 \rangle$ Cu	Goddard and
		$\|111\|$ Co $/\!/$ $\|111\|$ Cu	$\langle 1\bar{1}0 \rangle$ Co $/\!/$ $\langle 1\bar{1}0 \rangle$ Cu	Wright (1964)
bcc Cr	Ni	$\|110\|$ Cr $/\!/$ $\|001\|$ Ni	$\langle \bar{1}12 \rangle$ Cr $/\!/$ $\langle \bar{1}10 \rangle$ Ni	Cleghorn et al.
		$\|001\|$ Cr $/\!/$ $\|110\|$ Ni	$\langle \bar{1}10 \rangle$ Cr $/\!/$ $\langle \bar{1}12 \rangle$ Ni	(1968)
			or $\langle 010 \rangle$ Cr $/\!/$ $\langle \bar{1}12 \rangle$ Ni	
bcc Fe	β-Brass	$\|110\|$ Fe $/\!/$ $\|110\|$ brass	$\langle 001 \rangle$ Fe $/\!/$ $\langle 001 \rangle$ brass	Whyte (1965)
	Cu	$\|211\|$ Fe $/\!/$ $\|110\|$ Cu		Tanabe and
		$\|110\|$ Fe $/\!/$ $\|110\|$ Cu	$\langle 11 \rangle$ Fe $/\!/$ $\langle 110 \rangle$ Cu	Kamashi (1971)
		$\|110\|$ Fe $/\!/$ $\|111\|$ Cu		
fcc Fe	Cu	$\|110\|$ Fe $/\!/$ $\|110\|$ Cu	$\langle 001 \rangle$ Fe $/\!/$ $\langle 001 \rangle$ Cu	Wright (1971)
hcp Ni	α-Co	$\|10\bar{1}0\|$ Ni $/\!/$ $\|10\bar{1}0\|$ Co	$\langle 0001 \rangle$ Ni $/\!/$ $\langle 0001 \rangle$ Co	Wright and
				Goddard (1965)

[a] These systems are additions to the list given by Lawless (1967).

[b] The abbreviations used are bcc–body-centered cubic, fcc–face-centered cubic, and hcp–hexagonal close packed.

are of extremely good quality with the exception of the surface morphology. In particular the dislocation density may well be less than in vacuum deposits and has been observed to be as low as $10^6/\text{cm}^2$ (Bertocci and Bertocci, 1971).

Since all electrodeposited metal films must be prepared on metal bases, the most common forms of investigation are by reflection electron diffraction for structure and optical or replica electron microscopy for surface morphology. Several workers have attempted to prepare stripped layers for transmission electron microscopy and some degree of success has been obtained. The basic method of stripping is to remove the substrate chemically (Weil and Reid, 1950; Ogawa et al., 1957; Lawless, 1965) or to remove an intermediate deposited layer (Finch and Sun, 1936). This latter method has proved to be particularly successful for obtaining large areas (> 1 cm^2) of stripped single-crystal nickel films by preparation on copper covered with approximately 1000 Å of β-cobalt (Wright, 1972, 1974).

Although growth processes might at first sight appear to be quite different to growth by vacuum deposition due to the dense background of solution molecules, the general consensus of opinion is that the two growth processes

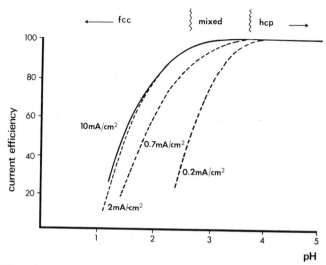

FIG. 6. Effect of solution pH on current efficiency at various current densities for cobalt from a sulfate bath. Values of 2 mA/cm², 0.7 mA/cm², and 0.2 mA/cm² are from Fukuda (1958), and 10 mA/cm² and structure of deposit are from Goddard and Wright (1964).

are similar. In electrodeposition, ions arrive at the cathode and are neutralized. The atoms then migrate on the surface until a suitable site is found, i.e., kink or growing nucleus. Hence growth is envisaged as taking place by nucleation and growth (see Chapter 4). However, due to the high deposition rates used (typically up to 200 Å/sec), far larger numbers of nuclei may be formed (although this does to some extent depend upon the plating parameters such as pH), and in some cases continuous layers are observed where the film thickness is less than 100 Å (Whyte, 1965; Lawless, 1965).

VI. Deposition of Epitaxial Alloy Layers

Although the deposition of binary alloys by electrodeposition has reached an advanced state in the area of industrial electroplating, almost no work has yet been performed on epitaxial layers. One of the most significant pieces of work (Lord, 1968) investigated in some detail the deposition of cobalt–nickel over a wide range of compositions.

At this point it is pertinent to discuss the problems of alloy deposition. It has been shown in Section II that because different ion species in a given solution will have different cathode potentials, one might expect them to deposit at different applied voltages. It is quite possible to obtain from a solution containing mostly species A with traces of species B, a deposit of only species B. Since the electrode potentials depend upon concentrations and in particular the polarizations, it is generally difficult to predict the behavior of a solution. The dynamic potentials in an alloy solution are often quite different from the separate static potentials of the components. It is, however, generally observed that metal pairs whose static potentials are within 0.1 V normally codeposit from simple acid baths. Examples are lead with tin, nickel with tin, nickel with cobalt, and copper with bismuth. Metals with widely differing potentials such as silver and zinc where $\Delta V > 1.5$ V do not readily codeposit. There are some exceptions to these generalizations, such as silver and palladium, whose static potentials are very close but which do not codeposit readily.

This would at first sight appear to limit the metals that may be codeposited. However, by the use of suitable chemical complexes the deposition potentials may be drastically altered as demonstrated in Table IV. Here copper and zinc, which have widely differing standard potentials, are easily codeposited to form brass from the cyanide solution.

In the case of the work by Lord on nickel–cobalt there were no significant electrode potential differences since cobalt and nickel are very close in the electrochemical series and have similar polarizations. However, this does not mean that the composition of the deposit will be the same as the composition of the solution. Figure 7 shows the relationship between these two quantities and demonstrates that the correspondence is by no means 1:1. The deposit composition was measured by X-ray fluorescence (Blavier et al., 1960), scanning probe analyzer, and chemical analysis (Burke and Yoe, 1962) with good agreement between the three methods.

One particular feature of this work was the measurement of composition at different positions on the sample. While the homogeneity was extremely good

TABLE IV

ILLUSTRATION OF HOW THE DIFFERENCE BETWEEN THE DYNAMIC ELECTRODE POTENTIALS OF COPPER AND ZINC, MEASURED AT A CURRENT DENSITY OF 1 MA/CM^2, MAY BE ADJUSTED BY SUITABLE COMPLEXING [a]

Metal complex		Dynamic potentials		
Cu	Zn	V_{Cu}	V_{Zn}	$V_{Cu} - V_{Zn}$
$CuSO_4$	$ZnSO_4$	0.261	−0.878	1.139
$Na_6Cu(P_2O_7)_2$	$Na_6Zn(P_2O_2)_2$	−1.102	−1.392	0.290
$Na_2Cu(CN)_3$	$Na_2Zn(CN)_3$	−1.448	−1.424	−0.024
K_2CuCl_3	$ZnSO_4$	−1.054	−0.878	−0.176

[a] After Stabrovsky (1952).

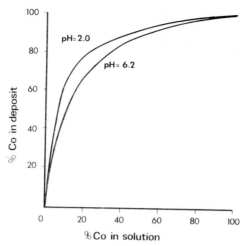

FIG. 7. Percentage of cobalt in nickel–cobalt deposits prepared from sulfate baths as a function of percentage cobalt in solution which contains 300 gm/liter of mixed hydrous sulfates of the two metals (after Lord, 1968).

over most of the sample (~ 1 cm^2 in area), an increase in thickness was observed at the edges, which was probably due to the bath geometry, cf. Section IV.

A second important measurement demonstrated that the film composition was constant with thickness. This is extremely important, since it is well known that in many thick alloy layers the components sometimes deposit alternately in striations, an effect that can be understood in terms of viscous layer depletion at the cathode boundary (Raub and Krause, 1944).

References

Benninghoff, H. (1970). *Galvanotechnik* **62**, 5.
Bertocci, U., and Bertocci, C. (1971). *J. Electrochem. Soc.* **118**, 1287.
Blavier, P., Hans, A., Tyou, P., and Houbert, I. (1960). *Cobalt* **7**, 33.
Brenner, A. (1963). "Electrodepostion of Alloys," Vols. I and II. Academic Press, New York.
Burke, R. W., and Yoe, J. H. (1962). *Anal. Chem.* **36**, 1378.
Chopra, K. L. (1969). *Phys. Status Solidi* **32**, 489.
Cleghorn, W. H., Warrington, D. H., and West, J. M. (1968). *Electrochim Acta* **13**, 331.
Cochrane, W. (1936). *Proc. Phys. Soc.* **48**, 723.
Damjanovic, A., Paunovic, M., Bockris, J. O'M. (1965). *Electrochim. Acta* **10**, 111.
Davies, C. W. (1967). "Electrochemistry." Newnes, London.
Der Tau Chin (1971). *J. Electrochem. Soc.* **118**, 818.
Dutta, P. K., and Clarke, M. (1968). *Trans. Inst. Metal Finish* **46**, 20.
Falkenhagen, H. (1932). "Electrolytes." Oxford Univ. Press, London and New York.
Finch, G. I., and Sun, C. H. (1936). *Trans. Faraday Soc.* **32**, 852.
Fukuda, S. (1958). *J. Appl. Phys. Japan* **27**, 236.
Garmon, L. B. (1960). M. Sc. Thesis, Univ. of Richmond, Richmond, Virginia.

Georgi, K. (1933). *Z. Electrochem.* **39**, 209.

Goddard, J., and Wright, J. G. (1964). *Brit. J. Appl. Phys.* **15**, 807.

Jacquet, P. A. (1956). *Met. Rev.* **1**, 157.

Lawless, K. R. (1965). *J. Vac. Sci. Technol.* **2**, 24.

Lawless, K. R. (1967). *In* "Physics of Thin Films" (G. Hass and R. E. Thun, eds.), Vol. 4, pp. 191–255. Academic Press, New York and London.

Lord, D. G. (1968). Ph.D. Thesis, Univ. of London.

Maja, M., and Spinelli, P. (1971). *J. Electrochem. Soc.* **118**, 1583.

Newman, R. C. (1956). *Proc. Phys. Soc.* **B64**, 432.

Ogawa, S., Mitzumo, J., Watanabe, D., and Fujita, F. E. (1957). *J. Phys. Soc. Japan* **12**, 999.

Pashley, D. W. (1965). *Advan. Phys.* **14**, 327.

Raub, E., and Krause, D. (1944). *Z. Electrochem.* **50**, 91.

Raub, E., and Müller, K. (1967). "Deposition of Metals." Elsevier, Amsterdam.

Raub, E., and Sautter, F. (1959). *Metalloberflache* **13**, 129.

Setty, T. H. V., and Wilman, H. (1955). *Trans. Faraday Soc.* **57**, 984.

Smith, D. P. (1948). "Hydrogen in Metals." Univ. of Chicago Press, Chicago, Illinois.

Stabrovsky, A. I. (1952). *Zhur. Fiz. Khim.* **26**, 949.

Tanabe, Y., and Kamashi, S. (1971). *J. Metal Finishing Soc. Japan* **22**, 54.

Takahashi, M. (1953). *J. Chemie. Phys.* **50**, 624.

Weil, R., and Reid, H. J. (1950). *J. Appl. Phys.* **21**, 1068.

Whyte, T. D. (1965). Ph. D. Thesis, Univ. of London.

Wood, W. A. (1931). *Proc. Phys. Soc.* **43**, 138.

Wood, W. A. (1935). *Trans. Faraday Soc.* **31**, 1248.

Wright, J. G., and Goddard, J. (1965). *Phil. Mag.* **11**, 485.

Wright, J. G. (1971). *Phil. Mag.* **24**, 217.

Wright, J. G. (1972). *Thin Solid Films* **9**, 309.

Wright, J. G. (1974). *Thin Solid Films* **24**, 265.

2.4 Chemical Vapor Deposition

Don W. Shaw

Texas Instruments, Inc.
Dallas, Texas

I. Introduction

Chemical vapor deposition (CVD) refers to the formation of a condensed phase from a gaseous medium of different chemical composition. It is distinguished from physical vapor deposition processes such as sublimation or sputtering, where condensation occurs in the absence of a chemical change. Chemical transport processes form a special class of CVD where initially a condensed phase (source) reacts with a transport agent to produce volatile species that, in another region under different conditions, undergo chemical reaction to re-form the condensed phase. Chemical transport processes require both a source and a transport agent, whereas other CVD processes employ only gaseous reactants, which are stable at room temperature and consequently require no separate condensed source.

89

The techniques of chemical vapor deposition have been applied to epitaxial growth with great success. In fact, in terms of commercial application, CVD is by far the most important method. Let us consider the reasons for its popularity and compare the CVD method with other techniques for epitaxial growth. CVD epitaxy is a very versatile process. In contrast to physical methods such as sublimation or sputtering, it is not limited in the choice of a starting material. Generally, a considerable number of different reactants are available, the choice being made in terms of purity, availability, ease of handling, thermodynamic stability, cost, and compatibility with other reagents. In comparison with physical methods, CVD techniques often permit growth under conditions that more closely approach equilibrium at the growing crystal surface and tend to produce layers with superior crystalline perfection. Epitaxial layers of materials that decompose on evaporation or sublimation can be grown. Also, CVD growth does not require the reduced pressure operation characteristic of most physical deposition processes and thus avoids the complications of associated vacuum equipment.

In comparison with electrodeposition, no ionized medium is required and the conductivity of the substrate is unimportant. Also, the higher deposition temperatures of most CVD processes lead to greater surface mobility of adsorbed species and improved crystal perfection of the layers. Epitaxial growth from solution, or liquid phase epitaxy, is plagued by the problem of solvent removal from the surface of the resulting layer, which leads to surfaces that are inferior to CVD layers with respect to smoothness and freedom from contamination. The ease of doping, and more important the ability to vary the dopant concentration during growth, is an exceptional feature of the CVD method, permitting growth of multiple-layer epitaxial structures and layers with controlled doping gradients. Graded layers or alloy deposits whose compositions vary as a function of thickness are routinely grown by CVD

TABLE I

CVD EPITAXY

Principal advantages	Principal disadvantages
Near equilibrium growth possible	Numerous control variables
Applicable to materials which decompose on melting or evaporation	Reactive atmosphere may attack substrates or apparatus
Atmospheric pressure operation	Apparatus complexity
Choice of reaction and starting materials	
Variable composition layers possible through gas composition control	
Insulating substrates may be used	
Adaptable to large-scale, multislice operation	

techniques. Finally, as has been amply demonstrated in the semiconductor industry, CVD epitaxial methods are readily adaptable to large-scale commercial processing.

Of course all techniques have disadvantages. The degree of freedom in selection of the growth conditions, which results in the versatility of CVD epitaxy, also demands careful control over a number of operating parameters, and for some materials no thermodynamically feasible CVD process is available. Some transport agents may act as detrimental impurities when incorporated into the layers or may attack the substrate. However, disadvantages such as these have not seriously interfered with the widespread application of CVD epitaxy to many materials. The principal advantages and disadvantages of the CVD method are summarized in Table I.

II. Fundamental Considerations

A. SYSTEM DESIGN

CVD epitaxy may be carried out either in a closed tube where free convection serves to transport the gaseous reactants from a source region to the substrate surface, or in an open-tube flow system where transport results from forced convection. Closed-tube approaches are discussed in considerable detail by Schaefer (1964), although the emphasis is placed on bulk, self-nucleated deposition rather than epitaxial growth. Closed-tube growth permits use of a wide range of pressures and is inherently simple in experimental execution as no complicated gas handling system is required. The source and substrate are located at opposite ends of a sealed tube containing the transport agent. By placing the tube in an appropriate temperature gradient, transport and epitaxial growth can be obtained. Closed-tube processes are limited in applicability, as only a small number of slices may be used and the temperature gradients may lead to nonuniformities. For these reasons, open-tube flow systems are most widely used and will here receive more attention.

A number of advantages are realized by use of flow systems in CVD epitaxy. These systems are usually operated at atmospheric pressure with a rather large excess of a carrier gas, which may be either reactive or inert. Reactant partial pressures are easily controlled externally, and the residence time of reactants in the deposition region may be varied by changing the total gas flow rate. Temperature gradients are not essential since forced convection of the flowing carrier gas provides for transport of fresh reactants into the growth region, thus maintaining the nonequilibrium reactant partial pressures which are the driving force for epitaxial growth. Minimum temperature gradients promote uniformity across the epitaxial layer and permit multiple-slice depositions. The impurity or dopant concentration in the gas flow may be varied to produce multiple-layer epitaxial structures, and alloy deposits whose compositions change with layer thickness (graded layers) are grown in open-tube processes with relative ease. Since the reactant partial pressures may be reduced to zero while the grown layer remains at the deposition temperature, cooling to room temperature in the absence of a reactive atmosphere is possible. This preserves good surfaces since no growth occurs except at the optimum deposition temperature.

The design and complexity of an epitaxial growth apparatus is determined to a large degree by the physical states of the reactants. The most straightforward and simplest apparatus results for processes whose only reactants are readily available as room-temperature gases. In this case no reservoirs are necessary for condensed reagents, and only gas handling equipment is required. A good example is epitaxial growth of silicon by pyrolysis of silane. Only two reagents are required, silane and the carrier gas. For epitaxial deposition these are simply passed over the heated substrate. Somewhat more complicated are systems where one of the reagents is a volatile liquid at room temperature. Here it is necessary to saturate the carrier gas by passing it over or through the liquid. However, the liquid reservoir is usually separate from the main deposition vessel so that its design remains relatively uncomplicated. More complex reactor designs are required when one or more gaseous reactants are unstable at low temperatures and must be generated in the main reaction vessel, typically by passing a reactive gas over a condensed source. Reaction then proceeds to generate the unstable reactants, which are then carried with the main flow into the substrate region where they undergo reaction to form the epitaxial layer.

A special type of chemical transport epitaxy is the close-spaced technique (Nicoll, 1963). The source, in a wafer form, is separated from the substrate by a very small distance, typically less than a millimeter. A transport agent is present in the ambient and, by imposing a suitable temperature gradient, transport occurs between the source and substrate wafers. The very narrow separation space confines the reaction within the region between the wafers and permits use of a very simple and compact apparatus.

B. THERMODYNAMIC FACTORS

Selection of a chemical process for epitaxial growth is based on a number of factors, including availability of suitably pure reactants, their compatibility with the substrate and growth apparatus, and the thermodynamics of the process. A thermodynamic analysis is essential not only to arrive at the optimum growth conditions but also for the proper apparatus design. A process with an overall negative enthalpy of reaction proceeds farther toward completion as the temperature is decreased. These processes require a reaction tube with hot walls such as may be obtained by resistance heating. For example, germanium may be epitaxially deposited by disproportionation of germanium diiodide according to:

$$2GeI_2(g) \leftrightharpoons Ge(s) + GeI_4(g) \qquad (1)$$

where $\Delta H \simeq -37$ kcal/mole (Jona et al., 1964). According to the van't Hoff expression,

$$d \ln K/dT = \Delta H/RT^2 \qquad (2)$$

where ΔH is the enthalpy of reaction, K the equilibrium constant, and T the absolute temperature, the equilibrium constant for reaction (1) decreases with increasing temperature, and from a thermodynamic point of view the direction of this reaction should shift to the right and the extent of germanium deposition should increase as the deposition temperature is lowered. This imposes a restriction on the design of an apparatus for germanium epitaxy using reaction (1). The growth process cannot take place in a cold wall, induction-heated vessel since most of the deposition would occur on the cooler tube walls rather than on the substrate. Instead the walls of the reaction vessel must be heated to a temperature equalling or exceeding that of the substrate, typically by an external resistance furnace.

Hot reactor tube walls are often suspected as a source of impurities that contaminate epitaxial deposits. In addition, they are subject to attack by the reactive atmosphere at high temperatures. These problems may be avoided by use of rf induction heating where only the substrates and their support or susceptor are heated while the walls remain relatively cool. In this case, according to Eq. (2), it is necessary to select a deposition process whose overall enthalpy of reaction is positive. Continuing with germanium as an example, the process

$$GeCl_4(g) + 2H_2(g) \rightarrow Ge(s) + 4HCl(g) \qquad (3)$$

is suitable for cool-wall deposition since $\Delta H = 34$ kcal/mole (Stull, 1965; Evans and Richards, 1952).

Thermodynamic analyses of potential CVD epitaxial processes also provide an important estimate of the extent of departure from equilibrium or the supersaturation of the gas stream in the growth region under various conditions. Here the contrast between the use of CVD for growth of epitaxial layers versus deposition of polycrystalline or amorphous coatings becomes apparent. Typically, CVD processes for coatings or similar applications operate with large departures from equilibrium in the deposition region in order to produce uniform coatings at maximum deposition rates. Epitaxial growth, on the other hand, must take place under relatively low gas-phase supersaturations so that the reactant partial pressures do not deviate excessively from the equilibrium values. If the arrival rate of reactant species greatly exceeds the nucleation and lattice incorporation rate at the crystal surface, poor crystal perfection will result. High fluxes of reactants to the surface increase the probability of misoriented two-dimensional nucleation and, in severe cases, result in polycrystalline layers. In addition, excessive super-saturations may lead to three-dimensional nucleation in the gas phase with the resultant nuclei falling onto the substrate surface to produce misoriented regions within the layers.

C. KINETIC FACTORS

Generally it is impractical to operate CVD epitaxial processes with flow rates sufficiently low to permit complete equilibration of the incoming reactants with the growing layer. Thus a portion of the reactants must pass by unreacted, and the overall epitaxial deposition rate may be limited by the rate of mass transfer of reactants from the main gas stream to the substrate surface where they undergo reaction. This is the classic diffusion or mass-transport controlled process. A convenient working model assumes a hypothetical boundary layer or stagnant gas region near the substrate surface where the gas stream velocity approaches zero. Outside this boundary layer the reactant concentrations are at their bulk or mainstream values. If the surface reactions are rapid, reactants reaching the substrate can react to produce the equilibrium concentration of reactants and products at the surface. In this case the driving force for epitaxial growth is a concentration gradient established throughout the boundary layer, which promotes diffusion of reactants and gaseous products. Actually material transport may result from either gas-phase diffusion or convection or a combination of both. In any case the rate is governed by a physical transport process. Increasing the mass transport rate, for example, by improving the flow dynamics, will increase the growth rate until the point is reached where the reactants are arriving at the surface at a rate faster than they can undergo reaction to form the epitaxial layer. The reactant concentration then increases at the surface, departing from the near equilibrium value and ultimately approaching the mainstream value. Now the overall epitaxial growth rate is limited or determined by the rate of a surface process and is said to be surface controlled or kinetically controlled with reference to the chemical reaction kinetics. The rate-determining surface process may be reactant adsorption, product desorption, or actual chemical reaction on the surface, all of which are chemical processes. Thus, depending on the gas-phase transport efficiency and the velocity of the epitaxial surface processes, the deposition rate may be limited by either chemical or physical steps.

Kinetically controlled processes exhibit growth rates that increase rapidly with increasing temperature due to the activated nature of chemical reactions Their rates are insensitive to variations in gas-stream velocity but are strongly influenced by the crystallographic orientation of the substrate surface. In contrast, diffusion-controlled processes are not so strongly temperature dependent, since gas-phase mass transport is not an activated process. Yet any factor that significantly influences the rate of mass transfer, such as the gas velocity or geometrical orientation of the crystal within the apparatus, will be expected to affect the overall growth rate of a mass transport controlled epitaxial process. A more detailed discussion of these factors may be found in a previous publication (Shaw, 1974).

It is difficult to decide a priori if a given CVD process will be mass-transport or kinetically controlled. This must be inferred from experimental growth-rate measurements. Nevertheless, such information is very important to the successful development of epitaxial processes, since it permits optimization of the operating variables. It has also been demonstrated that the rate-controlling step can influence the properties of the resultant layers (Silvestri, 1969; Dorfman and Belokon', 1969; Laukmanis and Feltyn, 1968).

D. BASIC CHEMICAL PROCESSES

Most of the basic classes of chemical reactions have been employed for chemical transport. Among those useful for CVD epitaxy are hydrolysis, synthesis, pyrolysis, oxidation, reduction, and disproportionation. These basic classifications will be discussed below along with several examples. Epitaxial growth of some II–VI compounds is possible by synthesis from the elements. For example,

$$Zn(g) + Te(g) \rightleftharpoons ZnTe(s) \qquad (4)$$

Either closed- or open-tube approaches may be used with separate elemental sources or a single source of the compound. The latter case represents sublimation, and even in an open-tube system where the source is sublimed into a flowing carrier gas for subsequent condensation to form an epitaxial layer it should not be classed as CVD epitaxy. It is vapor transport but not chemical vapor deposition, as no chemical reactions are involved. On the other hand, if the starting sources are elemental zinc and tellurium, their sublimation into a carrier gas stream with subsequent deposition might well be classified as CVD since the final solid phase is chemically different from the source.

Pyrolysis or thermal decomposition of gaseous compounds to form epitaxial layers is becoming increasingly popular. Generally it requires a relatively simple apparatus with a minimum number of reagents. In the simplest case only a single vapor species is decomposed to form the condensed epitaxial layer with the products being swept away by the carrier gas such as illustrated by the reaction

$$GeH_4(g) \rightarrow Ge(s) + 2H_2(g) \tag{5}$$

Compounds can also be deposited by pyrolysis of single gaseous reactants, as in Eqs. (6)–(8):

$$Zn(C_3H_5O_2)_2(g) \rightarrow ZnO(s) + \text{gaseous products} \tag{6}$$

(Korzo et al., 1969),

$$(CH_3)_2SiCl_2(g) \rightarrow SiC(s) + 2HCl(g) + CH_4(g) \tag{7}$$

(Rai-Choudhury and Formigoni, 1969),

$$GaBr_3 \cdot NH_3(g) \rightarrow GaN(s) + 3HBr(g) \tag{8}$$

(Chu, 1971). These reactions are usually irreversible, and low reactant partial pressures are required to produce reasonable supersaturations and prevent homogeneous decomposition with subsequent polycrystalline deposition. Compound deposits may also be formed by pyrolysis of multiple gaseous compounds containing the component elements. For example (Manasevit and Simpson, 1969, 1971),

$$(C_2H_5)_3Ga(g) + PH_3(g) \rightarrow GaP(s) + \text{gaseous products} \tag{9}$$

$$(C_2H_5)_2Zn(g) + (CH_3)_2Te(g) \rightarrow ZnTe(s) + \text{gaseous products} \tag{10}$$

The tendency to premature decomposition and the large departures from equilibrium during growth represent the greatest disadvantages of pyrolysis processes. In many cases these disadvantages are more than offset by the experimental simplicity and the absence of reactive products. Reactions using reactive transport species such as halogens must often be excluded because either the reactants or the products attack the substrate.

Hydrolysis reactions have been the method of choice for epitaxial growth of oxides. Examples include growth of simple oxides such as

$$2AlCl_3(g) + 3H_2O(g) \rightarrow Al_2O_3(s) + 6HCl(g) \tag{11}$$

(Messier and Wong, 1971),

$$VOCl_3(g) + H_2O(g) + \tfrac{1}{2}H_2(g) \rightarrow VO_2(s) + 3HCl(g) \tag{12}$$

(Takei, 1968) by direct hydrolysis, or in some cases the water vapor for hydrolysis can be generated by reaction of H_2 with NO or CO_2, e.g. (Nagai, 1969),

$$BeCl_2(g) + H_2(g) + CO_2(g) \rightarrow BeO(s) + 2HCl(g) + CO(g) \tag{13}$$

This approach can be useful to moderate rapid hydrolysis reactions to provide control and prevent premature deposition. Hydrolysis reactions can be employed to produce epitaxial layers of complex composition such as ferrites (Takei and Takasu, 1964) as

$$NiCl_2(g) + 2FeCl_3(g) + 4H_2O(g) \rightarrow NiFe_2O_4(s) + 8HCl(g) \tag{14}$$

or rare earth garnets (Mee et al., 1967) such as

$$6GdCl_3(g) + 10FeCl_2(g) + 19H_2O(g) + \tfrac{5}{2}O_2 \rightarrow 2Gd_3Fe_5O_{12}(s) + 38HCl(g) \tag{15}$$

The last reaction requires a more oxidizing atmosphere provided by addition of elemental oxygen. In some cases where hydrolysis is too rapid, the oxides may be formed by direct oxidation in an oxygen atmosphere, e.g., Ghoshtagore and Noreika (1970),

$$TiCl_4(g) + O_2(g) \rightarrow TiO_2(s) + 2Cl_2(g) \tag{16}$$

Here again it must be stressed that only a portion of the reactions used for general CVD can be applied to epitaxial growth because of substrate attack or large supersaturations that degrade the crystal perfection of the deposited layers.

Reduction of volatile halides, particularly by hydrogen, is another widespread approach. Typical examples are:

$$BCl_3(g) + \tfrac{3}{2}H_2(g) \rightarrow B(s) + 3HCl(g) \tag{17}$$

(Peters and Potter, 1966),

$$xMoF_6 + yWF_6(g) + 3(x+y)H_2(g) \rightarrow Mo_xW_y(s) + 6(x+y)HF(g) \tag{18}$$

(Gillardeau et al., 1971). Epitaxial growth of silicon by hydrogen reduction of silicon tetrachloride or trichlorosilane is a commercially successful epitaxial process in current use. Also, a number of III–V compound semiconductors are deposited by reduction of the group III halide by H_2 in the presence of the gaseous group V element as shown by the following reaction (Conrad, 1967):

$$xGaCl(g) + (1-x)InCl(g) + \tfrac{1}{4}As_4(g) + \tfrac{1}{2}H_2(g) \rightleftharpoons Ga_xIn_{1-x}As(s) + HCl(g) \tag{19}$$

Since many of these reactions are reversible in the temperature range of interest, they can be used as transport processes by passing the hydrogen halide over a condensed phase source, followed by a temperature change as the resulting mixture then passes into the growth region.

Disproportionation reactions are also useful for transport and epitaxial growth. A good example is growth of germanium by disproportionation of GeI_2 (Marinace, 1960) according to the reaction

$$2GeI_2(g) \rightarrow Ge(s) + GeI_4(g) \tag{20}$$

This classic deposition process will be discussed in more detail later. Other examples of disproportionation reactions for epitaxial growth include

$$2SiI_2(g) \rightarrow Si(s) + SiI_4(g) \tag{21}$$

(Newman and Wakefield, 1963),

$$3AlCl(g) + 2PH_3(g) \rightarrow 2AlP(s) + AlCl_3(g) + 3H_2(g) \qquad (22)$$

(Richman, 1968).

E. SUBSTRATES

As in any epitaxial growth technique, preparation of the substrate surface prior to growth is a critical step. The surface upon which epitaxial deposition is to occur must be relatively free of crystal imperfections and contamination. Prior to insertion in the growth apparatus the substrate is carefully polished both mechanically and chemically, if possible, to provide a surface with maximum perfection. However, external predeposition cleaning is not entirely successful with some materials because of the rapid growth of native oxides. Fortunately, with some CVD systems, particularly those with reversible deposition reactions, an additional in situ substrate cleaning can be obtained by vapor etching just prior to growth. With reversible deposition reactions the gas composition or temperature can be changed to effect a smooth transition from etch to growth conditions, thus contributing to the interfacial crystal perfection.

The substrate itself can influence the composition or phase of the deposited material. The orienting effect of the substrate during epitaxial growth can in principle produce metastable phases (Eversole, 1962) when these are crystallographically more compatible with the substrate. This effect has been used with success to prepare iron-deficient yttrium iron garnet films by CVD epitaxy where these films could not be obtained otherwise (Braginski, 1972).

III. Systems Illustrating the Principles of CVD Epitaxy

Application of the basic principles and chemical processes described in the preceding section can best be discussed with reference to a few well-established epitaxial processes. These processes were not chosen because of their commercial usage, but rather because they best illustrate the fundamentals and have been more thoroughly investigated than most other processes. The systems will be described in order of increasing complexity beginning with an elemental pyrolysis system, continuing through reduction and disproportionation processes to include deposition of stoichiometric compounds, and finally concluding with a process for heteroepitaxial growth of alloys.

A. SILICON FROM SILANE PYROLYSIS

Silane pyrolysis is the simplest and most straightforward process for silicon epitaxy. No bubbler or condensed reagent source is necessary—only the carrier

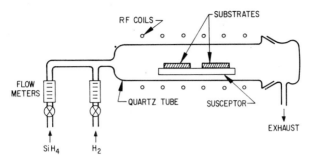

FIG. 1.   Silicon epitaxy by silane pyrolysis.

gas and silane, which is a gas at room temperature. The apparatus is shown schematically in Fig. 1. Although a horizontal reaction tube is illustrated, vertical tubes are also used. As the mixture of silane and carrier gas enters the vicinity of the heated substrates, decomposition occurs according to the reaction

$$SiH_4(g) \rightarrow Si(s) + 2H_2(g) \qquad (23)$$

In order to minimize contamination from the tube, the walls are usually water cooled. Substrate heating is provided by coupling rf energy into the susceptor, which also serves as a substrate support. At typical deposition temperatures silane has a Gibbs energy of formation of approximately 36 kcal/mole (Stull, 1965); hence, the decomposition reaction is essentially irreversible. The gas-phase supersaturation must then be controlled by the silane partial pressure alone. Also, the reverse reaction is not available for predeposition substrate etching, necessitating use of additional components for this process such as HCl and HI. Use of an inert carrier gas such as He in place of H_2 increases the deposition rate (Richman et al., 1970). At low temperatures the rate-limiting step for deposition is desorption of hydrogen from the surface (Joyce and Bradley, 1963). Note that hydrogen decreases the overall reaction rate by influencing a surface process rather than by shifting the point of equilibrium in reaction (23), since the process is so irreversible as to be essentially insensitive to the hydrogen partial pressure under ordinary conditions.

Since the growth rate decreases with decreasing temperature (Gupta, 1971), a practical minimum temperature exists. However, by use of helium as a carrier gas the inhibiting effect of hydrogen on the deposition rate is avoided, permitting reduction of the deposition temperature from the usual 1000–1050°C down to around 900°C. This is an excellent example of the application of fundamental reaction mechanism data to solution of a practical problem. Low epitaxial-growth temperatures promote purer deposits and diminish out-diffusion of impurities from the substrate into the epitaxial layer. These features are especially appealing for growth of epitaxial layers for electronics application and are responsible for the increasing use of the silane process.

Pyrolysis processes are particularly useful for heteroepitaxial growth. The hydrogen formed as a byproduct of reaction (23) is relatively unreactive toward substrates of interest in silicon heteroepitaxy. This permits growth of silicon layers on insulating substrates such as sapphire or spinel, which would be attacked by the HCl formed as a byproduct in the $SiCl_4$ process discussed below.

B. Silicon from Hydrogen Reduction of Silicon Halides

Although this approach to silicon epitaxy requires one more reactant and is chemically more complex than the simple silane decomposition process, it was the first to achieve commercial acceptance and remains the workhorse of the electronics industry. Silane is pyrophoric and, consequently, special handling techniques are required. On the other hand, silicon tetrachloride represents a convenient and easily purified source of silicon. Both silicon tetrachloride and trichlorosilane have been used with the respective reduction reactions

$$SiCl_4(g) + H_2(g) \rightarrow Si(s) + 4HCl(g) \tag{24}$$

$$SiHCl_3(g) + H_2(g) \rightarrow Si(s) + 3HCl(g) \tag{25}$$

Since these reactions are partially reversible, the reverse reaction can lead to varying degrees of substrate etching. Silicon tetrachloride is by far the most popular of the halide growth processes. Experimentally, the epitaxial apparatus does not differ greatly from that used for epitaxial growth from silane. However, a condensed source ($SiCl_4$) is present, and provision must be made for its introduction into the main reaction region. Figure 2 illustrates a typical deposition system. Silicon tetrachloride is transported into the growth region by bubbling hydrogen through the liquid. The input partial pressures may be

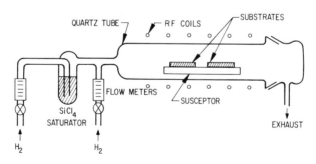

Fig. 2. Silicon epitaxy by silicon tetrachloride reduction.

controlled by regulation of the temperature of the $SiCl_4$ saturator or by dilution of the gas stream with hydrogen. Only two reagents are required, since hydrogen serves not only as the carrier gas, but also as one of the reactants. The reduction process is endothermic, permitting use of rf heating in conjunction with cool tube walls.

Since the reduction reactions are reversible, they can be used for growth under conditions that are closer to equilibrium than the irreversible silane process. This can lead to more selective deposition under certain conditions, with growth in some crystallographic directions proceeding at rates up to 300 times faster than others, while no rate differences are found under similar conditions with the silane process (Nishizawa *et al.*, 1972). Of course, such large rate variations would not be expected under all conditions even with a halide reduction system. The reversible nature of halide reduction also makes this method less susceptible to homogeneous nucleation of silicon crystallites, allowing more degrees of freedom in controlling the gas-phase supersaturation. For this purpose both the deposition temperature and the $SiCl_4$ partial pressure may be varied. However, for good deposition the temperature is usually around 1200 to 1250°C, and this represents the principal disadvantage of the halide reduction process. High growth temperatures promote solid-state diffusion of impurities from the substrate into the growing layer, and the HCl produced as a product of reactions (24) or (25) may attack the substrate locally to release impurities that later become incorporated into the layer. These two methods of silicon epitaxy—pyrolysis and halide reduction—are good examples of modification of the fundamental chemistry of a CVD process to produce the desired results.

C. GERMANIUM FROM GERMANIUM DIIODIDE DISPROPORTIONATION

Germanium can be epitaxially deposited at very low temperatures by CVD epitaxy. The fundamental reaction is disproportionation of germanium diiodide

$$GeI_2(g) \leftrightharpoons Ge(s) + GeI_4(g) \tag{26}$$

Although the reaction appears very simple, the experimental apparatus is rather complicated because GeI_2 is unstable at room temperature and must be generated at higher temperatures. Figure 3 gives a diagram of the deposition system. Note that there are two condensed sources, each of which must be maintained at a specific temperature. In the first zone iodine is sublimed into the flowing carrier gas. As the resulting mixture passes over a solid germanium source, the following reaction occurs:

$$I_2(g) + Ge(s) \rightarrow GeI_2(g) \tag{27}$$

The carrier gas together with GeI_2 formed in the second zone then enters the

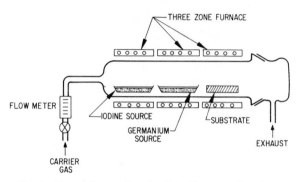

FIG. 3. Germanium epitaxy by GeI_2 disproportionation.

third or growth zone where, at a relatively low temperature, disproportionation with subsequent formation of epitaxial germanium occurs. In general, the total gas flow rates for a CVD epitaxial system with solid sources are relatively low because of the importance of controlling the extent of the source reaction. If high flow rates are required, it becomes necessary to use packed beds or other approaches to increase the overall reaction efficiency between the solid and gas phases. It is possible to reduce the complexity of the main reaction tube by eliminating the solid iodine source (and its temperature-controlled zone) and replacing it with gaseous HI. Then the reaction at the germanium source becomes

$$2HI(g) + Ge(s) \rightarrow GeI_2(g) + H_2(g) \tag{28}$$

Gaseous HI is not readily available in a high-purity state but may be generated externally by reaction of H_2 with solid iodine using a Pt catalyst (Reisman and Berkenblit, 1965).

The germanium diiodide disproportionation reaction is reversible in the temperature range of interest and has been subjected to detailed thermo-dynamic analysis (Reisman and Alyanakan, 1964). The deposition reaction has a negative enthalpy and illustrates a process where hot reactor tube walls are essential. Germanium diiodide formed at the source becomes increasingly unstable with respect to germanium at lower temperatures, and if the tube walls were at a lower temperature than the substrate, most of the deposition would occur there. Substrate vapor etching can be easily achieved by raising the substrate temperature. Note that in reaction vessels such as these care must be taken to ensure that no temperature minima are present between the source and the growth region, or premature deposition and depletion of the gas stream will result.

Germanium epitaxial layers with a high degree of crystal perfection may be obtained at very low temperatures ($\sim 400°C$) with this process. As a rule, low temperatures lead to relatively imperfect deposits, probably because of the decreased mobility of the surface-diffusing species. The reversible nature of the disproportionation process may be responsible for its successful low-temperature operation. Perhaps the diffusing species are adsorbed reactants such as GeI_2, which may be easily desorbed if no favorable or low-energy site is available. Certainly, in a process such as evaporation onto a substrate at a temperature where the vapor pressure is very low, most of the species must remain on the surface where they will have a high probability of being "frozen in" at incorrect sites or forming misoriented, two-dimensional nuclei, all of which would result in reduced crystal perfection.

D. GALLIUM ARSENIDE FROM A Ga–AsCl$_3$–H$_2$ SYSTEM

Previous examples of epitaxial systems were limited to elemental materials. As expected, epitaxial growth of a compound is a more complex process. Not only the supersaturation but also the stoichiometry of the gas passing over the substrate is a matter of concern. Gallium arsenide is a good example for illustration of CVD epitaxy of compounds because it has been the object of a number of theoretical and experimental investigations.

Gallium arsenide dissociates at high temperatures, and at the melting point (1230°C) an arsenic pressure of ~ 0.9 atm is in equilibrium with the melt. This greatly complicates crystal growth from the melt and has prevented the use of bulk GaAs in many promising electronic applications. However, CVD epitaxial techniques permit growth of layers at temperatures well below the melting point where dissociation is negligible. A number of volatile arsenic compounds are available for CVD epitaxy, but the choice of gallium compounds is limited. Thus, the various epitaxial systems differ with respect to the arsenic-containing reactant, but are alike in the transport of gallium as a monohalide. (One exception is a system using alkyl gallium compounds, but this has not gained widespread application.) The monohalides, GaCl and GaI, are unstable at low temperatures and must be formed in situ by reaction of a halogen or volatile halide with liquid gallium. After formation the monohalides are carried into a lower temperature deposition region where they react with arsenic to form gallium arsenide according to the reactions

$$GaX(g) + \tfrac{1}{4}As_4(g) + \tfrac{1}{2}H_2(g) \rightarrow GaAs(s) + HX(g) \tag{29}$$

or

$$3GaX(g) + \tfrac{1}{2}As_4(g) \rightarrow 2GaAs(s) + GaX_3(g) \tag{30}$$

There has been some controversy as to which reaction, reduction (29) or disproportionation (30), is most important. Recent work (Boucher and Hollan, 1970; Shaw, 1971), however, indicates that the reduction process predominates because of the large excess of H_2, which also serves as the carrier gas. Both reactions are exothermic and require a deposition region with hot walls.

Since many applications for epitaxial GaAs require high-purity layers, the chemical processes and apparatus must be selected with great care. Hydrogen is attractive as a carrier gas and reactant because it can be easily purified to a high degree by diffusion through palladium. Gallium is also readily available in high purities. However, there are several possible choices for the halide transport agent and the arsenic source. Associated with high purity is simplicity of apparatus with a minimum number of sources of contamination and the use of a limited number of reagents. These considerations lead to the use of $AsCl_3$, which is easily purified by distillation and serves both as the transport agent for gallium and the source of arsenic. Thus only three reagents are required: Ga, $AsCl_3$, and H_2. This system has resulted in very high-purity gallium arsenide layers (Maruyama *et al.*, 1969; DiLorenzo and Machala, 1971). The experimental apparatus is diagrammed in Fig. 4.

Purified hydrogen is bubbled into the $AsCl_3$ saturator, and the resulting mixture passes into a heated source region. At high temperatures $AsCl_3$ is unstable in hydrogen, and the following reaction occurs immediately:

$$4AsCl_3(g) + 6H_2(g) \rightarrow As_4(g) + 12HCl(g) \tag{31}$$

The HCl formed in (31) then reacts with gallium to form the transportable gallium monochloride

$$Ga(l) + HCl(g) \rightarrow GaCl(g) + \tfrac{1}{2}H_2(g) \tag{32}$$

Actually no growth occurs initially with a fresh gallium source because the arsenic dissolves in the source until saturation is complete. After saturation gallium arsenide precipitates and floats on the gallium surface. Thus, although initially the source is liquid gallium, later only the GaAs floating on the surface is exposed to the gas stream. For this reason in thermodynamic analyses the source is considered to be solid GaAs, and the source reaction is best described

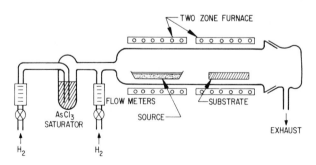

FIG. 4. Ga–$AsCl_3$–H_2 process for gallium arsenide epitaxy.

as the reverse of reaction (29). In this process we have an example of preparation of a source in situ prior to epitaxial growth. After leaving the source the mixture of GaCl, As$_4$, and H$_2$ enters the growth region, which is maintained at a temperature lower than the source, and epitaxial growth occurs via the forward portion of reaction (29).

Gallium arsenide can be grown from a wide range of gas-phase Ga:As ratios; however, the electrical properties of the resulting layers will be a function of the ratio (Shaw *et al.*, 1967). An interesting feature is the absence of GaAs deposition on the tube walls or the quartz substrate holder, even though they are at a temperature close to that of the substrate. This indicates that nucleation of GaAs on fused silica is a slow or difficult process. From another point of view, since GaAs deposition is thermodynamically favorable throughout the deposition region, any GaAs present acts as an autocatalyst for additional deposition. Depending on the temperature and gas-phase concentration, either kinetic or mass-transfer rate limitations have been identified. As expected in the kinetic region, the rate of growth depends strongly on the crystallographic orientation of the surface and the deposition temperature but is independent of the flow dynamics (Shaw, 1968).

From examination of the Ga–AsCl$_3$–H$_2$ epitaxial system it is evident that reactions occurring at condensed sources require as much attention for successful epitaxy as those occurring at the substrate. In addition, it demonstrates that careful selection of a chemical process permits growth of layers of complex composition from a relatively simple apparatus.

E. GALLIUM ARSENIDE PHOSPHIDE ALLOY DEPOSITS FROM A Ga–HCl–AsH$_3$–PH$_3$–H$_2$ SYSTEM

Gallium arsenide and gallium phosphide form a complete series of solid solutions. Most applications of epitaxial gallium arsenide phosphide alloys require only moderate purity levels, but necessitate careful control over the deposit composition and crystal perfection. These requirements result in use of a different chemical system from that described above for GaAs. Since these layers are usually grown heteroepitaxially on GaAs substrates with different crystal-lattice parameters, lattice mismatch and interfacial dislocations are encountered. Reduction of these problems may be approached by initially growing phosphorus-free epitaxial GaAs on the substrate and then gradually increasing the phosphorus concentration in the vapor until the desired steady-state layer composition is obtained. Thus the principal epitaxial GaAs$_x$P$_{1-x}$ layer is isolated from the GaAs substrate by a "graded" region where the composition varies. Such growth requires good control over the gas-phase composition, which may be achieved by the apparatus depicted in Fig. 5. Although more complicated than the simple GaAs system previously described,

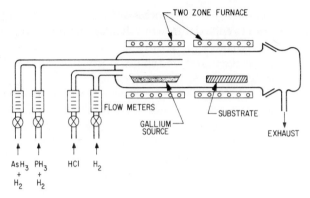

FIG. 5. Ga–HCl–AsH₃–PH₃–H₂ process for gallium arsenide phosphide epitaxy.

it is very versatile with independent control over the arsenic and phosphorus composition as well as control over the ratio of the group III to group V vapor phase species. The chemical reactions at the gallium source and in the growth region are similar to those in the previous GaAs system; however, HCl is not formed within the main reaction tube but is externally supplied as a starting reagent. In comparison with the previous process for homoepitaxial GaAs growth, this system is more complicated and uses reagents that are more difficult to purify. Thus simplicity and, to some degree, layer purity are sacrificed in favor of versatility and independent control over the process parameters, which are critical for growth of variable composition epitaxial layers.

   In summary, these examples of CVD epitaxial systems illustrate many basic principles of general applicability. Examination of these processes should provide insight that is useful for design and development of CVD processes for epitaxial growth of other materials.

References

Boucher, A., and Hollan, L. (1970). *J. Electrochem. Soc.* **117**, 932.
Braginski, A. I., Oeffinger, T. R., and Takii, W. J. (1972). *Mater. Res. Bull.* **7**, 627.
Chu, T. L. (1971). *J. Electrochem. Soc.* **118**, 1200.
Conrad, R. W., Hoyt, P. L., and Martin, D. D. (1967). *J. Electrochem. Soc.* **114**, 164.
DiLorenzo, J. V., and Machala, A. E., (1971). *J. Electrochem. Soc.* **118**, 1516.
Dorfman, V. F., and Belokon', M. S. (1961). "Growth of Crystals," Vol. 8, p. 128. Consultant Bureau, New York.
Evans, D. F., and Richards, R. E. (1952). *J. Chem. Soc.* 1292.
Eversole, W. G. (1962). U.S. Patent Nos. 3030187 and 3030188.
Ghoshtagore, R. N., and Noreika, A. J. (1970). *J. Electrochem. Soc.* **117**, 1310.
Gillardeau, J., Faron, R., Bargues, M., Hasson, R., Dejachy, G., and Durand, J. P. (1971). *J. Crystal Growth* **9**, 255.

Gupta, D. (1971). *Solid State Technol.* **14 (10)**, 33.

Jona, F., Lever, R. F., and Wendt, H. R. (1964). *J. Electrochem. Soc.* **111**, 413.

Joyce, B. A., and Bradley, R. R. (1963). *J. Electrochem. Soc.* **110**, 1235.

Korzo, V. F., Kiriev, P. F., and Lyashchenko, G. A. (1969). *Inorg. Mater.* **5**, 304.

Laukmanis, L. A., and Feltyn, I. A. (1968). *Izv. Akad. Nauk. SSSR Neorg. Mater.* **4**, 1275.

Manasevit, H. M., and Simpson, W. I. (1969). *J. Electrochem. Soc.* **116**, 1725.

Manasevit, H. M., and Simpson, W. I. (1971). *J. Electrochem. Soc.* **118**, 645.

Marinace, J. C. (1960). *IBM J. Res. Develop.* **4**, 248.

Maruyama, M., Kikuchi, S., and Mizuno, O. (1969). *J. Electrochem. Soc.* **116**, 413.

Mee, J. E., Archer, J. L., Meade, R. H., and Hamilton, T. N. (1967). *Appl. Phys. Lett.* **10**, 289.

Messier, D. R., and Wong, P. (1971). *J. Electrochem. Soc.* **118**, 771.

Nagai, H. (1969). *Jap. J. Appl. Phys.* **8**, 1221.

Newman, R. C., and Wakefield, J. (1963). *J. Electrochem. Soc.* **110**, 1068.

Nicoll, F. H. (1963). *J. Electrochem. Soc.* **110**, 1165.

Nishizawa, J., Terasake, T., and Shimbo, M. (1972). *J. Crystal Growth* **13/14**, 297.

Peters, E. T., and Potter, W. D. (1966). *Trans. Met. Soc. AIME* **236**, 473.

Rai-Choudhury, P., and Formigoni, N. P. (1969). *J. Electrochem. Soc.* **116**, 1440.

Reisman, A., and Alyanakan, S. A. (1964). *J. Electrochem. Soc.* **111**, 1134.

Reisman, A., and Berkenblit, M. (1965). *J. Electrochem. Soc.* **112**, 812.

Richman, D. (1968). *J. Electrochem. Soc.* **115** 945.

Richman, D., Chiang, Y. S., and Robinson, P. H. (1970). *RCA Rev.* **31**, 613.

Schaefer, H. (1964). "Chemical Transport Reactions." Academic Press, New York.

Shaw, D. W. (1968). *J. Electrochem. Soc.* **115**, 405.

Shaw, D. W. (1971). *J. Crystal Growth* **8**, 117.

Shaw, D. W. (1974). *In* "Crystal Growth-Theory and Techniques" (C. H. L. Goodman, ed.). Plenum Press, New York.

Shaw, D. W., Conrad, R. W., Mehal, E. W., and Wilson, O. W. (1967). *Proc. Int. Symp. Gallium Arsenide, 1st* p. 10. Inst. Phys. Soc., London.

Silvestri, V. J. (1969). *J. Electrochem. Soc.* **116**, 81.

Stull, D. R. (ed.) (1965). JANAF Thermochemical Tables. Dow Chem., Midland, Michigan.

Takai, H. (1968). *Jap. J. Appl. Phys.* **7**, 827.

Takai, H., and Takasu, (1964). *Jap. J. Appl. Phys.* **3**, 175.

2.5 Growth of Epitaxial Films by Sputtering

M. H. Francombe

Westinghouse Electric Corporation
Research and Development Center
Pittsburgh, Pennsylvania

I. Introduction

In comparison with the numerous studies reported on the growth of epitaxial layers by vacuum evaporation and chemical-vapor deposition, relatively little work has yet been done on the epitaxy of films deposited by sputtering. There have been several reasons for this. Those workers interested in the fundamental mechanisms of nucleation and epitaxial growth have tended to prefer simpler evaporation approaches for depositing pure metals or alkali halides, where the correlation of structural features with deposition conditions is more amenable to analysis. For the formation of epitaxial semi-conductor layers and junctions, chemical methods have dominated the field, since they offer high rates of deposition and are more appropriate to component manufacturing technology. In consequence, those studies of epitaxy by sputtering that have been made to date have been motivated by basic scientific interest or by the special need for film compositions or orientations not immediately accessible via the more common evaporation and chemical approaches.

Another factor that should not be discounted is that clean rf sputtering techniques, applicable both to chemically reactive metallic and to insulating materials, are still relatively new and unfamiliar. Their development has been prompted in large part to meet the growing needs of the electronics industry for large area, multilayer resistor–conductor patterns, or for uniform capacitor dielectric or passivating coatings. In this work the tendency, understandably, has been to achieve standardized, reproducible deposition conditions for simple metallic and dielectric films, there being no interest or need to produce these in oriented or epitaxial form.

Probably the main driving force in exploring the potentialities of sputtering for epitaxial film growth has been generated by interest in novel applications of semiconductor, transducer, and magneto- or electrooptic films. Several important advantages were anticipated for sputtering, as compared to other techniques, in the preparation of such films. First, on the basis of work with sputtered metal alloys, it was expected that in general the desired film composi-tion could be achieved simply by sputtering from a target of the same composition. Next, from studies on nonreactive metal films such as gold and silver, it appeared likely that lower epitaxy temperatures might be achieved in sputtering. This has obvious attraction in the growth of abrupt semiconductor junctions and in satisfying low-temperature processing needs in hybrid integrated circuits. Unlike chemical deposition methods, sputtering does not involve side reactions with the substrate material and thus permits greater flexibility in the choice of substrate crystals.

In this chapter we begin by discussing the mechanisms involved in cathode disintegration, transfer of sputtered material, and formation of the deposit at the substrate. Next we consider the main experimental sputtering approaches available for growing epitaxial layers. The available literature on sputtered epitaxial films of metals, semiconductors, and insulators is reviewed next, and finally some comments are included on future research directions and applications.

II. Sputtering and Film-Deposition Processes

A. SPUTTERING PROCESS

Cathodic sputtering, i.e., the process by which the surface of a negatively biased target disintegrates under the action of bombardment by positive ions of inert gas, has formed the topic of much experimental research and theoretical speculation for over a century. Technological applications for the process were found well before the basic mechanisms were properly studied or understood. Largely due to the experimental investigations of Wehner (1955) and Wehner and Anderson (1970), factors affecting the sputtering yield from pure elements have now carefully been documented. The characteristic sputtering yields for the elements and for several compounds have been measured and have been found to increase (initially sharply) with ion bombarding energy to a maximum (usually at several kilovolts), and then saturate—or at higher energies slowly decrease. The yields in atoms per ion range typically from 2 to 8 for elements under bombardment by Ar^+ ions at energies of about 10 kV. Sputtering yield is a function of the mass of the bombarding ion and can be increased considerably by using heavier ions such as Kr^+ or Hg^+. The angle of incidence of the arriving ions with respect to the target surface also affects sensitively the rate of sputtering. Increases in the range of 6 to 10 are found in changing from normal incidence (90°) to an incidence angle of 30°. The yield is also strongly influenced by the crystallographic orientation of the target, sputtered material being emitted at higher rates along close-packed directions of the lattice.

Current theories (Pease, 1960) visualize the sputtering process as one of momentum transfer in which the arriving ion produces a primary knock-on atom at the surface of the target. This knock-on atom then moves up to several atomic layers deep into the target, losing energy at each collision until it is brought to rest. In the process a target atom may be ejected. By suitably modifying the assumption made concerning interactions between the arriving ion and the electron screening cloud for the target atoms, the theory can be made to explain changes of yield over a wide range of ion energies.

B. NATURE AND ENERGY OF SPUTTERED PARTICLES

Mass spectrographic evidence on the nature of particles sputtered from metals such as Cu (Woodyard and Cooper, 1964) shows these in general to be atoms or groups of atoms, of which only a small fraction ($\sim 1\%$) are electrically charged. At higher bombardment energies increasing numbers of polyatomic aggregates are ejected, as evidenced by the work of Herzog et al. (1967) on aluminum. Results for multicomponent targets such as semiconductor compounds are conflicting in that, for the case of GaSb, Wolsky et al. (1962)

state that sputtering occurs via the removal of molecular groups, while for GaAs, Comas and Cooper (1967) report that the sputtered material comprises mainly ($\sim 99.4\%$) neutral Ga and As atoms. The latter data were based upon direct mass spectrographic evidence, however, while the GaSb results were inferred from the yield data and from analysis of the collected deposit.

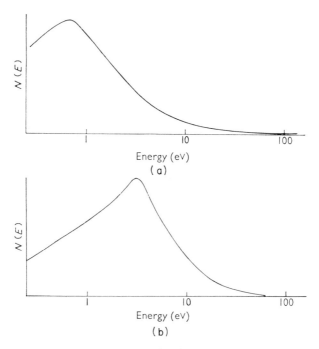

FIG. 1. Energy spectra of gold sputtered in different ejection directions from a single-crystal target: (a) ⟨100⟩ direction; (b) ⟨110⟩ direction (after Thomson, 1968).

Studies by several authors (Stuart and Wehner, 1962; Thomson, 1968) indicate that sputtered particles possess much higher energies than particles generated by thermal evaporation. For a thermal source operating at 2000°K the average energy of the emitted atoms is about 0.2 eV. However, for sputtered atoms, depending upon the energy of the bombarding ions and upon the crystallographic direction of ejection, mean energies as high as 100 eV may be developed. Results obtained by Thomson (1968) on energies of ejection from gold bombarded by 41-keV argon ions are shown in Fig. 1 for two different crystal directions. Energies along the more close-packed ⟨110⟩ directions are significantly higher, giving a mean value of 93.5 eV compared with 22.7 eV for the ⟨100⟩ directions.

C. FILM-DEPOSITION PROCESS

A wide variety of sputtering techniques has been developed for meeting the deposition needs peculiar to different materials and technological applications. The way these influence the conditions of film deposition will become clearer in our discussion in the following section on experimental methods. However, some general comments can be made at this point concerning major differences in the conditions of growth for sputtered as opposed, for example, to evaporated films.

Depending upon the method used to sustain the discharge during sputtering, the surface of the substrate and the growing film will be subjected to bombardment to a varying degree by electrons and neutral and charged gas atoms (Koenig and Maissel, 1970; Maissel, 1970). Associated with these differences will be wide variations in the incident energy and directions of atoms arriving from the cathode. We may illustrate this by considering the two cases of an unsupported, diode-type, glow discharge operated typically at pressures in the range from 20 to 100 mTorr, and thermionically and magnetically assisted discharge at pressures of 1 to 5 mTorr such as would be achieved with a triode-type system (described below). In the first case, the secondary electrons required to sustain the supply of positive ions are generated during ion bombardment of the target, and to maintain a supply of electrons sufficient to keep the discharge going, relatively high target voltages (2–5 kV) and gas pressures are needed. In the second case, the electrons are furnished by a thermionic (filament) source, and their ionizing effect is enhanced by the magnetic field, which increases their path length. In this situation the gas pressure can be relatively low, and a fairly high sputtering yield is achieved with modest target voltages (500–1000 V).

Under high-pressure, glow-discharge conditions, the sputtered atoms undergo many collisions with gas atoms before reaching the substrate, and indeed a high proportion may be back-sputtered onto the target surface. These collisions attenuate the initial high ejection energy and cause the target atoms to arrive at the substrate over a wide range of incident angles, giving rise to the well-known "back-coating" effect. As a result of electron bombardment and energy dissipated at the substrate surface by ion–electron recombination, the surface temperature (of an insulating substrate) rises rapidly within a few seconds from the initiation of the discharge (or from exposure to the negative glow) to reach values as high as several hundred degrees centigrade (Francombe and Noreika, 1961a) (see Fig. 2). Although the temperature rise may be suppressed by lowering the cathode potential, this leads to a reduction in sputtering rate to values where the relative rate of arrival of background impurities becomes significant. These effects have a significant bearing upon conditions chosen for epitaxial growth. Substrate temperatures chosen for

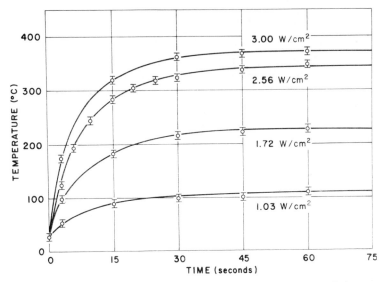

FIG. 2. Temperature rise at substrate surface during glow-discharge diode sputtering (after Francombe and Noreika, 1961a).

epitaxy and the conditions of sputtering should be adjusted so that the discharge causes no further appreciable rise from the initial-growth temperature value.

Using low-pressure, assisted discharge conditions of operation, the sputtered atoms, although possibly ejected at lower velocities (due to the smaller values of target voltage), undergo relatively few collisions during transfer to the substrate and consequently arrive with high energies and at more or less normal incidence. The high arrival energy is sufficient to cause significant lattice penetration and can induce surface damage (Francombe, 1966a) and even alloying or chemical reaction (Mattox and McDonald, 1963) with the substrate surface. Under the higher substrate temperature conditions used in epitaxial growth, the damage is probably minimized by annealing effects. An important difference produced by operating at low pressures is that the substrate heating effect due to the discharge is considerably reduced. This is illustrated in Fig. 3, which shows the temperature measured at the surface of a vitreous silica substrate during triode sputtering from an uncooled platinum target at an argon pressure of 5 mTorr. In this case the temperature rises very slowly during sputtering and, on turning off the discharge, cooling occurs at a comparable rate. Measurements of the target temperature during the same period showed that it attained a maximum temperature of about 400°C. There seems little doubt, therefore, that the substrate heating under these conditions

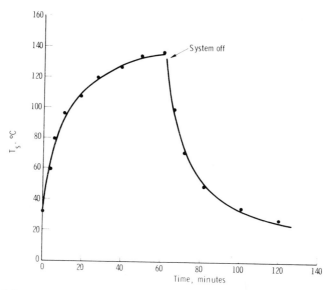

FIG. 3. Substrate surface temperature during low-pressure triode sputtering of platinum on quartz, where the target potential is 925 V, the target current is 39 mA, and the area is 65 cm².

can be attributed primarily to radiation from the target. Water cooling of the target, as is done routinely in low-pressure rf sputtering, is found to reduce substrate heating considerably (Lau and Mills, 1972).

Several of the sputtering techniques to be discussed tend to introduce additional factors that are important in influencing the film nucleation and growth processes. In particular, bias sputtering under both dc and rf conditions, through reemission effects, can cause major changes in both the structure and physical and electrical properties of the growing film. These effects will be considered briefly in the next section and, where appropriate, will be related in later discussion to specific examples of epitaxial growth.

III. Experimental Techniques

Several detailed review articles have been written on sputtering and on techniques for film deposition, and the reader is referred to these (Holland, 1960; Maissel, 1966, 1970; Francombe, 1963) for more complete information on experimental arrangements. In particular, Holland (1960) and, more recently, Maissel (1970) have fully discussed the literature on experimental approaches as applied primarily to the preparation of nonepitaxial films. The present discussion is confined chiefly to those aspects that are known or may be expected to be relevant to the growth and compositional control of epitaxial layers.

A. dc METHODS

1. *Unassisted Glow-Discharge Sputtering*

This is the simplest, historically the oldest, and, until recently, most widely used of the sputtering methods. Both apparatus design aspects and a wealth of empirical data relating experimental factors such as gas pressure, composition, electrode spacing and geometry, cathode potential, and discharge current to rate of sputtering have been published over the years, and are reviewed by Holland (1960). In this arrangement an "abnormal" glow-discharge condition is commonly used, with the cathode completely covered by the cathode glow, and with the boundary separating the cathode dark space from the negative glow positioned at approximately halfway between the electrodes. The exact location of this boundary is not critical and, at a given cathode potential, will vary with gas pressure, moving farther from the cathode as the pressure is reduced. In equilibrium, an electron emitted at the cathode will be accelerated across the cathode fall region (in the cathode dark space) and must, by ionization, produce sufficient positive inert gas ions to eject another electron from the cathode. If either the pressure is too low or the electrode spacing too small, the anode surface enters the cathode dark space, and ionization is reduced to a point where the discharge is quenched. This fact is exploited in preventing sputtering from the back surface of the cathode by positioning a grounded shield close to the surface.

We have already pointed out that glow-discharge sputtering tends to produce significant heating of both the cathode and the substrate surface. Other surfaces in the vacuum chamber exposed to the discharge also become heated and are subjected to ion and electron bombardment, with the result that severe outgassing may occur and seriously contaminate the growing film. To ensure optimum purity, particularly in sputtered films of reactive metals and semiconductors, several precautions must be taken. A well-trapped, high-speed oil-diffusion pump system capable of base pressures in the low 10^{-7} Torr range is preferred, operated under conditions where the sputtering gas is admitted continuously through an adjustable leak and is pumped with the high-vacuum valve throttled. The target should be free of pores or gas occlusions and should be well cooled during operation, preferably directly with flowing water. The walls of the chamber exposed to the discharge should also be cooled in a similar fashion, and for this reason steel bell jars are preferred. The highest-purity sputtering gas should be employed and, if necessary, the sputtering stmosphere should be purified further by passing the gas over heated copper or titanium or through a gettering discharge [cf. Theurer and Hauser (1964)].

The main advantages of high-pressure, glow-discharge systems are their extreme simplicity and their ability, by using large-area cathodes, to produce deposits uniform in thickness and physical properties over very large substrate areas. Disadvantages arise from the fact that, in order to obtain high currents, high pressures are needed. Unfortunately, although the sputtering rate increases with the current, the sputtering efficiency is reduced due to back-sputtering of atoms ejected at the cathode, and by the diffusion-limited transfer of these atoms through the ionized plasma. Also, occlusion of gas in the deposit is increased, and the film grows in a high-temperature glow region under the influence on intense electron bombardment. This presents problems in the measurements and control of temperature during epitaxial growth.

2.   *Assisted Discharge Sputtering*

With the aim of improving sputtering efficiency and of minimizing structural and chemical changes induced in films grown in a glow-discharge environment, several approaches have been developed in which ionization is enhanced at low gas pressures, either by increasing the effective path lengths of the ionizing electrons in the cathode fall region, or increasing the number of electrons by using a secondary source. The first approach is achieved most effectively by applying a magnetic field transverse to the electron path; but this displaces the plasma between the cathode and anode and leads to nonuniform sputtering. A magnetic field parallel to the potential gradient is reasonably effective, since most of the electrons have a transverse velocity component. However, Kay (1963) has reported that by using a quadrupole magnetic field (produced by two opposed coils coaxial with the direction of the electrical field) a radially symmetric transverse magnetic field is produced, optimizing the desired influence on electron path length. As much as an eightfold increase in sputtering rate is thereby achieved.

A more generally accepted arrangement that employs both a thermionic electron source and an assisting magnetic field has been developed for use in bell-jar systems by Nickerson and Moseson (1964) and is illustrated schematically in a form suitable for epitaxial growth in Fig. 4. The basic concept is due to Ivanov *et al.* (1961), who employed a horizontal electron-gun system. A system of the type shown in Fig. 4 has been used extensively in the author's laboratory for epitaxial growth of a wide range of metals and oxides. Typical operating conditions for the epitaxial growth of Pt on CaF_2 are: argon pressure, 5 mTorr; substrate temperature, 530°C; anode potential, 65 V; anode current, 3 A; target potential, 1200 V; target current, 2 mA/cm^2; deposition rate, 200 Å/min.

One of the advantages of such a low-pressure system lies in the fact that,

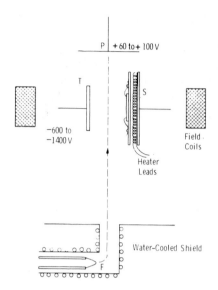

Fig. 4. Schematic diagram of low-pressure triode sputtering system for epitaxial growth.

once a few calibration runs have been made, the sputtering rate as a function of voltage and ion mass can be predicted for different metals with confidence from the yield data in the literature (Wehner, 1955; Wehner and Anderson, 1970).

3. *Bias Sputtering*

This version of dc sputtering emerged from some work on asymmetric ac sputtering conducted by Frerichs (1962) for the deposition of superconducting films. In these experiments a diode-resistive bridge was used, which permitted the application of the full negative voltage to the cathode on one half-cycle and of a greatly reduced negative voltage (limited by a resistor) to the anode on the next. Maissel and Schaible (1965) developed a more efficient means of producing the effect under dc glow-discharge conditions, in which the cathode was operated in the normal mode and a much smaller variable dc voltage was applied to the growing (conductive) film. The objective of this approach is to sputter selectively impurities normally adsorbed on and incorporated into the growing film back into the vapor phase. For the approach to be successful, the strength of the chemical bond between the element being deposited, e.g., Ta, Nb, Ni, etc., and the reactive contaminant, e.g., O_2, N_2, should be significantly weaker than the bonding between constituent atoms of the growing film. The efficiency of bias sputtering in reducing impurities in superconducting and magnetic films has been convincingly demonstrated by relating measurements of impurity-sensitive parameters such as resistivity (Maissel and Schaible, 1965) and coercive force (Griest and Flur, 1967) to the magnitude of the bias voltage. In general, as the bias is increased resistivity and coercive force are found to decrease.

Recent work by Winters and Kay (1972) has highlighted the need for understanding the nature of chemical interactions occurring between impurities and the sputtered deposit in order to optimize bias conditions for achieving purer films. Based upon their studies of the reactive sputtering of metals such as Ni, W, and Au in N_2–Ar, these authors conclude that the important controlling mechanisms are: (a) the sputtering coefficient for impurities relative to that of the main film material, (b) the tendency for chemisorption and compound formation to occur, and (c) the trapping of ionized impurities during growth. In films such as Ni and W, bias sputtering reduces contamination since impurity back-sputtering occurs at a much higher rate than sorption by mechanism (c). With Au, however, the reverse is true, and bias sputtering, by increasing the arrival rate of charge impurities, enhances the incorporation of contaminants.

Thus far, most evaluations of bias sputtering have been made on nominally unheated substrates. However, since it is well known that impurities may influence strongly both the epitaxial temperature and crystalline perfection of single-crystal films of metals and semiconductors (Khan, 1970), the technique is clearly important to the theme under discussion here. To a limited extent this has been recognized in the work of Wehner (1962) and Haq (1965) on the epitaxial growth of germanium films.

B. rf METHODS

One of the biggest limitations experienced with dc sputtering has been the difficulty of applying this method to the deposition of insulating materials such as oxides and high-resistivity semiconductors. Application of a high negative potential to insulating targets leads to positive ion bombardment and, within a short time, to the buildup of a compensating positive surface charge that brings further ion bombardment to a halt. Attempts to remove this charge using surface leakage grids or supplementary electron bombardment (Wehner and Anderson, 1970) have proven somewhat cumbersome and unsatisfactory. It was not until the work of Anderson et al. (1962), who used rf voltages successfully to clean the walls of glass discharge tubes, that an effective solution to this problem was found. Davidse and Maissel (1965) showed that practical systems for the rapid deposition of insulators, based upon the use of rf voltages, could be built, and a new and possibly the most fruitful chapter in the field of sputtering began.

1. Standard rf Sputtering Approach

The basic concepts of rf sputtering and details of apparatus suitable for the deposition of insulators such as SiO_2 have been described and reviewed extensively (Koenig and Maissel, 1970; Maissel, 1970; Davidse and Maissel, 1965). It has been shown that maintenance of ion bombardment at the surface of an insulator-covered electrode may be achieved through the marked difference in electron and ion mobilities occurring at higher rf frequencies. In operation, a glow region is developed that is separated from the target and grounded substrate electrodes by dark spaces. The more efficient ionization in the glow region caused by rf excitation of electrons enables the discharge to be maintained at low pressures (e.g., 5–10 mTorr) without the aid of thermionic or magnetic support. The glow acquires an appreciable positive potential relative to the electrodes due to electron depletion via the adjoining dark spaces. These more mobile electrons are collected by the electrodes when the latter become positive with respect to the glow. As a consequence, most of the rf current through the dark spaces is in the form of electron displacement current. The less mobile ions take several rf cycles to travel from the glow region to the electrodes and consitute a small ion conduction current. The rectifying action resulting from the different mobilities of electrons and ions sustains, in the absence of a further capacitor in the external circuit, equal dc biases between the glow and the two electrodes. However, it may be shown that if such a capacitor is inserted (asymmetric system) the respective biases divide in inverse proportion to the fourth power of the electrode areas (Koenig and Maissel, 1970). Since the grounded electrode includes not only the substrate support but also the surrounding baseplate shielding and bell-jar walls, the bias to the substrate surface is relatively small. Nevertheless, for small systems employing large targets it may still be significant and will lead to appreciable ion bombardment and reemission of material from the growing film.

Maissel and co-workers have studied the species bombarding the substrate during sputtering (Koenig and Maissel, 1970) and the influence of reemission on the reduction of voids (Maissel et al., 1970) in vitreous SiO_2 films. They report that the growing film is subjected to bombardment by energetic negative ions and neutral atoms, as well as by positive ions and by electrons originating at the target surface. Reemission coefficients as high as 0.85 have been measured. The reemission coefficient in the growth of SiO_2 films increases with temperature, changing typically by 25% for a rise from room temperature to 300°C. High reemission coefficients are found to reduce voids and increase the density of the SiO_2 films. Unfortunately, virtually nothing is yet known about the possible influence of such factors on the growth of crystalline or epitaxial films. However, it seems highly likely that they will have an important bearing upon structural quality and possibly upon physical properties.

Details of experimental rf systems and of matching networks for connecting a typical 13.56-MHz rf generator to a sputtering target have been reviewed (Maissel, 1970), and some of the effects specific to target design, e.g., ground shield spacing and effective water cooling, are discussed fully in the literature.

We have previously mentioned the operating features of an asymmetric system, in which an additional capacitor is inserted in the external circuit and the substrate support forms one of the electrodes. An alternative arrangement that dispenses with the external capacitor and prevents sputtering of the substrate electrode is the symmetrical (Kloss and Herte, 1967) system in which sputtering targets of equal area are made the electrodes and can be disposed so as to achieve more uniform depositions at higher rates.

The construction and mounting of ceramic targets for rf sputtering presents special problems. High-density and refractory vitreous or crystalline materials can be cut thin and held by means of solder or epoxy in intimate thermal contact with a well-cooled metal electrode. Certain ceramics such as ferro-electrics, ferrites, or garnets may require special forming (e.g., hot pressing), lapping, and finishing to obtain a thin flat sample that can be cooled effectively. Semiconductors are often commercially available as sputtering targets in sintered form. For growth of epitaxial films, where the objective is to achieve bulk-type semiconducting properties, such targets should be avoided in favor of vacuum-melted material free from occluded gas.

Figures 5a and 5b illustrate schematically two substrate heater designs that

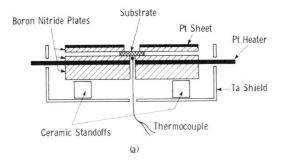

(a)

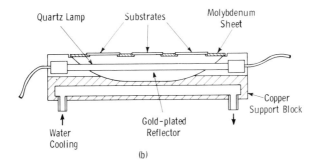

(b)

Fig. 5. Substrate heater arrangements for rf sputtering systems: (a) resistive; (b) radiant (courtesy of J. J. Cuomo, IBM Yorktown Heights).

have been used successfully for the epitaxial growth of ferroelectric and magnetic oxides, respectively. The resistively heated version has been used for
growth over periods of many hours at temperatures in the range of 700 to
800°C. The radiation-heated system is capable of operation at temperatures in
excess of 1000°C (using suitably refractory substrate supports) and generates
less heat in the bell jar.

2. rf Bias Sputtering

We have indicated above that in the asymmetric system a significant bias
may exist between the glow space and the substrate surface, resulting in sputtering and reemission of the growing film. A certain amount of reemission is
desirable in order to improve the density and structure of the film, but too much
results in a needless reduction in deposition rate. In practice, the level of substrate sputtering can be adjusted so as to optimize film properties. Logan
(1970) has described a substrate tuning circuit suitable for this purpose, in
which a variable inductance is inserted to cancel partially the capacitive
reactance of the substrate sheath and hence lower the net impedance between
the plasma and ground through the substrate holder. This increases the
substrate component of the rf current relative to the component flowing to the
walls and shielding of the system. Consequently, the effective bias at the
substrate and the amount of reemission are increased.

Logan's data show that depending upon the strength of the magnetic field
superimposed during sputtering, substrate bias voltages of 100 V (negative) or
greater could be achieved. A bias in the range of -30 to -40 V was found to
be adequate to optimize the density of SiO_2 films (as gauged by a lowering in
their etch rate in a standard buffered HF nitric solution).

C. REACTIVE SPUTTERING

Prior to the development of rf sputtering, the deposition of insulators such
as oxides, nitrides, sulfides, etc., was performed primarily by reactive sputtering. In this method the cathode, in a simple glow-discharge, triode, or other
system, comprises one or more elements, e.g., Ta, Nb, Ti, Si, $NiFe_2$, PbTi, etc.,
and usually the ionized gas consists of argon with a small content of reactive
gas such as oxygen, nitrogen, ammonia, or H_2S. In the case of an argon–
oxygen mixture the objective would be to obtain as a film product Ta_2O_5,
Nb_2O_5, TiO_2, SiO_2, $NiFe_2O_4$, and $PbTiO_3$, respectively. Much of the earlier
work on oxides was performed with pure oxygen or with oxygen-rich argon
mixtures (Holland, 1960). Although this tended to ensure complete conversion
of the sputtered material to the oxide, it also limited the deposition rate
severely, due probably to the rapid buildup of an electrically insulating oxide
layer on the target surface. This effect, for the case of reactively sputtered
SiO_2, is illustrated in Fig. 6, taken from the more recent study of Valletta *et al.*

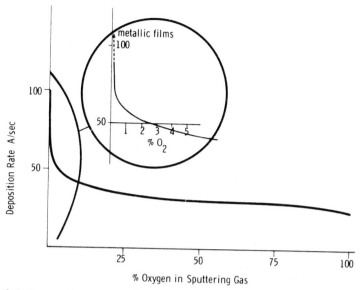

FIG. 6. Influence of percentage of oxygen in argon on growth rate of reactively sputtered SiO$_2$ films (after Valletta *et al.*, 1966).

(1966). Although in certain cases, e.g., in the reactive sputtering of Si$_3$N$_4$ in nitrogen, high partial pressures of the reactive gas are conducive to better film quality, fully reacted films with quite acceptable properties, and grown at far higher rates, can in general be obtained using very low (1–2%) partial pressures (Maissel, 1970).

In most cases systems used for reactive sputtering, except for the provision of facilities for premixing and measuring varying proportions of the reactive and inert gas components, are identical to the normal physical sputtering arrangements. Any of the systems referred to above, including rf sputtering, can in fact be used in the reactive mode. The thermionically and magnetically assisted triode system (Fig. 4) does present some problems in this case, since the tungsten filaments normally used tend to erode rapidly due to the formation and rapid evaporation of volatile tungsten oxides. Also, care must be taken in the choice of substrate heater elements and other structural parts used at elevated temperature for epitaxial growth. rf reactive sputtering offers considerable advantages over the normally used dc methods in that any insulating layer tending to form on the surface of the target is sputtered away. Thus, the deposition rate is far less sensitive to partial pressure of the reactive gas than in the case of dc sputtering.

There has been considerable speculation (Maissel, 1970; Perny, 1966) as to the nature of the reactions involved in reactive sputtering. In particular, for any given situation it is not known whether the compound to be deposited is formed (a) at the target surface and then transported to the substrate, (b) in the plasma, or (c) at the substrate surface by combination of atoms of the target material with impinging atoms of the reactive gas. Much of the available evidence seems to suggest that the main process involved is (c), i.e., reaction at the substrate surface, or alternatively in a zone very close to this surface (Perny, 1966). To the limited extent that this speculation applies to examples of epitaxial growth, we shall discuss it further in the following sections. For a more complete discussion of the models advanced to explain reactive sputtering of nonepitaxial films, the reader is referred to Maissel's review (1970) and to the references therein.

D. MULTICOMPONENT SPUTTERING

One of the main advantages often claimed for sputtering is its ability to transfer the composition of a multicomponent target unchanged to the growing film. Evidence in support of this claim has been obtained for metal alloys (Wehner and Anderson, 1970; Maissel, 1970), semiconductor compounds (Wolsky et al., 1962), and mixed oxides. When sputtering is initiated with such materials, the fastest sputtering component leaves the surface first causing a thin region of altered composition to be formed. The effective exposed area of this component (relative to that of the slower sputtering components) is reduced to a level that compensates for its higher sputtering rate, and after a short period the composition sputtered is essentially that of the bulk target. In some cases diffusion can play a significant role, and the altered layer can be relatively thick. Thus, Gillam (1959) in his studies on $AuCu_3$ found that copper sputtered initially much faster than gold, leaving a gold-rich layer some 40 Å thick. Diffusion of Cu through this layer occurred subsequently at a sufficient rate to ensure that the metal atoms were ejected in proportion to the bulk alloy composition. For nonrefractory systems containing elements of high volatility or high chemical reactivity toward background gases such as oxygen, the composition of the cathode may not be preserved in the growing film. Maissel (1970) has listed three key mechanisms that may influence film composition.

(1) Overheating of the cathode. This may enhance diffusion through the altered layer, and if one of the components has a relatively high vapor pressure, selective loss will occur.

(2) Chemical reaction effects. With metal alloys in particular, residual surface oxide may be present, so that the relative sputtering rates will be those of the oxides, not of the elements. In the case of permalloy films it has been shown (Francombe and Noreika, 1961a) that this can lead to initial imbalance in the Ni:Fe metal–atom ratio and hence in the magnetic properties of the films.

(3) *Selective resputtering.* One of the components may be selectively removed from the growing film due to resputtering. The highest-yield component is removed preferentially, and its loss may not be compensated completely by the arrival of fresh material from the cathode.

While these three mechanisms may account for the main causes of compositional change, the detailed effects may become quite complicated, especially in the case of multiphase target structures. Such structures are used, for example, when reactive sputtering from physically separate nonmiscible, metal sources is employed to obtain a mixed-oxide film, or when excess of a volatile phase is used in a sintered target to compensate for loss due to the high substrate temperature needed for epitaxial growth, The latter situation will be discussed in Section VI in reviewing work on mixed-oxide epitaxial films. Typical experimental structures representative of two-component metal target sources are shown schematically in Fig. 7. Figure 7a shows an arrangement used

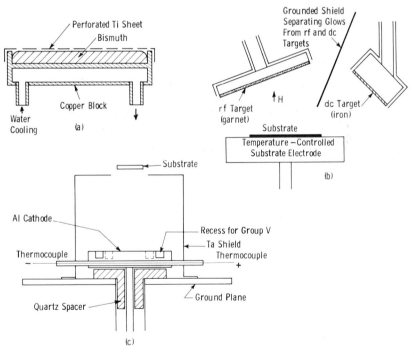

FIG. 7. Schematic diagrams of some two-component sputtering systems: (a) for rf reactive sputtering of $Bi_4Ti_3O_{12}$ (after Takel *et al.*, 1970); (b) for combined rf and dc reactive sputtering of GdIG and Fe_2O_3, respectively (after Sawatzky and Kay, 1969a); and (c) for the dc sputtering of AlSb (after Noreika *et al.*, 1969).

by Takei *et al.* (1970) for the epitaxial growth of bismuth titanate $Bi_4Ti_3O_{12}$ using rf reactive sputtering. The target comprises a water-cooled recessed copper block upon which a layer of Bi is spread by melting. Over this is placed a titanium sheet containing a uniform array of holes, the proportion of the hole area determining the Bi–Ti surface area ratio. The approach is reminiscent of an earlier scheme used by Bickley and Campbell (1962) for reactive sputtering of Pb–Ti to form $PbTiO_3$. In their case, the target comprised a titanium sheet on the surface of which was photoprinted a pattern of Pb dots. Figure 7b shows a combined rf and dc sputtering scheme used by Sawatzky and Kay (1969a) for cosputtering garnet and iron oxide. The aim was to compensate for iron cation deficiencies occurring in films sputtered from a single garnet source. Due to nonuniform deposit thicknesses produced by this arrangement, exact compensation could be achieved only at one position on the substrate. Figure 7c shows an arrangement used for sputtering epitaxial AlSb films (Noreika *et al.*, 1969), using heat generated in sputtering aluminum to sublime excess Sb vapor onto the heated substrate.

E. Ion-Beam and High-Rate Sputtering

1. *Ion-Beam Sputtering*

In the assisted-ionization sputtering schemes described above, the lowest pressure at which a supply of ions adequate to maintain sputtering at reasonable rates is obtained is in the range of 10^{-3} Torr. There has been some interest in sputtering at lower pressures than this under conditions such that the background pressure of impurities is reduced and the substrate is maintained in a

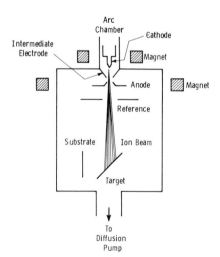

FIG. 8. Schematic of the duoplasmatron argon ion source used for vacuum sputtering of thin films (after Chopra *et al.*, 1963).

plasma-free region. An effective way of achieving these aims is to generate the ions in a high-pressure chamber and, using suitable electron and ion optics, extract them in the form of a beam into a lower-pressure ($\sim 10^{-5}$ Torr) chamber. Such a "duoplasmatron" arrangement was originally proposed by von Ardenne (1956) and has been developed both for sputtering of films and more recently for the large-area etching and polishing of surfaces (Spencer and Schmidt, 1971). Chopra and Randlett (1967) designed a duoplasmation source (shown schematically in Fig. 8) yielding ion-beam current of about 500 mA for a beam cross section of about 1 cm^2. More recent commercial systems (Spencer and Schmidt, 1971) utilize many "beamlets" in order to sputter from larger target areas. In the system used by Chopra and Randlett (1967) typical deposition rates onto a substrate placed 8 cm away from the target are 40 Å min^{-1} for quartz and 500 Å min^{-1} for Ag, using 50 mA cm^{-2} current density, and an accelerating potential of 5 kV.

2. High-Rate Sputtering

One of the criticisms often leveled at sputtering is that the deposition rates attainable are too low for the method to be competitive with chemical-vapor deposition for the growth of thick coatings. As a case in point, the growth of epitaxial semiconductors such as Ge and Si is carried out at several microns per minute by the pyrolytic decomposition of the appropriate hydride or halide, whereas in sputtering, deposition rates even in rf systems are usually limited to tenths of a micron per minute at most. Grantham et al. (1970) have pointed out that a limitation on sputtering rate is imposed by the breakdown of the low-pressure gas between the sputtering electrode and the ground shield in proximity to it. Typically, for this spacing to be just less than the cathode dark space, its value at 10^{-2} Torr is about 0.6 cm. For rf power densities at the target beyond about 50 W cm^{-2}, electrical arc breakdown occurs across the gap to the ground shield.

From Paschen's law (Cobine, 1958) it can be seen that the gaseous break-down voltage passes through a minimum value and then increases rapidly as the pd product decreases, where p is the gas pressure and d the spacing from the target to the ground shield. To obtain a higher breakdown capability through reduction of p, Grantham et al. (1970) designed a differentially pumped, water-cooled, rf target assembly shown schematically in Fig. 9. Among the design features are: (1) a clamping ring to apply even pressure to form the vacuum seal at the target face, (2) direct bonding of the target to the water-cooled electrode for best thermal contact and cooling, (3) flexible couplings in the cooling lines to prevent stressing through misalignment, and (4) provision of a water-cooled electromagnet capable of producing about 100 Oe. With the target assembly pumped to about 10^{-5} Torr and an argon pressure of 2.8×10^{-2} Torr, power levels up to 120 W cm^{-2} could be applied

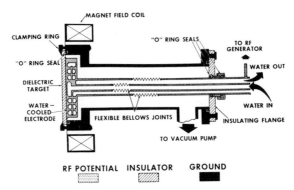

FIG. 9. Sectional view of target designed for high-rate rf sputtering (after Grantham *et al.*, 1970).

to an Al_2O_3 target, yielding deposition rates of 6000 Å min^{-1}. Rates in excess of 10,000 Å min^{-1} were achieved for SiO_2 and Cu targets. The results indicate that thermal problems may limit further increases in deposition rate, particularly for insulating materials, but the high rates obtained are certainly comparable with those achieved chemically using low-temperature pyrolysis.

IV. Epitaxial Films of Metals and Alloys

Although relatively few data have appeared there seems little doubt, from those results available, that virtually all metals and alloys that have been epitaxed by evaporation can similarly be grown by means of sputtering. Indeed, claims have often been made that epitaxy is more easily achieved by sputtering than by evaporation, although the precise experimental conditions required to optimize epitaxial growth and convincing descriptions of the mechanisms involved are still not available.

Certainly, one of the greatest advantages of sputtering is that it has enabled the list of metals and alloys epitaxed by evaporation to be extended to include the more refractory materials, e.g., Pt, Nb, Zr, etc., not readily amenable to containment as molten sources.

A. NONREACTIVE METALS: Au, Ag, Pt

One of the most systematic studies to date of the initial stages of condensation and epitaxial growth of sputtered gold was carried out by Chapman and Campbell (1969), who examined nucleation and growth on rock salt. Using a high-energy beam of argon ions produced by a linear accelerator, they bombarded a gold crystal along $\langle 110 \rangle$ axes in such a way as to produce atom ejection along $\langle 100 \rangle$ (see Fig. 10). The mean energy of the ejected atoms under

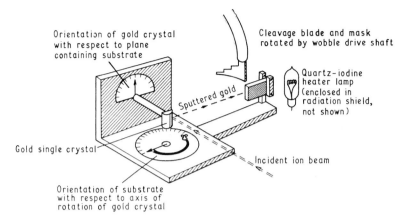

Fig. 10. Apparatus used for growth of gold in selected crystallographic directions from a gold single crystal onto vacuum-cleaved rocksalt (after Chapman and Campbell, 1969).

these conditions (Fig. 1) was about 20 eV with an incidence rate on the freshly cleaved rock salt crystal face of about 0.5 Å min^{-1}. Efforts were made to compare the saturation island density N_s (i.e., density of groups of atoms) formed in the initial stages of growth, and also the epitaxial orientation of deposits sputtered in this fashion, with those observed primarily for thermal evaporation sources yielding much lower incidence energies, in the range of 0.2 eV atom^{-1}.

The energy parameters controlling nucleation and growth are E_a, the adsorption energy between an adatom and substrate; E_d, the surface diffusion energy required for transfer of an atom to an adjacent adsorption site; and E_b, the cohesion energy between adjacent atoms on the substrate. Earlier work by Lewis and Campbell (1967), in which log N_s vs $1/T$ plots were made for gold evaporated on rock salt, showed that such plots could display two regions of different slope. At low temperatures the saturation island density is determined by surface diffusion, and the slope of the straight-line plot obtained gives the surface diffusion energy E_d. At higher temperature reevaporation plays an important role, and the slope of the line is governed by the adsorption energy E_a as well as E_d. According to Lewis and Campbell (1967) in the low-temperature region when pairs of atoms are stable, the saturation island density N_s is proportional to the square root of the incidence rate R:

$$N_s = (N_0 R/v_0)^{1/2} \exp(E_d/2kT) \tag{1}$$

Assuming this relationship to hold true for the high-energy conditions, the results obtained by sputtering were recalculated by Chapman and Campbell for a deposition rate equivalent to that used in the earlier study on evaporated

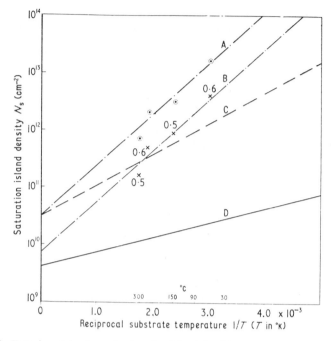

FIG. 11. Experimental values of saturation island densities of gold on rocksalt. Curve **A**, sputtered (results modified to a rate of 8.4 Å min^{-1}); curve B, sputtered (high energy); curve C, evaporated (electron bombardment), electron flux approximately 10^{15} cm^{-2} sec^{-1} (240 eV); curve D, evaporated. Incidence rates: gold 0.14 Å sec^{-1} except where shown by numbers at $\times$; these rates are in angstroms per minute (after Chapman and Campbell, 1969).

gold (8.4 Å min^{-1}). The data thus obtained are shown in Fig. 11 together with the thermal evaporation data and also some results obtained by Jordan and Stirland (see Chapman and Campbell, 1969) in which electron bombardment was carried out during thermal evaporation. It is apparent from Fig. 11 that both high-energy sputtering and electron bombardment produce large and comparable increases in N_s and in the intercept at $1/T = 0$, when referred to the thermal evaporation results. There is also an increase in the slope of the straight-line plot, this being far more significant for the sputtered deposits. The increase in gradient is consistent with a change in surface diffusion energy from 0.10 eV for the evaporation case to a new value of 0.36 eV for sputtered gold. These results were essentially unaffected by changing the sputtering conditions. Thus, ejected atom beams along $\langle 100 \rangle$ and $\langle 110 \rangle$ produced by bombardment with xenon ions at 50 keV along $\langle 111 \rangle$ gave about the same values of N_s at equivalent substrate temperatures. It was concluded, therefore, that the effects shown in Fig. 11 must be produced at energies lower than those employed in the experiments, and that increase in the energy above some lower threshold effected little further change in the log N_s vs. $1/T$ plot.

Of greater relevance to the present discussion were the epitaxial orientation effects observed for the sputtered Au deposits grown at 150°C. Usually, epitaxial films of Au grown on (100) rock salt display mixed orientations, i.e., a parallel orientation component with $(100)_{Au} \| (100)_{NaCl}$ and a second orientation component with $(111)_{Au} \| (100)_{NaCl}$. In the case of the sputtered deposits grown by Chapman and Campbell (1969), the (111) component was completely absent. The authors suggest that both the increase in N_s values and effective surface-diffusion energy, and also the preference for the single (100) orientation, might be attributed to the formation of special nucleation sites by the high-energy atom beam, although the actual mechanism involved is not known.

As indicated previously, it has been speculated that on the grounds that the higher-energy atoms produced by sputtering should diffuse more readily on the substrate surface, much lower epitaxy temperatures might be achieved than in the case of evaporated films. Investigations by Campbell and Stirland (1964), using a diode glow-discharge system, have in fact demonstrated that large reductions in epitaxy temperature for growth on (100) rock salt of Au and Ag are produced by sputtering the metals at low rates. The conditions for Au have been examined systematically for a wide range of sputtering pressures and rates by Francombe and Schlacter (1964). It was found that the use of low sputtering pressures, which should lead to the arrival of highly energetic atoms at the substrate, does not give rise to oriented growth. Indeed, an amorphous film is developed under these conditions possibly arising from damage to the rock salt surface. As the sputtering pressure increases, however, a crystalline orientation develops, becoming much more perfect, and at pressures of 300 mTorr and rates of 0.2 Å min^{-1} epitaxial growth could be obtained reproducibly at 30°C or lower. Diffraction patterns produced from single and folded regions of a thin film made under these conditions are illustrated in Fig. 12, and a micrograph indicating their high degree of structural continuity is shown in Fig. 13. It is interesting to note that a single parallel (100) orientation is produced here, as was the case with the more energetic atom beam conditions used by Chapman and Campbell (1969). Epitaxial growth at such low temperatures implies a high surface mobility for the atoms and initial nuclei. It seems likely that this arises from the "scrubbing" action due to collision of gas atoms with the substrate.

An interesting effect caused by a transverse electric field on the growth sequence for epitaxial Au films formed on mica was reported by Chopra et al. (1963). In this study, attempts were made to interpret electrical size effects in single-crystal films by measuring their equivalent resistivity as a function of thickness. Epitaxial growth of Au with the (111) planes parallel to the cleavage

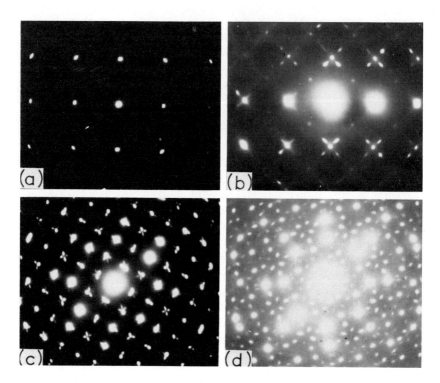

FIG. 12. Electron diffraction patterns from epitaxial gold film (75 Å thick) sputtered onto rock salt crystal, deposition rate 0.2 Å min⁻¹, argon pressure 300 mTorr: (a) unfolded region; (b) twinned region showing double diffraction; (c) and (d) multiple diffraction due to double and triple layers, respectively, of folded film (after Francombe and Schlacter, 1964).

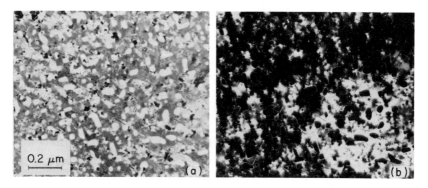

FIG. 13. Electron micrograph of unfolded region of film shown in Fig. 12 with $t = 75$ Å, 1.25 Å/min, 310 mTorr: (a) bright field; (b) dark field.

face of mica was achieved (using relatively high deposition rates, 50–250 Å min^{-1}) in a conventional glow-discharge system, at about 280°C. Measurements of resistance were made in situ both continuously during deposition and also at discrete time intervals, corresponding to definite film-thickness increments (as judged from previous calibration runs). The results are depicted in Fig. 14 and show that under both conditions of measurement the equivalent film resistivity (calculated on the assumption of a structurally continuous deposit) drops rapidly toward the bulk value for pure Au at a critical thickness. However, this effect occurs at a considerably greater thickness (300–350 Å) when the measuring voltage is applied continuously than when it is interrupted (160–240 Å). Suplementary electron-microscopy studies revealed that the effect of the continuous measuring voltage was to delay linking of the epitaxial islands during growth [cf. Bassett and Pashley (1959)], possibly due to electrostatic repulsion. The fact that bulk resistivity was achieved at thickness significantly less than the mean free path for conduction electrons (~ 500

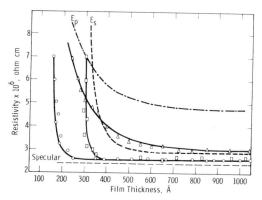

FIG. 14. Resistivity of gold films sputtered on mica as a function of film thickness: △—polycrystalline film on cold substrate; □—single-crystal films with measuring voltage applied continuously; ○—single-crystal films with measuring voltage interrupted (after Chopra *et al.*, 1969).

Å) was taken as evidence that specular reflection of these electrons was occurring at the atomically smooth surfaces of the epitaxial film.

Platinum, in common with gold and silver, is useful in epitaxial film form as an electrode material for the study of electrical properties of epitaxial films of insulators such as oxides (Francombe *et al.*, 1967). Epitaxial layers of Pt have been grown in the author's laboratory by means of diode and triode (Fig. 4) sputtering on CaF_2 and MgO substrates. The conditions for epitaxy and orientation relationships obtained are listed in Table I. Experimentally, such films are quite easily deposited using Pt foil cathodes wrapped around a metal

TABLE I SPUTTERED EPITAXIAL FILMS OF METALS AND ALLOYS

Metal or alloy	Deposition method	Substrate	Epitaxy temp (°C)	Orientation dep ∥ subst	Reference	Remarks
Au	Ion-beam	NaCl(001)	150	Parallel	Chapman and Campbell (1969)	Influence of Au particle energy on nucleation and orientation
Au	Glow-discharge	NaCl(001)	<30	Parallel	Campbell and Stirland (1964)	Low-temperature epitaxy achieved at low deposition rate
Au	Glow-discharge	NaCl(001)	<30	Parallel	Francombe and Schlacter (1964)	Low rates and high argon pressure
Au	Glow-discharge	Mica(00.1)	280	$(111) \parallel (00.1)$ $[1\bar{1}0] \parallel [11.0]$	Chopra et al. (1963)	Electrical study of size effects
Ag	Glow-discharge	Mica(00.1)	<30	Parallel	Campbell and Stirland (1964)	As above for Au
Pt	Glow-discharge and triode	$CaF_2(111)$	530	Parallel	Francombe et al. (1967)	Epitaxial electrode used for oxide epitaxy
		MgO(001)	550	Parallel	Francombe et al. (1967), Takei et al. (1970)	
Ti	Triode	$CaF_2(111)$	410	$(00.1) \parallel (111)$ $[11.0] \parallel [1\bar{1}0]$	Francombe et al. (1967)	Study of oxide overgrowth
Zr	Triode	NaCl(111)	450	$(00.1) \parallel (111)$	Francombe et al. (1967)	Study of oxide overgrowth
		$CaF_2(111)$	450	$(00.1) \parallel (111)$ $[11.0] \parallel [1\bar{1}0]$		
Nb	Triode	MgO(001)	450	Parallel	Francombe et al. (1967)	—
Ta, Zr, Mo, W	Ion-beam	NaCl(001)	250–400	Parallel	Chopra et al. (1967)	Epitaxy of metastable fcc phase
$Ni_{81}Fe_{19}$	Glow-discharge	NaCl(001)	420	Parallel	Francombe (1966b)	—
$NiFe_2$	Glow-discharge	NaCl(001)	420	Parallel	Francombe (1966b)	Oriented conversion to $NiFe_2O_4$
Fe	Glow-discharge	MgO(001)	420	$(001) \parallel (001)$ $[100] \parallel [110]$	Francombe (1966b)	Oriented oxidation and interaction to form $MgFe_2O_4$
$Au_{85}Ni_{15}$	Glow-discharge	NaCl(001)	400	Parallel	Khan and Francombe (1965)	Oriented overgrowth of NiO
$Ag_{80}Mg_{20}$	Glow-discharge	NaCl(001)	340	Parallel	Francombe et al. (1967)	Unsuccessful attempt to form oriented overgrowth of MgO and Al_2O_3
$Ag_{80}Al_{20}$	Glow-discharge	NaCl(001)	340	Parallel	Francombe et al. (1967)	

backing plate and spot-welded at the folded corners. Figures 15a and 15b show, respectively, a transmission electron diffraction pattern and electron micrograph obtained from an epitaxial film grown on the (111) cleavage plane of CaF_2.

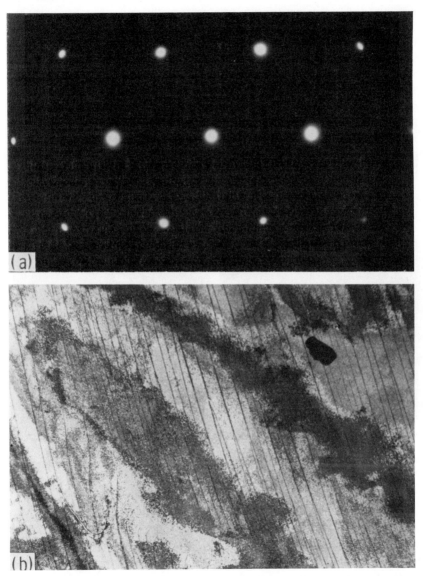

FIG. 15. Electron diffraction pattern (a) and micrograph (b) ($\times$ 3600) of an epitaxial Pt film grown on CaF_2 at 600°C (after Francombe *et al.*, 1967).

B. Refractory Metals: Ti, Zr, Nb, Ta, Mo, W

Relatively little work appears to have been done on the epitaxial growth of the more chemically reactive, refractory metals. This fact may be due to the ease with which impurities are known to be incorporated into such films and the uncertainty as to whether the film product is really a pure metal or perhaps a compound containing small amounts of oxygen, nitrogen, or carbon. Films of satisfactory purity can in fact be produced from high-density cathodes using well-trapped vacuum systems. Epitaxy of hcp Ti and Zr and bcc Nb has been achieved in the author's laboratory (Francombe et al., 1967) using a triode sputtering system with high-purity sheet-metal cathodes. The results are summarized in Table I. It was found possible to duplicate by sputtering the results of Arntz and Chernow (1964–1965) for evaporated Ti on CaF_2 and those of Denoux and Trillat (1964) for evaporated Zr on NaCl and CaF_2. In agreement with these latter authors, considerably better orientation was obtained on CaF_2 crystals than on crystals of NaCl (see Fig. 16). Films of Nb were found to grow epitaxially not only on MgO crystals [as reported by Hutchinson (1965) for evaporated films] but also on NaCl.

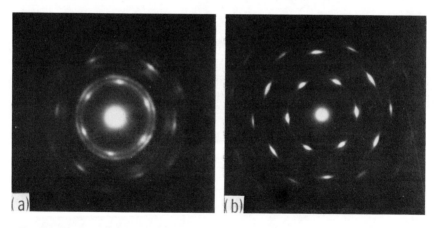

Fig. 16. Electron diffraction patterns from sputtered Zr films deposited at 400°C: (a) on NaCl (111) surface; (b) on CaF_2 (111) surface (after Francombe et al., 1967).

Epitaxial films of thickness greater than about 1500 Å in some cases showed wrinkling after cooling to room temperature. To a large extent this effect appears to be attributable to mismatch in thermal-expansion coefficients of the film and substrate. However, the fact that, in thinner films, wrinkling is initiated or accentuated by exposure to the atmosphere suggests also that surface oxidation plays an important role. Thus, the volume mismatch between the surface oxide and the metal film may (together with the thermal stress present in the film) generate sufficient stress to break the bond between film and substrate surface.

As indicated above, the triode-sputtering approach led to the growth of epitaxial films of Ti, Zr, and Nb possessing the normal crystal structure of the bulk metals. However, it has been reported by Chopra *et al.* (1967) that ion-beam sputtering of Ta, Mo, W, Re, Hf, and Zr can at lower growth temperature lead to the formation of a metastable fcc structure, which at larger film thicknesses or higher growth temperatures reverts to the normal bulk crystal structure. In cases where the transition temperature is reasonably high, e.g., Ta, Mo, W, and Zr ($\sim 400°C$), epitaxial growth of this fcc phase was achieved on the (100) face of rock salt at temperatures lying between 250 (for Zr) and 400°C (for Ta). In contrast to the data presented above, it is stated that epitaxy of the normal bulk structures is not obtained at temperatures below 600°C. However, few data are given in this reference on the precise conditions for epitaxy. Consideration was given by Chopra *et al.* (1967) to the possible role of impurities such as nitrogen in producing the fcc structure, and this possibility was finally discounted.

C. MAGNETIC METALS AND ALLOYS

Virtually all studies on epitaxial films of magnetic metals and alloys to date have been performed on samples prepared by evaporation. However, the limited sputtering data available suggest that epitaxy of such materials should present no difficulty in clean vacuum conditions, and that in the case of alloys with constituents differing significantly in vapor pressure, sputtering should offer important advantages in helping to control film composition. In connection with work on sputtered permalloy films for memory-matrix development (Francombe and Noreika, 1961a; Francombe, 1963; Francombe and Noreika, 1961b) and on the growth of epitaxial ferrite films (Francombe, 1966b) some preliminary data were obtained on epitaxial layers of $Ni_{81}Fe_{19}$, $NiFe_2$, and Fe. These films were produced by conventional glow-discharge sputtering using high-density targets fabricated by vacuum melting. The results are summarized in Table I. In the case of the permalloy and iron film the epitaxial orientations are the same as those reported, for example, by Heavens (1964) and Sato *et al.* (1964), respectively, for evaporated films. The $NiFe_2$ epitaxial layers, contrary to the situation found in the bulk phase diagram (Hansen, 1958d), were single-phase with the γ fcc structure. It is probable that thermal annealing would cause this to revert to the normal two-phase structure expected for the $NiFe_2$ composition. Results on the oxidation of the $NiFe_2$ and Fe films will be discussed further in our treatment of epitaxial oxide films (Section 6).

D. OTHER ALLOYS: Au–Ni, Ag–Mg, Ag–Al

In the course of studies aimed at producing thin epitaxial surface oxides of NiO, MgO, and Al_2O_3, Khan and Francombe (1965) and Francombe *et al.* (1967) investigated the sputtering of epitaxial films of the alloys Au–15% Ni, Ag–20% Mg, and Ag–20% Al. Epitaxial layers of the Au–15% Ni alloy (Khan and Francombe, 1965) were grown successfully by glow-discharge sputtering from a vacuum-melted cathode of the same composition, on the (100) cleavage face of rock salt (see Table I). At the temperature of growth ($\sim 400°C$) the

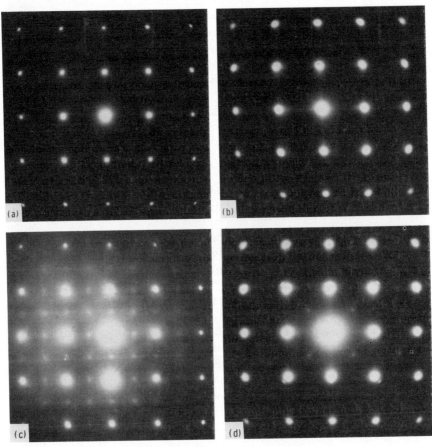

FIG. 17. Electron diffraction patterns showing development of superlattice and precipitation effects on Au–15% atomic Ni alloy film (thickness 700 Å) epitaxed by sputtering onto cleaved rock salt: (a) Annealed in vacuo at 400°C; (b) annealed at 500°C; (c) annealed at 520°C; and (d) annealed at 650°C. Outer-satellite spots due to Ni-rich precipitate (after Khan and Francombe, 1965).

alloy is supersaturated in Ni but a single-phase condition is quenched in. On annealing above 500°C, preprecipitation effects, typified by local enrichment of the alloy matrix in Ni were observed in the electron-diffraction patterns (Fig. 17). These are associated with the formation, in succession, of the superlattice structures $NiAu_3$ and NiAu (isostructural, respectively, with the cubic $CuAu_3$ and tetragonal CuAu superlattices), leading eventually to the precipitation of Ni-rich alloy at temperatures above 600°C. It is interesting that these super-lattice structures were not observed in the work of Fukano (1961), who produced similar epitaxial alloys by evaporation followed by homogenization at 700°C, and then annealing at lower temperatures (~ 500°C). Apparently, the low epitaxy temperature used for the sputtered alloy ensured a condition of atomic disorder, so that subsequent annealing promoted rapid recrystalliza-tion and diffusion, leading to superlattice ordering. The ordering and precipita-tion sequence can be completed in a matter of minutes, in contrast to the annealing periods of hours or weeks often used (Hansen, 1958b) for bulk alloys prepared from the melt.

Epitaxial films of the alloy Ag–20% Mg were grown by triode sputtering (Table I) taking special precautions to water-cool the target adequately. Use of an uncooled target led to rapid loss of the high vapor pressure Mg. It was found possible to epitax the alloy on the (100) face of rock salt (Table I) at temperatures slightly above 340°C. Substrate temperature appeared to be a very critical factor in controlling film composition. At temperatures above 360°C rapid loss of Mg from the alloy occurred (or possibly the sticking co-efficient of Mg was greatly reduced), and the resulting films were found to comprise Ag only. The loss of Mg was detected from X-ray diffractometer measurements of the lattice parameters of thicker films, using the bulk-lattice parameter data for alloys ranging from pure Ag ($a_0 = 4.0862$ Å) to Ag–20% Mg ($a_0 = 4.1058$ Å). Single-phase films of the alloy Ag–20% Al were epitaxed successfully using a conventional diode glow-discharge system (Table I). No signs of the low-temperature μ phase (Hansen, 1958a) (composition variable around 23% Al) were detected, indicating that a metastable film structure had formed with the a_{Ag} phase slightly supersaturated in Al.

V. Epitaxial Semiconductor Films

Some of the most complete and systematic studies of epitaxy by sputtering have been performed on films of semiconductors, in particular, Ge and GaAs. A fairly wide range of materials has now been investigated including elements, sulfides, narrow band-gap tellurides and wide band-gap nitrides (e.g., AlN and BN). Much of this effort has in recent years been stimulated by the need for microelectronic circuits and for large-area optical sensors in the ultra-violet, visible, and infrared, and was prompted by the fact that facilities for the growth of large homogeneous bulk crystals of such materials were not available or were prohibitively expensive to develop. Some of the earlier published work on epitaxial semiconductor films was covered by Francombe and Johnson (1969) in a recent review.

A. Elements: Ge, Si

In the case of elemental semiconductors, by far the major effort has been concentrated on germanium. Main emphasis has been placed on homoepitaxy on (111) Ge substrates, although some work (Krikorian, 1964) on cleaved (111) faces of CaF_2 has been described. Probably the first serious attempt made at producing epitaxial semiconducting films by the diode technique was that of Reizman and Basseches (1962), who grew epitaxial junctions of Ge on Ge substrates at temperatures up to 840°C using both p- and n-type cathodes. In all cases the deposited film was p-type and could only be converted to n-type by prolonged treatment with phosphorus vapor. The large number of faults revealed in these films by chemical etching led the authors to suggest that the high concentration of acceptors could be attributed to structural defects.

The results of Krikorian (1964) and Krikorian and Sneed (1966) on using a glow-discharge diode system were among the first to provide a more detailed description of the conditions for epitaxial growth. These results contained some interesting features (Fig. 18) that generated a certain amount of con-

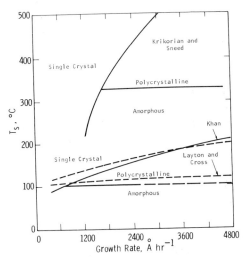

Fig. 18. Effect of film growth rate and substrate temperature T_s on the structure of Ge films sputter deposited on Ge (111) substrates (after Krikorian and Sneed, 1966; Layton and Cross, 1967; and Khan, 1973).

troversy and led to several subsequent studies both by sputtering and evaporation. Krikorian and Sneed found that at higher deposition rates (>1200 Å hr^{-1}) the structure of the Ge deposit may be amorphous, polycrystalline (randomly oriented), or single-crystal (more correctly polycrystalline-epitaxial), depending upon the deposition temperature. The amorphous-crystalline transformation seems unaffected by deposition rate in the range above 1200 Å hr^{-1}, but at lower rates this transformation temperature falls steeply, continuing the trend shown by the polycrystalline–epitaxial transition for higher rates. Films prepared at the triple-point of this "phase diagram" appear to have heterogeneous structures containing proportions of both amorphous and polycrystalline (epitaxial and random) phases. The admixture of crystalline and amorphous components in films prepared near the transformation temperature is not surprising in view of the fact that Sloope and Tiller (1962, 1963) report the same situation in evaporated films. A more intriguing aspect of these results is the direct transformation at low deposition rates from an amorphous to an epitaxial structure. This presumably implies that the amorphous phase itself may possess some structural orientation that varies with deposition rate and temperature.

Low-pressure sputtering of Ge was studied by Wolsky et al. (1965), who prepared their films in an ultrahigh vacuum system, with a thermionically assisted discharge, at rates of about 300 to 420 Å hr^{-1}. Under these conditions, the epitaxy temperature was about 150°C compared with the value of 120°C that might be predicted by extending the data of Krikorian and Sneed (1964) shown in Fig. 18. The results obtained included data on the amorphous–crystalline transition, the conditions for epitaxial growth, and the variation of electrical properties such as carrier mobility and density with growth conditions, film thickness, and depth within the film.

Further investigations of the dependence of structure of homoepitaxial Ge films on deposition rate and temperature were made by Layton and Cross (1967) with triode sputtering using technical vacuum conditions (background pressure of 10^{-6} Torr), and recently by Khan (1973) with diode sputtering using UHV conditions (residual background pressure of 2×10^{-10} Torr). Layton and Cross (1967) obtained phase transition data similar in form to those described by Krikorian and Sneed (1964) but defined regions within which amorphous, polycrystalline, single-crystal twinned, and single-crystal untwinned were grown. Their polycrystalline region is much narrower than that shown by Krikorian and Sneed, and the phase transitions lie at considerably lower temperatures (Fig. 18). Also a well-defined triple-point was not detected even at the lowest deposition rates used. The lowering of the phase transition and epitaxial temperatures was attributed by Layton and Cross (1967) to a substrate cleaning action of the arriving high-energy atoms.

Khan (1973) has recently reexamined the homoepitaxial growth of Ge by diode sputtering in a UHV system and evaluated the influence of substrate cleanliness on the conditions for epitaxy. LEED and Auger spectroscopy were employed in a separate system to determine the contaminants present on the initial chemically cleaned surface and the heat treatment needed to remove these. Contamination by sulfur, carbon, and oxygen was observed, but annealing at 600°C in UHV was sufficient to remove these completely. It was proposed that sulfur was removed via sublimation of volatile GeS, and oxygen through decomposition of GeO_2 to GeO, followed by sublimation of the lower oxide. Carbon was presumed to react with oxygen to form CO. It is interesting to note that the phase transition temperatures determined in this study lie very close to those found by Layton and Cross (1967) using triode sputtering under conventional vacuum conditions (Fig. 18), but that Khan's data do confirm the existence of the triple-point claimed by Krikorian and Sneed (1964). Khan found that the epitaxial temperature on an "unclean" Ge surface was considerably higher (> 150°C) than for a surface cleaned by thermal regeneration, and noted that surface oxide (Adamsky et al., 1969) probably played the key role in raising the epitaxy temperature. In the presence of deliberately added oxygen levels of 10^{-9} to 10^{-7} Torr, the epitaxial transition temperatures were observed to increase by 75 to 100°C.

As a final note on the sputtering of homoepitaxial Ge films we reference a study made by Haq (1965), using asymmetric ac sputtering, which not only confirms the findings relating to the effect of deposition rate and impurities but also sheds further light on the origin of acceptor states in Ge films. As indicated in our discussion in Section III, in asymmetric ac sputtering (Frerichs, 1962), bias sputter cleaning of the growing film is effected by applying a negative "cleaning" potential (lower than the main sputtering potential) to the substrate on each alternate half-cycle. With low values of cleaning potential Haq (1965) found, as did Reizman and Basseches (1962) for normal diode sputtering, that even when films were sputtered from an n-type target the epitaxial film product was strongly p-type. However, at higher cleaning-potential values, using an n-type cathode doped with $2-3 \times 10^{17}$ atoms cm^{-3} of antimony, it was possible to grow epitaxial n-type films close in doping level to the cathode. Table II shows values of sputtering and cleaning voltages needed to obtain n-type conductivity in such layers. Structures formed by growth on p-type substrates yielded electrical data consistent with an ideal abrupt junction with the epitaxial n-type layer possessing near-bulk values of mobility.

These asymmetric ac sputtering results appear to indicate that the impurities produced in outgassing during sputtering are mainly responsible for the tendency to form p-type films. Thus, careful shielding of the growth region from the walls of the vacuum chamber and an increase in the cleaning potential relative to the sputtering potential reduce the tendency. However, this still

TABLE II

Highest Sputtering Voltages at Which n-Type Ge Films Were Obtained on p-Type (111) Oriented Substrates at Various Substrate Temperatures and Cleaning Voltages

Substrate temp (°C)	Cleaning voltage	Sputtering voltage	Deposition rate (μm/hr)	Crystallinity
500	500	3700	1.0	Single
400	250	2000	0.32	Single
400	600	3000	1.1	Single
350	600	3000	0.9	Single
325	600	3000	1.5	Poly
300	600	3000	1.4	Poly

does not rule out the possible role of defects that may be generated by chemical impurities, e.g., oxygen, as the epitaxial film grows.

In comparison with the studies on Ge, relatively little published work has appeared on sputtered epitaxial films of Si. Wolsky et al. (1965) made brief reference to the epitaxial growth of Si in their study on the epitaxy of Ge films, and Clark and Alibozek (1968) published a brief note claiming the growth of single-crystal films of Si on Si (under conventional vacuum conditions) at 200°C by triode sputtering through a moving mask. The aim in this latter study was to favor lateral growth preferentially from selected nuclei by moving the mask across the substrate surface slowly during deposition. In view of the fact that the lowest epitaxial temperature reported for evaporated Si on (111) Si even under conditions of stringent cleanliness is about 400°C (Thomas and Francombe, 1971) (deposits at lower temperatures being amorphous), the claim of epitaxy at 200°C by Clark and Alibozek (1968) appears open to some question.

Two investigations have been made to date of the epitaxial growth of Si by ion-beam sputtering. The first, made by Unvala and Pearmain (1970), describes homoepitaxy of Si on (111) Si in a bakeable, helium cryopumped system. An ion beam of from 3 to 8 mA at 12 keV energy was used to bombard an inclined silicon slice, in a background pressure of about 10^{-4} Torr. The silicon substrate was heated radiantly and was held at a potential of 9 kV to avoid bombardment by positive ions. With a deposition rate of 400 Å min^{-1} single-crystal layers, free from crystallographic faults, were grown successfully at temperatures above 730°C. Also, epitaxial diodes displaying sharp rectification characteristics were formed by epitaxy of n-type layers on p-type substrates and of p-type layers on n-type substrates.

The second ion-beam study was performed by Weissmantel et al. (1972) and was directed toward heteroepitaxial growth of Si on spinel. A residual system

TABLE III   SPUTTERED EPITAXIAL FILMS OF SEMICONDUCTOR ELEMENTS AND COMPOUNDS

Element or compound	Deposition method	Substrate	Epitaxy temp (°C)	Orientation dep ‖ subst	Reference	Remarks
Ge	Glow-discharge	Ge(111)	<840	Parallel	Reizman and Basseches (1962)	Attempt to produce p–n junctions
Ge	Glow-discharge	Ge(111) CaF₂(111)	200–500	Parallel	Heavens (1964), Krikorian and Sneed (1966)	Effect of rate on film crystallinity and epitaxial temp.
Ge	Thermionically assisted	Ge(111)	150	Parallel	Wolsky et al. (1965), Wallis et al. (1965)	Low-deposition rate, residual UHV conditions
Ge	Triode	Ge(111)	120–200	Parallel	Layton and Cross (1967)	Effect of rate on crystallinity, epitaxial temperature, and twinning
Ge	Glow-discharge	Ge(111)	100–200	Parallel	Khan (1973)	As Krikorian (1964), Krikorian and Sneed (1966), and Layton and Cross (1967). Also effect of impurities. Residual UHV conditions
Ge	Asymmetric ac	Ge(111)	300–500	Parallel	Haq (1965)	Successful growth of p–n junctions. Electrical measurements
Si	Triode	Si(111)	200	Parallel	Clark and Alibozek (1968)	Growth through moving mask
Si	Ion-beam	Si(111)	730	Parallel	Unvala and Pearmain (1970)	Fault-free epitaxial diodes grown
Si	Ion-beam	MgAl₂O₄(001)	750	Parallel	Weissmantel et al. (1972)	High-mobility p-type films
Si	Rf	Si(111)	720	Parallel	Noreika (to be published)	Resistivity profile on n-type films
Bi₂Te₃	Glow-discharge	NaCl(001) NaCl(111)	<250 >300	(10.5)‖(001) (00.1)‖(111) [11.0]‖[1̄10]	Francombe et al. (1962), Francombe (1964)	Study of film structure and decomposition effects

144

Material	Method	Substrate	Temperature (°C)	Orientation	Reference	Remarks
$Cd_xHg_{1-x}Te$	Triode	NaCl(001)	50	(111)∥(001) $[\bar{1}10],[\bar{1}10]$∥[110] As at 50°C with parallel orient. comp.	Cohen-Solal et al. (1970), Zozime et al. (1972)	Structural and optical studies. Mercury vapor used for sputtering to reduce loss of Hg from films
		Mica(00.1)	260	(111)∥(00.1) [110]∥[11.0] Parallel		
PbTe	Glow-discharge	CdTe	50	Parallel	Francombe et al. (1962)	—
$Pb_{1-x}Sn_xTe$ $0.15 \leqslant x \leqslant 0.32$	Triode	NaCl(001)	50	Parallel	Krikorian et al. (1972)	Influence of rate on epitaxial temperature. Structural optical and electrical studies
		CaF$_2$(111)	250	Parallel		
		CaF$_2$(001)	200–300	Parallel		
CdS	Rf	BaF$_2$(111)	170	Parallel	Lichtensteiger et al. (1969)	Deposited p-type film to form junctions
ZnS	Rf	CdS(00.1)	RT	Parallel (with poly.)	Bunton and Day (1972)	—
InSb	Glow-discharge	NaCl(001)	250	Parallel and (110)∥(001)	Francombe et al. (1962)	—
		NaCl(001)	325	$[1\bar{1}1]$∥[110] Parallel		
InSb	Glow-discharge	NaCl(110)	325	Parallel	Francombe and Schlacter (1964)	—
InSb	Glow-discharge	NaCl(001)	>250	Parallel	Khan (1968)	In situ electron diffraction study. Excess indium in initial deposit
GaAs	Glow-discharge	Ge(111)	500–570	Parallel	Molnar et al. (1965)	Loss of As above 570°C
GaAs	Glow-discharge	NaCl(001)	125–400	Parallel	Evans and Noreika (1966)	Epitaxial temperature sensitive to gaseous impurities
GaAs	Rf	NaCl(001)	236	Parallel	Bunton and Day (1972)	—
AlSb	Glow-discharge reactive	CaF$_2$(111)	650	Parallel	Noreika et al. (1969)	—
AlN	Glow-discharge reactive	Si(111)	1000	(00.1)∥(111) [11.0]∥$[1\bar{1}0]$	Noreika et al. (1969)	Residual UHV conditions
		SiC(00.1)	1300	Parallel		

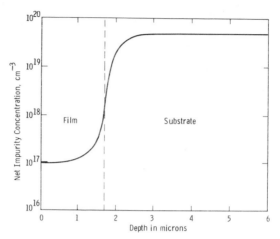

FIG. 19. Carrier concentration profile for homoepitaxial n-type (111) Si film grown by rf sputtering on n$^+$ substrate [courtesy of A. J. Noreika (to be published)].

pressure below 1×10^{-6} Torr was used, and the pressure in the deposition chamber during operation, with a 15-mA ion beam at 10 keV, did not exceed 6×10^{-6} Torr. Good quality, parallel orientation, epitaxy on clean (100) spinel was achieved at 750°C with deposition rates in the range of 50 to 100 Å min^{-1}. Films of thicknesses of 1 to 2 μm showing p-type conductivity with a carrier concentration of 10^{16} cm^{-3} and mobility $\mu_p = 120$ cm^2 V^{-1} sec^{-1} (near bulk values) were successfully grown from a weakly p-type target.

The feasibility of growing homoepitaxial films of Si by rf sputtering has also been demonstrated recently by Noreika (to be published). Deposition from an n-type target at a rate of about 200 Å min^{-1} resulted in n-type films. Typically, growth from a target of carrier concentration $N_D = 2 \times 10^{17}$ cm^{-3} of a layer 2 μm thick at 720°C, resulted in good epitaxial quality with $N_D = 1 \times 10^{17}$ cm^{-3}. A carrier concentration profile, obtained from resistivity measurements, from a film grown on an n$^+$ substrate is shown in Fig. 19. Data on sputtered epitaxial films of semiconductor elements and compounds are summarized in Table III.

B. Compound Semiconductors

To date, a wide variety of semiconductor compounds has been deposited epitaxially by means of sputtering. These range from high-conductivity materials with near-metallic properties, e.g., Bi$_2$Te$_3$, HgCdTe, etc., to near-insulating materials, e.g., CdS, AlN, BN, etc. While these studies are often superficial and lacking in detail, they provide useful guidelines as to the optimum choice of sputtering techniques needed for compositional control. Also, in many cases they enable the researcher to evaluate prospects of using the sputtering approach (relative to evaporation or chemical deposition) as a means of producing films suitable for technological applications.

1. *Tellurides:* Bi_2Te_3, HgTe–CdTe, PbTe, PbSnTe

The fibered and epitaxial growth of Bi_2Te_3 films deposited by diode sputtering from a Bi_2Te_3 cathode have been investigated extensively by Francombe and co-workers (1962, 1964). Preliminary studies of growth on amorphous substrates showed a sequence of fiber orientations with changing substrate temperature. These film textures are peculiar to the pseudocubic atomic packing of the bismuth and tellurium atoms in the Bi_2Te_3 structure and show a strong correlation with the temperature conditions required for epitaxial growth on different faces of rock salt. Thus, at temperatures below 250°C, the fiber texture involves parallelism between the {10.5} planes of the hexagonal telluride structure and the substrate surface. These planes correspond to the faces of a primitive pseudocubic cell unit, which would represent the repeat unit for a lattice in which the Bi and Te atoms are randomly distributed. Epitaxy within this temperature range is readily achieved on the (100) face of rock salt and the resulting deposit comprises doubly positioned crystallites symmetrically arranged with respect to the cube axes of the substrate (Fig. 20a). At higher temperatures, above 300°C, the fiber texture on amorphous substrates is of the {00.1} hexagonal type, corresponding to a set of the {111} planes of the pseudocubic packing unit for the disordered structure. The epitaxial structure most strongly favored in this temperature range is obtained on the (111) face of rock salt (see Fig. 20b).

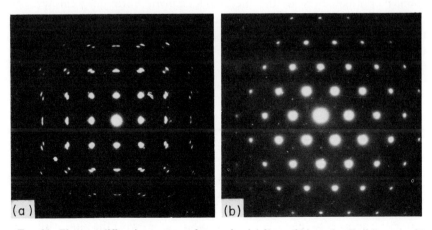

FIG. 20. Electron diffraction patterns from epitaxial films of bismuth telluride grown by sputtering onto faces of rock salt crystal: (a) (100) deposited at 200°C; (b) (111) deposited at 320°C (after Francombe, 1964).

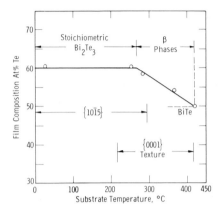

Fig. 21. Composition of sputtered bismuth telluride films as a function of substrate temperature, showing temperature ranges for fiber textures (prepared from data in Francombe, 1964).

Interpretation of both the fiber and epitaxial structures is complicated by the fact that as the temperature is raised above 250°C, a progressive loss of tellurium from the film occurs. This loss is not readily evident since it is not accompanied either by a discontinuous change in film structure, or in the appearance of a second bismuth phase, which might reasonably be expected from the equilibrium phase diagram (Hansen, 1958c). The film structures produced at temperatures between 280 and 420°C in fact comprise the metastable β-phase (Brown and Lewis, 1962) and are isostructural with Bi_2Te_3 but differ from it in that the hexagonal "a" and "c" parameters increase and decrease, respectively, as the tellurium content is reduced (see Fig. 21). The limiting composition BiTe is obtained in films deposited at 420°C and corresponds approximately to the metastable mineral form of wehrlite. This analysis was in fact confirmed by annealing the tellurium deficient films at a temperature within the liquidus range for the β phase, whereupon they transformed to two-phase structures comprising metallic bismuth and stoichiometric Bi_2Te_3.

Considerable interest has developed in recent years in the synthesis of crystals based on the solid-solution system $Cd_xHg_{1-x}Te$. Materials of composition $0.18 \leqslant x \leqslant 0.35$ possess small band gaps and have important potential application for infrared detectors and imaging systems (Willardson and Beer, 1970), operating in the wavelength range of 4 to 20 μm. For imaging devices, large uniform arrays of photovoltaic elements comprising junctions in the pseudobinary telluride systems $Hg_{1-x}Cd_xTe$ or $Pb_{1-x}Sn_xTe$ are being considered in conjunction with integrated-circuit address schemes. The epitaxial films provide a highly suitable geometry for such devices, and sputtering appears to offer a convenient means of growing the films with good compositional control and uniformity over large substrate areas. Sella and his co-workers (Cohen-Solal et al., 1970; Zozime et al., 1972) have explored the feasibility of growing epitaxial films of compounds in the system $Cd_xHg_{1-x}Te$ using triode sputtering and were able to obtain epitaxy at low temperatures of both the pure tellurides and their mutual solid solutions (Table III). Substrates used in these studies were cleaved rock salt, mica, and the (111) faces of CdTe. In their earlier work (Cohen-Solal *et al.*, 1970) triode sputtering was carried out in an argon atmosphere (at low pressure, 10^{-4}–10^{-3} Torr). It is interesting to note that these materials, which crystallize with the cubic zinc blende structure, have a strong tendency to deposit with the {111} planes parallel to the substrate surface. This results in nonparallel-type epitaxy in the case of the (001) NaCl substrate, simpler single-orientation epitaxy on cleaved (00.1) mica, and parallel orientation growth on (111) CdTe. These results are summarized in Table III. It seems possible that parallel orientation might be achieved on NaCl at higher substrate temperatures. However, the rapid loss of Hg occurring at higher temperatures makes it difficult to grow films with any but the lowest Hg content. Another disadvantage of growth under these conditions is the tendency for argon to be trapped in the epitaxial layers, probably accentuated by the low growth temperatures needed to reduce Hg loss. This gas occlusion affects adversely the electrical properties of the grown layers.

In an attempt to avoid argon trappings and reduce Hg loss, Zozime *et al.* (1972) modified the triode sputtering approach, using a mercury diffusion pump to evacuate the steel vacuum chamber and performing sputtering in a Hg vapor atmosphere supplied from a small tank source. The substrate was held outside the plasma, which was confined in the center of the chamber by means of a magnetic field and a cylindrical electrostatic screen. With this system it was possible to grow films at higher temperatures while retaining most of the Hg in the original target composition. At lower rates of deposition (40 Å min^{-1}) films deposited at 200°C on NaCl showed a pronounced (111) fiber texture with some tendency toward the double-positioned epitaxy found for argon sputtering (Table III). However, at 260°C two orientations developed, a main parallel epitaxial component and a weaker double-positioned orientation with (111) parallel to the (001) NaCl plane (Table III). By increasing the deposition rate to 400 Å min^{-1} the double-positioned (111) epitaxial orientation alone was achieved without heating the substrate.

Preliminary studies using diode sputtering, by Francombe *et al.* (1964), demonstrated the feasibility of growing epitaxial layers of PbTe in parallel orientation on the (001) face of NaCl at 250°C. Of greater technological interest, however, is the epitaxial growth of solid solutions in the system $Pb_{1-x}Sn_xTe$, which, as indicated above, are of considerable importance for infrared detection. Extensive studies on sputtered films in this system have been carried out by Krikorian *et al.* (1972), but unfortunately are available only in government reports. Krikorian *et al.* (1972) investigated epitaxial growth from targets with compositions $x = 0.15$, 0.20, 0.25, and 0.32, using triode, ion-beam, and diode sputtering. Substrates comprised (111) oriented CaF_2 and BaF_2 and (001) CaF_2, and were prepared, depending on the orientation of the crystal rod, by cleaving or by cutting and polishing. Structural (X-ray and electron-diffraction), optical transmission, and electrical (Hall effect) measurements were used to characterize the film composition, energy gap, type, carrier concentration, and mobility. Conditions for epitaxy of the rock salt-type solid solution structures were explored systematically by varying the deposition rate and substrate temperature over wide ranges. As reported previously for epitaxial Ge films (Krikorian, 1964), an epitaxial transition temperature could be defined and was found to vary with deposition rate, increasing typically (for a target with $x = 0.20$) from about 220°C at 0.5 $\mu m \ hr^{-1}$ to about 310°C at about 3.0 $\mu m \ hr^{-1}$. Films grown below this temperature were polycrystalline. An upper limiting temperature (in the vicinity

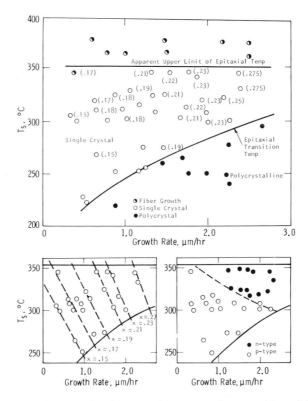

FIG. 22. Characterization of epitaxy, carrier type, and composition of $Pb_{1-x}Sn_xTe$ sputtered thin films as functions of deposition conditions; target composition $Pb_{0.8}Sn_{0.2}Te$ (after Krikorian et al., 1972).

of 360°C) was also defined, above which the film lost its structural integrity, due apparently to chemical decomposition.

Representative data for films grown from a target of composition $x = 0.20$ are presented graphically in Fig. 22 and show the growth rate and temperature conditions needed for epitaxial growth and for producing films of p- and n-type conductivity. The surprising feature of these results is the fact that, depending upon the growth conditions, epitaxial films can be formed possessing a range of compositions extending each side of the x value for the starting target material. Moreover, lines of constant composition can be drawn (Fig. 22b) corresponding to combinations of values of growth rate and substrate temperature. At present, it would be premature to attempt an explanation of these effects, since they probably are influenced by many factors such as target decomposition, differing volatility, and sticking coefficients of various element and compound species arriving at the substrate, and differing energies of formation for the constituents PbTe and SnTe. Suffice to say that to date the results serve to indicate the flexibility and also the complexity of the growth conditions in producing films of desired composition and carrier type.

It has been a common experience with crystals of $Pb_{1-x}Sn_xTe$ that optimum electrical properties can be arrived at by annealing the solid solutions in the presence of a slightly metal-rich solid solution possessing essentially the same Pb:Sn ratio. The effect of such annealing in controlling electrical type and carrier concentration may be illustrated by reference to Fig. 23, which is a magnified reproduction of the $Pb_{1-x}Sn_xTe$ phase diagram in the vicinity of the stoichiometric composition. In interpreting this diagram it should be borne in mind that, in the lead salts, excess lead and excess nonmetal introduce, respectively, donor and acceptor levels (Willardson and Beer, 1970). Thus, isothermal, metal-rich, saturation annealing will have opposite effects depending upon whether the annealing temperature lies above T_x (say at T_1) or below T_x (say at T_2). At T_1, metal saturation of the solid solution leads to a metal-deficient condition (relative to the stoichiometric compound) and hence

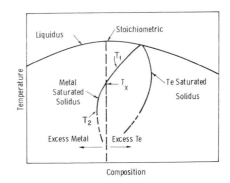

FIG. 23. Schematic of the equilibrium phase diagram of $Pb_{1-x}Sn_xTe$ alloy near the stoichiometric composition.

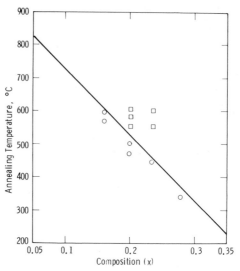

FIG. 24. Annealing temperature versus composition for the characterization of carrier type in $Pb_{1-x}Sn_xTe$ annealed in metal-rich charge (after Krikorian et al., 1972). The solid line is the bulk crossover temperature of metal-rich boundary of solidus field and stoichiometric composition. The points are for annealed p-type films to n-type (O) and remaining p-type (□).

to p-type conductivity, while at T_2 it leads to a metal-excess condition, and hence to n-type conductivity. Krikorian et al. (1972) have annealed sputtered epitaxial films with various x values in metal-rich charges and found that the transition temperatures from p- to n-type agree closely with those for bulk crystals (Fig. 24).

2. Sulfides: CdS, ZnS

Little effort appears thus far to have been devoted to investigation of the epitaxy of sulfide films by sputtering. Some information is available on epitaxial growth CdS and ZnS by rf sputtering, but this provides only sparse details on conditions of epitaxy and film structure. Lichtensteiger et al. (1969) grew films of CdS from a high-purity CdS cathode in an argon atmosphere (10–30 mTorr) containing small additions of phosphine. For a substrate temperature of about 170°C, films grown on glass showed a strong basal plane (0001) fiber orientation, while those grown on CdS single-crystal substrates were epitaxial. In all cases, the films were found to be p-type with carrier concentrations in the range of 1.1×10^{13} to 4.8×10^{15} cm^{-3}, and mobilities in the range of 6 to 15 cm^2 V^{-1} sec^{-1}. Rectifying junctions were produced successfully by sputtering p-type films on high-resistivity n-type substrates. These results are significant in regard to the potential exploitation of CdS in semiconductor device applications. Previously, attempts to make p-type CdS have only been successful through conversion of n-type CdS using ion implantation.

Bunton and Day (1972) have reported the epitaxial growth of rf- sputtered ZnS films on (100) NaCl. Parallel orientation was achieved on unheated substrates; however, the films contained a proportion of polycrystalline and twinned material. This represents a considerable lowering of the epitaxial temperature as compared to electron-beam evaporated films, in which epitaxy occurred at 225°C (Unvala *et al.*, 1968).

3. III–V *Compounds:* InSb, GaAs, AlSb, AlN

Although epitaxial growth by sputtering has been investigated for only a few of the many known III–V compounds, the data available at present suggest that epitaxy of these compounds by one or more of the several techniques now in vogue should present no significant problems. The four compositions to be discussed differ widely in properties, ranging from the narrow band gap ($E_G = 0.17$ eV) highly conductive, near-metallic InSb to the wide-gap ($E_G = 5.9$ eV) near-insulating AlN. While none of the materials described here is representative of the problems to be encountered for the class of compounds as a whole, experience with their epitaxial growth provides a useful guide in optimizing deposition conditions for other members of the class (e.g., InAs, GaAs, InP, GaP, etc.).

Thin epitaxial films of InSb were grown on air-cleaved NaCl by Francombe *et al.* (1962) using glow-discharge diode sputtering. Films deposited at 250°C were found to comprise a polycrystalline randomly oriented component together with two epitaxial components, as shown in Table III. The secondary epitaxial component displayed parallelism between the (110) planes and the rock salt substrate. This observation is satisfactorily explained on the basis of structural studies carried out on films deposited on glass surfaces. At lower temperatures these films show a strong $\langle 110 \rangle$ fiber texture. This situation is reminiscent of a tendency noted by Cohen-Solal *et al.* (1970) for films of $Cd_{1-x}Hg_xTe$ deposited on (100) NaCl to grow at lower temperatures with (111) planes oriented parallel to the substrate surface. For InSb it was found that raising the substrate temperature to 320°C led to the formation of a single parallel-oriented epitaxial structure (Fig. 25). In a subsequent study Francombe and Schlacter (1964) studied the various stages of growth for epitaxial InSb films grown at 325°C on (110) NaCl and obtained the electron microscope data shown in Figs. 26 and 27. Figure 26 shows in transmission the progression, with increasing nominal thickness, from the initial-island stage to the stage at which intergrowth between the islands is almost complete. Evidence of twinning is apparent within some of the islands at all stages of growth. The surface topology of the film is brought out clearly in Fig. 27,

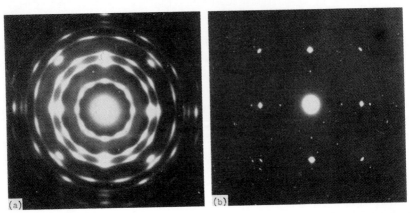

FIG. 25. Electron diffraction patterns from InSb films grown on (100) rock salt by dc glow-discharge sputtering: (a) partial epitaxy, substrate temperature 250°C; (b) complete epitaxy, substrate temperature 320°C.

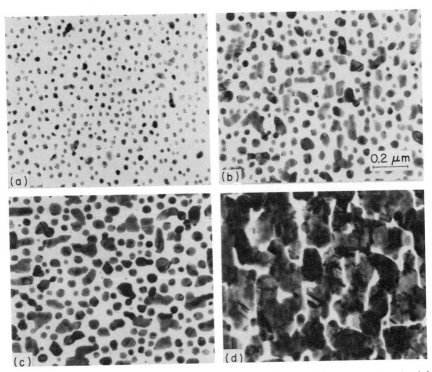

FIG. 26. Transmission electron micrographs showing various thickness stages in epitaxial growth of InSb deposited by glow-discharge sputtering on (110) NaCl: (a) 30 Å; (b) 100 Å; (c) 150 Å; (d) 300 Å (after Francombe and Schlacter, 1964).

154

FIG. 27. Electron microscope replicas of growth stages depicted in Fig. 26: (a) 50 Å; (b) 100 Å; (c) 150 Å; (d) 300 Å.

which shows a series of electron replicas corresponding to the film thicknesses indicated in Fig. 26. An unexpected feature of Fig. 27 is the rounded, granular appearance of the islands, which seems inconsistent with the impression of platelet growth conveyed by the transmission data.

A detailed in situ electron diffraction study of the epitaxy of diode-sputtered InSb on (100) NaCl was made by Khan (1968), who reported that epitaxy with a single parallel orientation occurred at temperatures above 250°C, but he did not observe any tendency to form other orientations at lower temperatures. In the initial stages of growth a thin (~ 20 Å) layer of epitaxial indium was formed, followed at greater thicknesses (> 32 Å) by epitaxial InSb. The apparent discrepancy between Khan's observations and those described above (Francombe et al., 1962; Francombe and Schlacter, 1964) in which no free indium was detected, is difficult to account for. Khan's reflection electron-diffraction technique would be more sensitive to small amounts of indium than the transmission diffraction approach used by Francombe et al. However, one would certainly expect to detect a 20 Å layer of indium, even in transmission, at the early stages of growth. Alternatively, if the InSb target became

.erheated in any presputtering treatment used to clean it, loss of Sb could well occur, leading to initial deposition of an indium-rich layer.

Epitaxial growth of GaAs films has been examined by several workers (Bunton and Day, 1972; Molnar et al., 1965; Evans and Noreika, 1966). Molnar et al. (1965) have examined the structural characteristics of gallium arsenide films deposited on substrates of vitreous silica and single-crystal germanium at growth rates in the range of 60 to 200 Å min^{-1} and substrate temperatures between 30 and 700°C. The crystallinity and orientation of the films were found to depend strongly upon whether they were grown in the positive glow region of the discharge or in the Faraday dark space, and the observed differences could be explained satisfactorily by taking into account the wall-heating effect at the substrate surface due to the glow. In the absence of wall heating, the film structure on silica was amorphous at temperatures below about 400°C, and at higher temperatures—up to 560°C—fiber orientations, first $\langle 110 \rangle$ and then, above 520°C, $\langle 111 \rangle$ were produced. Decomposition due to loss of arsenic occurred above 580°C.

Epitaxial growth was observed in films grown on germanium only within a limited temperature range between 500 and 570°C, and for lower temperatures in this range the structures were invariably twinned on {111} planes. At 570°C it was frequently possible to obtain films showing little or no twinning, but it was difficult to optimize temperature conditions for untwinned growth because of the strong tendency for the epitaxial orientation to degenerate, due to arsenic loss, if the temperature rose slightly during deposition. The surfaces of epitaxial gallium arsenide films grown in this way were usually rough and faceted, particularly at higher values of film thickness (Fig. 28).

Growth of GaAs on the cleavage plane on NaCl has also been studied by Evans and Noreika (1966) and by Bunton and Day (1972). Evans and Noreika carried out their studies in a UHV system capable of vacua of 3×10^{-10} Torr, using glow-discharge diode sputtering. The effect of residual gases such as H_2O, CO_2 on the quality of epitaxial growth at various temperatures between 27 and 400°C was carefully investigated under conditions in which the substrate was cleaved in situ during sputtering and in air prior to sputtering. In general, cleavage of the substrate in air and the presence of small partial pressures of water vapor (approximately 10^{-7} Torr) during growth was found to hinder the development of crystalline or epitaxial growth at lower substrate temperatures. However, careful exclusion of water vapor, and the use of a freshly cleaved surface yielded epitaxial growth at temperatures as low as 120°C, using initial vacua of the order of 10^{-9} Torr.

Imperfect epitaxial growth of rf sputtered GaAs films on air-cleaved NaCl was achieved by Bunton and Day (1972) using a conventional vacuum system. They found that polycrystalline growth developed at temperatures above 130°C, and partial epitaxy at higher temperatures, up to about 236°C, the

FIG. 28. Electron micrographs of gallium arsenide films grown epitaxially by dc glow-discharge sputtering onto (111) germanium at 560°C: (a) thickness 0.5 μm; (b) thickness 3 μm (after Molnar *et al.*, 1965).

highest substrate temperature examined. Their electron diffraction results for an epitaxial film grown at 236°C show rather extensive twinning, and are comparable to the results obtained by Evans and Noreika (1966) on a vacuum-cleaved substrate at 120°C.

Some preliminary observations have been described on the epitaxial growth of AlSb films by dc reactive sputtering in a UHV ion-pumped system by Noreika et al. (1969). The target arrangement used is illustrated in Fig. 7c and provides for the generation of antimony vapor during the sputtering of Al by using the heat generated during sputtering to sublime the solid antimony.

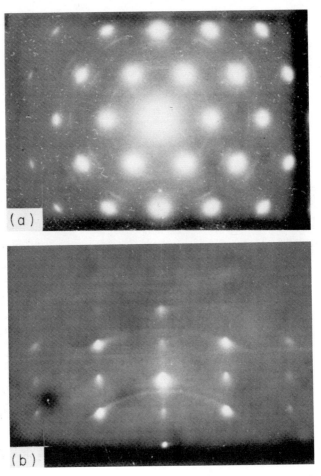

FIG. 29. Electron diffraction patterns of AlN films grown epitaxially by dc glow-discharge reactive sputtering: (a) on (111) Si at 925°C; (b) on the (00.1) face of 6H-type SiC at 1300°C [electron beam along (11.0)] (after Noreika et al., 1969).

With a cathode potential of 2500 V, current density of 1 mA cm^{-2}, and argon pressure of 80 mTorr, a temperature of about 435°C was reached by the Al cathode, sufficient to produce an Sb vapor pressure of about 10^{-4} Torr. This should provide an arrival rate of about 150 monolayers min^{-1} at the substrate surface placed about 4 cm away. This represents a significant excess over the deposition rate for Al (approximately 30 Å min^{-1}). Films were deposited on cleaved CaF_2 (111) substrates heated in the range of 400 to 700°C. The temperature was confined in this range to prevent both condensation of excess Sb (Günther, 1966) and thermal etching of the substrate. Relatively slow deposition rates (approximately 50 Å min^{-1}) were achieved, and the backened appearance of the target suggested that this was due to the formation of AlSb on its surface, so that subsequent sputtering was impeded by the high resistivity of this layer. Epitaxial growth was obtained at 650°C with the orientation relationship as indicated in Table III.

Epitaxial growth of diode-sputtered AlN films was examined during the same study, using the same basic experimental system. In this case pure nitrogen was leaked into the chamber at a rate sufficient to balance its consumption by the reactive sputtering of Al. Using low growth rates, again limited by the formation of high-resistivity AlN on the target surface, epitaxial growth was found to occur both on (111) Si and (00.1) SiC. High substrate temperatures were required in both cases in order to achieve epitaxial growth, about 1000°C in the case of Si and 1300°C for SiC. Corresponding electron diffraction patterns for these deposits are shown in Fig. 29, and the epitaxial relationships are listed in Table III.

VI. Epitaxial Oxide Films

Some of the most comprehensive and detailed studies of epitaxial-sputtered films have been performed within the past few years on simple and mixed oxide systems. In most cases the films have been epitaxed by direct deposition on single-crystal substrates of refractory halides or oxides. However, a number of indirect methods have been employed, involving conversion of an oriented metal or metal–alloy film to the oxide or recrystallization into an epitaxial structure of a film initially amorphous in character. Epitaxial data for oxide films are summarized in Table V (see Section C).

A. ORIENTED OXIDE GROWTH ON METALS OR ALLOYS

Oriented overgrowth of thin oxide films on metals such as Ni, Cu, Fe, Zn, Cd (Pashley, 1956), and Mg (Schwoebel, 1963), etc., is a well-known effect and provides one simple means of obtaining epitaxial oxides, where direct growth is more difficult than epitaxial growth of the parent metal. Oxidation of suitable alloys, e.g., $NiFe_2$, $CoFe_2$, and $MnFe_2$, has been explored as a means of producing epitaxial ferrite films (Francombe, 1966) and some limited success has been achieved. Figure 30, for example, shows the oriented oxidation achieved for a film of $NiFe_2$ deposited epitaxially by diode sputtering on NaCl (Table I). The film was detached from the substrate before oxidizing, and this fact probably accounts for the multiple orientations generated in the resulting ferrite film (Fig. 30b). Better success has been obtained by oxidation of a metal film in situ on a refractory oxide substrate. In this case an epitaxial iron film was grown by diode sputtering onto single-crystal MgO, and oxidation resulted both in conversion of the metal to an oxide and in interdiffusion with

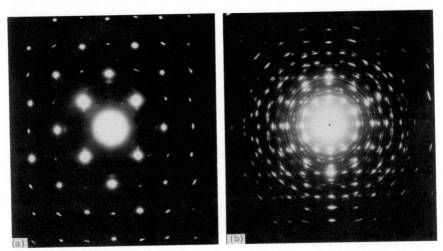

FIG. 30. Electron diffraction patterns illustrating oxidation of epitaxial film of alloy $NiFe_2$ grown on (100) rock salt: (a) alloy as grown at 450°C; (b) after oxidation at 700°C.

the substrate to form a magnesium ferrite layer (see Fig. 31). Some preliminary magnetic measurements of anisotropy and saturation magnetization (Francombe, 1966) suggest that they are well-oriented $MgFe_2O_4$ structures possessing essentially similar properties to the bulk material. This is demonstrated by data summarized in Table IV, which compares some of the magnetization and anisotropy values measured from the ferrite films with those for ceramic bodies.

Work by Kahn and Francombe (1965) also has demonstrated that selective oriented oxidation of a metal in a noble-metal matrix, such as Au, can lead to good quality epitaxial overgrowth. In this case, as discussed in Section IV, a Au–Ni alloy containing about 15% Ni was grown epitaxially by sputtering on NaCl. After homogenizing the single-crystal alloy film for several hours at

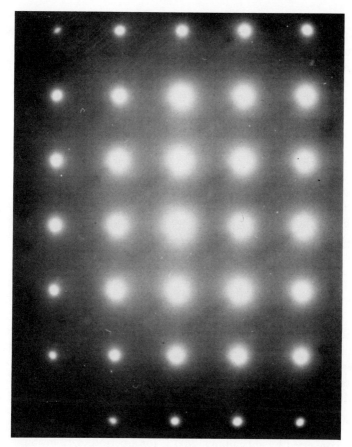

FIG. 31. Electron diffraction pattern of epitaxial $MgFe_2O_4$ film (thickness 1500°C) produced by oxidizing iron film (epitaxed on MgO by dc glow-discharge sputtering) at 950°C.

TABLE IV

MAGNETIC PROPERTIES OF EPITAXIAL FILMS OF $MgFe_2O_4$[a]

Quenching temp (°C)	σ_s (emu/gm) at 25°C		K_1 (10^4 ergs/cm^3)	
	Film	Bulk	Film	Bulk
1000	47.5	54.5	−5.2	2.5
400	23.9	25.2	−4.5	—

[a] Values taken at $t = 4500$ Å (after Francombe, 1966).

390°C, oxidation was performed at the same temperature in air or oxygen. This resulted in the formation of a double-layered structure: (100) NiO on (100) Au. The parallelism between the oxide and gold lattices was confirmed by electron diffraction, which showed that a secondary satellite spot pattern had developed with slightly streaked spots attributable to NiO ($a_0 = 4.14$ Å), positioned inside the main spots due to residual Au ($a_0 = 4.078$ Å).

B. ORIENTED CRYSTALLIZATION OF AMORPHOUS FILM

Earlier studies by Sawatzky and Kay (1968) demonstrated the feasibility of producing films of gadolinium iron garnet, GdIG, by rf (and also by dc) sputtering using ceramic or metal targets in a reactive sputtering mode. It was found that films sputtered from a GdIG ($Gd_3Fe_5O_{12}$) target were deficient in iron when deposited at elevated substrate temperatures. To retain the stoichiometric composition it was necessary to maintain the substrate temperature at a level below 50°C. Films grown at temperatures less than 500°C were found to be amorphous and could only be converted to a crystalline form by annealing at temperatures above 700°C. It was discovered that if the amorphous films were grown on a nonmagnetic garnet substrate, they could be converted by careful thermal cycling into a single-crystal layer in parallel epitaxial orientation to the substrate (Sawatzky and Kay, 1969b). Typical heat treatment involved slow heating (at 600°C hr^{-1}) to above 700°C and annealing for several hours before cooling at a similar rate.

Subsequently, the same basic approach was adopted by Cuomo et al. (1972) to produce high-quality epitaxial films of Gd:Ga:YIG ($Gd_{0.3}Y_{2.7}(Fe_{3.8})$ $Ga_{1.2}O_{12}$) on nonmagnetic gadolinium gallium garnet, GGG, substrates. In this case the deposited amorphous film had a thickness typically of 3.65 μm and was converted by annealing at 900°C for a period of 12 hr. Slow heating and cooling rates (50°C hr^{-1}) were essential to avoid cracking in the annealed films. It was also remarked that sputtering initially from an unreacted mixed-component target conveyed certain preparative advantages, in that a higher deposition rate was possible and that stoichiometric films could be grown over wider ranges of rf power density than was possible in growth from fully reacted ceramic targets. Strip and bubble domains were clearly visible in these films.

C. DIRECT DEPOSITION OF EPITAXIAL FILMS

1. Simple Oxides: Ta_2O_5, Al_2O_3, ZrO_2, ZnO, VO_2

A variety of sputtering methods has been explored for the epitaxial growth of oxide films by direct deposition on single-crystal substrates. These include dc glow-discharge reactive sputtering, triode sputtering, and rf reactive sputtering. Some of the earliest studies were concerned with the epitaxy of oxides of Ta, Al, and Zr (Krikorian and Sneed, 1967; Francombe *et al.*, 1967). Krikorian and Sneed (1967) carried out a systematic investigation of the conditions required for epitaxial growth of Ta_2O_5 and Al_2O_3 using both reactive sputtering and evaporation. They found that two of the key factors influencing epitaxy of these oxides on substrates such as Al_2O_3 (sapphire) and CaF_2 were the partial pressure of oxygen and the deposition rate. Some of the data obtained for sputtered Al_2O_3 grown on sapphire substrates are shown in Figs. 32 and 33. The results obtained for epitaxial Ta_2O_5 films were closely similar. For partial pressures of oxygen less than about 10^{-6} Torr (or 10^{-5} Torr, depending upon the sputtering conditions used), the deposited films

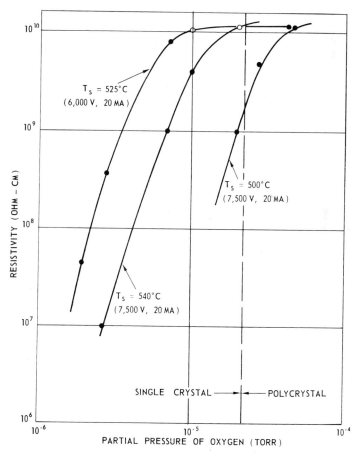

Fig. 32. Formation conditions for epitaxial Al_2O_3 on sapphire grown by reactive, supported discharge sputtering from Al target (after Krikorian and Sneed, 1967).

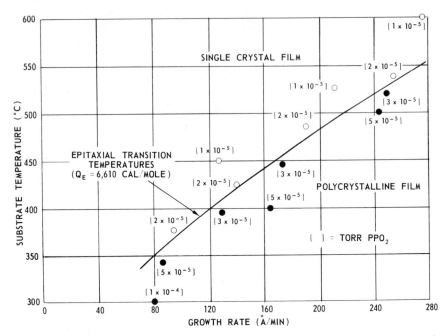

FIG. 33. Single-crystal formation conditions for Al_2O_3 on sapphire by reactive, supported discharge sputtering (after Krikorian and Sneed, 1967).

were essentially metallic, possessing low resistivity. With increasing oxygen pressure, however, the resistivity rose rapidly to a value consistent with that for insulating Al_2O_3.

Two critical oxygen pressures were defined by Krikorian and Sneed (1967): (a) that corresponding to the beginning of the high-resistivity plateau—the stoichiometric pressure, and (b) that corresponding (for elevated growth temperatures) to the transition from an epitaxial to a polycrystalline oxide film structure—the epitaxial pressure. Epitaxial growth was observed only for oxygen pressures lying between these two values. It was stated that the value of the epitaxial pressure varied with the deposition technique used, but for each technique it remained effectively constant over the range of critical growth rates and temperatures explored. However, the value of the stoichiometric pressure was sensitive to substrate temperature and to growth rate, decreasing with increasing temperature and decreasing growth rate. Krikorian and Sneed (1967) report that in those cases where, due to the growth rate being too high or the substrate temperature too low, the stoichiometric pressure rises above the epitaxial pressure, single-crystal growth cannot be achieved. The interdependence of the epitaxial transition temperature and growth rate for reactive sputtering of Al_2O_3 is illustrated in Fig. 33, and is qualitatively similar to that reported by Krikorian and her co-workers for the growth of sputtered Ge (Krikorian, 1964; Krikorian and Sneed, 1966) and $Pb_{1-x}Sn_xTe$ (Krikorian et al., 1972) films.

Krikorian and Sneed (1967) reported that the Ta_2O_5 epitaxial structure grown epitaxially on sapphire, MgO, mica, and CaF_2 by sputtering was in each case the normal low-temperature β-phase. An interesting difference was found between Al_2O_3 epitaxial film deposited, respectively, on sapphire and MgO. In the case of growth on sapphire the high temperature rhombohedral phase of α-Al_2O_3 was observed, whereas epitaxial growth upon cubic MgO gave rise to the low temperature cubic spinel type γ-Al_2O_3 form.

Subsequent studies by Francombe et al. (1967) using triode reactive sputtering exclusively revealed no critical dependence of epitaxial growth condition for Ta_2O_5 upon partial pressure of oxygen. Epitaxial films of Ta_2O_5 were grown on CaF_2, MgO, and epitaxial (111) Pt, using oxygen partial pressures as high as 10^{-3} Torr. Figure 34 shows electron diffraction patterns

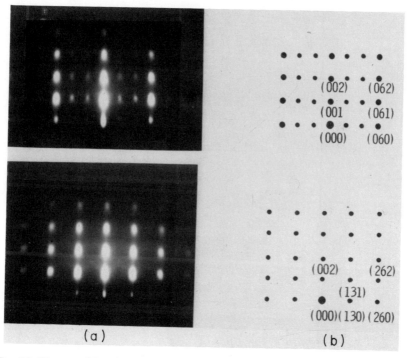

(a)   (b)

FIG. 34. Electron diffraction patterns (a) and indexed reciprocal lattices (b) of β-Ta_2O_5 (disordered form) grown by triode reactive sputtering on epitaxial Pt on (111) CaF_2 (after Francombe et al., 1967).

TABLE V

SPUTTERED EPITAXIAL FILMS OF OXIDES

Oxide	Deposition method	Substrate	Epitaxy temp (°C)	Orientation dep ∥ subst	Reference	Remarks
$NiFe_2O_4$	Glow discharge for metal or alloy	NaCl(001)	420 ($NiFe_2$)	Parallel with other components	Francombe (1966)	Oriented oxidation of $NiFe_2$ epitaxial films
$MgFe_2O_4$		MgO(001)	420 (Fe)	Parallel	Francombe (1966)	Oriented oxidation and interaction to form $MgFe_2O_4$
NiO		NaCl(001)	400 ($Au_{85}Ni_{15}$)	Parallel	Khan and Francombe (1965)	Oriented overgrowth of NiO
GdIG	Rf reactive	YAG (various)	>700	Parallel	Sawatzky and Kay (1968, 1969b)	Oriented crystallization of amorphous film
Gd:Ga:YIG	Rf reactive	GGG(111)	900	Parallel	Cuomo et al. (1972)	Oriented crystallization of amorphous film
β-Ta_2O_5	Glow- and supported-discharge reactive	α-Al_2O_3(00.1) CaF_2(111) MgO(001)	680–800 600–750	Not stated	Krikorian and Sneed (1967)	Epitaxial temperature increased with increasing growth rate
β-Ta_2O_5	Triode reactive	CaF_2(111) Pt–CaF_2(111)	600 650	(001)∥(111) [010]∥[110]	Francombe et al. (1967)	Disordered β-Ta_2O_5 form showing no superlattice along b-axis
Al_2O_3	Rf-supported discharge	α-Al_2O_3(00.1) CaF_2(111) MgO(001)	350–550	Parallel Not stated γ-Al_2O_3 phase, parallel	Krikorian and Sneed (1967)	Epitaxial temperature increased with increasing growth rate

Material	Deposition method	Substrate	Temperature	Orientation relationship	Reference	Remarks
ZrO_2	Triode reactive	$CaF_2(111)$	600	$\{111\} \parallel (111)$ $\langle 1\bar{1}0 \rangle \parallel [1\bar{1}0]$	Francombe et al. (1967)	Multiple positioning of the pseudocubic (monoclinic) axes
ZnO	Glow-discharge reactive	CdS(00.1)	300–550	Parallel	Rozgonyi and Polito (1969)	Structural, optical, and electrical studies
VO_2	Glow-discharge reactive	α-Al_2O_3(00.1)	300–450	Parallel	Rozgonyi and Hensler (1968)	No data given on azimuthal orientations of epitaxial films
		α-Al_2O_3(00.1) TiO_2(001) (100) (110)	400	$(100) \parallel (00.1)$ $(011) \parallel (001)$ $(100) \parallel (100)$ $(100) \parallel (110)$		
GdIG	Rf reactive	YAG (various)	500	Parallel	Sawatzky and Kay (1969b)	Garnet epitaxial films iron-deficient
$TbFeO_3$	Rf–dc reactive	α-Al_2O_3(01.2)	>350	Not stated. Presumably $(001) \parallel (01.2)$	Sosniak (1971)	Iron deficiency compensated by using iron wire grid on target
		$YAlO_3$(001)		Not stated. Presumably parallel		
$Bi_4Ti_3O_{12}$	Rf reactive	Pt–MgO(001)	500–700	$(001) \parallel (001)$ $[100,010] \parallel [110]$	Takei et al. (1970), Takei et al. (1969), Wu et al. (1972), Francombe et al. (1972)	Structural, optical, and electrical studies of films for electrooptic display applications
		Pt–MgO(110)	650	$(001) \parallel (001)$ $[100] \parallel [1\bar{1}0]$		
		MgO(001)	650	$(110) \parallel (001)$ $[001,110] \parallel [100]$		
		MgO(110)	650	$(100,010) \parallel (110)$ $[001] \parallel [001]$		
		$MgAl_2O_4$(110)	725	$[001] \parallel [001]$ $(010) \parallel (110)$ $[001] \parallel [001]$		

and corresponding reciprocal lattices for β-Ta_2O_5 grown on epitaxial Pt. The orthorhombic lattice parameters for the oxide film structure are $a = 6.18$ Å, $b = 10.25$ Å, $c = 3.89$ Å. The structure is essentially that described by Zaslavskii et al. (1955) for the bulk form, but differs from it in that the "b" dimension is approximately one-quarter the value measured for the bulk structure. In the case of epitaxial growth on (111) oriented substrates of CaF_2 and Pt an orientation was adopted in which the short orthorhombic "c" axis of Ta_2O_5 lay parallel to the substrate normal (Fig. 34 and Table V) and the "b" axis (i.e., [010] direction) parallel with the [110] axis in the substrate surface.

Films of ZrO_2 (Francombe et al., 1967) were grown epitaxially by triode sputtering on cleaved CaF_2 substrates, developing orientations in which the {111} planes of the monoclinic (pseudocubic) structure lay parallel with the substrate surface. Due to the equivalence of the {111} planes in this structure, a multiple positioning of the ZrO_2 crystallites comprising the epitaxial film was observed, and this is recognized in the notation of Fig. 35.

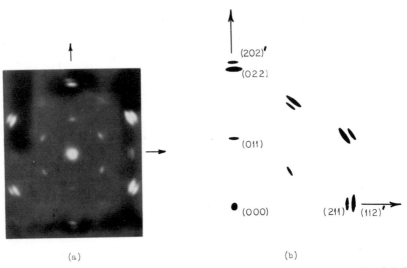

(a)   (b)

FIG. 35. Electron diffraction pattern (a) and reciprocal lattice quadrant (b) of ZrO_2 epitaxial film grown by triode reactive sputtering. Indexing in terms of a monoclinic cell, $a = 5.12$, $b = 5.20$, $c = 5.30$ Å, $\beta = 99.18°$ (primed indices denote an alternative orientation) (after Francombe et al., 1967).

More recently, Rozgonyi and Polito (1969) and Rozgonyi and Hensler (1968) studied the epitaxial growth of ZnO and VO_2 films by dc reactive sputtering. Depending upon the conditions of growth, ZnO films can be deposited in oriented form with either semiconducting or semi-insulating properties. Typical applications utilizing such films are acoustoelectric light scanners (Hakki, 1967) (involving a thin film, high mobility semiconductor with a large electromechanical coupling coefficient), semiconductor heterojunctions used for example for generating electroluminescence (Foster *et al.*, 1968; Foster and Rozgonyi, 1966), and piezoelectric transducers (Foster *et al.*, 1968; Foster and Rozgonyi, 1966). Epitaxial growth of ZnO was achieved on CdS and sapphire substrates using both ceramic and single-crystal ZnO sputtering targets. Film structures were evaluated by means of X-ray and electron diffraction, and from the X-ray diffraction spot shapes the films were classified as moderately, highly, and very highly oriented, or epitaxial. Figure 36 shows the conditions of growth rate and substrate temperature giving rise to oriented and epitaxial growth on (00.1) oriented CdS substrates. As reported above for sputtering studies on Ge (Krikorian, 1964), $Pb_{1-x}Sn_xTe$ (Krikorian *et al.*, 1972), Ta_2O_5, and Al_2O_3 (Krikorian and Sneed, 1967), the epitaxial transition temperature decreases with decreasing growth rate.

The quality of epitaxial growth on CdS was found to depend sensitively upon the method of polishing the CdS substrates, and upon the polarity of the surface [i.e., (00.1) or (00.$\bar{1}$)]. Syton polishing of the substrate was found to give rise to high quality on both the (00.1) and (00.$\bar{1}$) faces, whereas polishing in an HCl solution led to higher quality epitaxy on the S-rich face than on the Cd-rich face. In the case of growth on (00.1) sapphire substrates, the epitaxial quality was found to degrade at higher substrate temperatures, and the transition from oriented to epitaxial growth was far less well defined than for films prepared on CdS. No amorphous structures were produced within the ranges of substrate temperature and growth rate explored.

Films of VO_2 were deposited by Rozgonyi and Hensler (1968) using dc reactive sputtering from a high-purity vanadium metal target. Interest in this material stems from the fact that it displays an abrupt phase transition at about 67°C, changing from a monoclinic, semiconducting phase below this temperature to a tetragonal (rutile-type) metallic phase above. The phase transition is accompanied by a refractive-index change so that the material has potential application in thermistor bolometry and infrared imaging. Substrates, comprising glass and single-crystal sapphire and rutile TiO_2, were heated at 400°C during deposition, and sputtering was carried out in an argon–oxygen mixture (total pressure 25 mTorr) in the ratio 100:1, giving a deposition rate of about 15 Å min^{-1}. The films obtained on (00.1) sapphire were highly oriented, near epitaxial, while those grown on the (001), (100), and (110) faces of TiO_2 were fully epitaxial. It should be stressed that at the temperature of growth the VO_2 films possessed the tetragonal structure, reverting to monoclinic (pseudotetragonal) on cooling to room temperature. Referring (presumably) to the high-temperature tetragonal form, Rozgonyi and Hensler give the orientations for their VO_2 films as follows (see also Table V): On

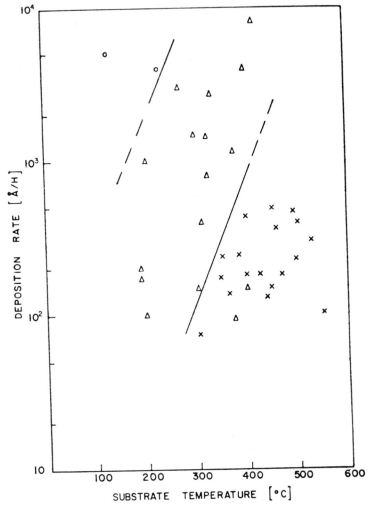

FIG. 36. Effect of deposition rate and substrate temperature on structure of ZnO films on CdS substrates: ○—moderately oriented; △—highly oriented; ×—epitaxial (after Rozgonyi and Polito, 1969).

sapphire the (100) VO_2 plane lay parallel with the (00.1) plane of Al_2O_3, while on rutile the relationships were stated as (011) VO_2 on (001) TiO_2 and (200) VO_2 on (100) and (110) TiO_2. No data were presented on azimuthal relationships or on multiple positioning arising from transition to the monoclinic phase on cooling. Electrical measurements showed that the oriented VO_2 films grown on sapphire and rutile displayed a 40 times larger change in resistivity at the transition temperature than did the polycrystalline films grown on glass. Also the films grown on TiO_2 possessed a lower transition temperature (by about 9°C) than those grown on the other substrates. It is not yet known whether this lowering of the transition temperature is caused by interdiffusion between the film and the substrate or by residual stress produced on cooling from the growth temperature.

2. Mixed Oxides: GdIG, TbFeO₃, Bi₄Ti₃O₁₂

2. *Mixed Oxides:* $GdIG$, $TbFeO_3$, $Bi_4Ti_3O_{12}$

In the previous discussion of indirect methods of epitaxial oxide growth we referred to work by Sawatzky and Kay (1968) and by Cuomo *et al.* (1972) on the growth of epitaxial films of magnetic garnets. In their work on sputtered films of GdIG, Sawatzky and Kay (1969b) also demonstrated that epitaxy on YAG (yttrium aluminum garnet) was obtained by direct growth at temperatures at or above 500°C. However, under these growth conditions severe loss of iron occurred from the films. Somewhat surprisingly, despite the iron deficiency, a GdIG-type, single-phase epitaxial structure still was developed. It was postulated that this structure contained cation vacancies and had been stabilized by growth on the isostructural garnet substrate. (Films deposited at the same temperature on glass substrates displayed iron deficiencies of about 6% and developed a two-phase structure, i.e., GdIG, or $Gd_3Fe_5O_{12}$, and $GdFeO_3$.)

In a later study Sosniak (1971) using combined rf–dc sputtering with ceramic targets, investigated the epitaxial growth of $TbFeO_3$, one of the members of the orthorhombic orthoferrite class. As with the Gd:Ga:YIG garnet films grown by Cuomo *et al.* (1972) the motivation was to produce single-crystal layers displaying high-mobility magnetic bubble domains suitable for memory storage applications. Thick films (up to 35 μm) were grown at rates of 80 to 120 Å min^{-1} on single-crystal substrates of (001) $YAlO_3$ and (01$\bar{1}$2) sapphire. Under the growth conditions used (nominal heater temperature of 350°C) a deficiency of some 20% in iron occurred, and to compensate for this loss Sosniak used an iron-wire grid to overlay the ceramic $TbFeO_3$ target. Final film compositions could thus be controlled to within 2% of the desired stoichiometric condition. Good quality epitaxial layers with smooth surfaces were achieved on $YAlO_3$ substrates, but structures grown on sapphire were cracked, presumably due to the thermal-expansion mismatch between the film and substrate. Magnetic measurements on these films indicated that layers grown on sapphire possessed significantly higher coercivity (by a factor of 3) than those on $YAlO_3$. Films on $YAlO_3$ showed magnetic domain structure, but the coercivity was too high for bubble domains to be developed.

One of the most detailed studies of the epitaxial growth of mixed oxides by sputtering was carried out by Takei *et al.* (1970) on the ferroelectric compound $Bi_4Ti_3O_{12}$. Single crystals of this material possess unusual electrooptic

properties in that reversal of a small component of the spontaneous polarization along one crystal axis results in a rotation of the extinction position,
between crossed polarizers, of approximately 50° (Cummins and Cross, 1968).
This provides a near-optimum change of transmitted light intensity and is well
suited to the needs of electrooptic displays or memories. The crystal structure
is monoclinic (Dorrian *et al.*, 1971) with axial lengths of 5.448, 5.410, and
32.84 Å for *a*, *b*, and *c*, respectively, and a small deviation of β from 90°. Bulk
crystals grown from a Bi_2O_3 flux possess a (001) tabular habit and, as might be
expected from the near-equality of the *a*- and *b*-axes, twinning often occurs,
resulting in an interchange of these two directions within the plane of the
crystal. The polarization lies in the *ac*-plane, at 4.5° to the *a*-axis, and switching
of the *c*-component within this plane results in the large optical contrast effect
(Fig. 37). Unfortunately, the normal crystal habit makes it difficult to produce
large *ac* face areas suitable for optical devices by mechanical sectioning.

   In an attempt to "force" a titanate crystal morphology in which an *ac* face
of large area was developed, efforts were made to explore sputtering conditions
leading to the epitaxial growth of (010) oriented films. The investigation
involved four phases, i.e., development of techniques favoring the growth of
the correct stoichiometric $Bi_4Ti_3O_{12}$ composition, evaluation of epitaxial

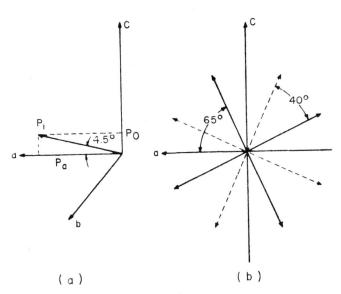

(a)   (b)

FIG. 37. Electrooptic switching in the (010) plane of single-crystal $Bi_4Ti_3O_{12}$. (a) Orientation of resultant polarization P_r with respect to crystallographic axes, *c*-component P_0
exaggerated for clarity. (b) Orientation of optical indicatrix when crystal viewed along *b*.
Solid arrows show orientation for one polarization state. When *c*-component is switched
indicatrix rotates to position indicated by dotted arrows.

growth on a variety of single-crystal oxide substrates [and also (001) epitaxial Pt layers], optimization of growth conditions for the epitaxy of thick films with untwinned (010) orientations, and correlation of electrical and electrooptic properties of the epitaxial titanate with film structure and comparison with bulk crystal properties. Only a brief summary of these four phases is presented here. For more complete details the interested reader is referred to the original published work (Takei *et al.*, 1969, 1970; Wu *et al.*, 1972; Francombe *et al.*, 1972).

Takei *et al.* (1970) explored three sputtering approaches for the growth of $Bi_4Ti_3O_{12}$ films, viz., reactive sputtering with a mixed Bi–Ti target (Fig. 7b), rf sputtering with ceramic targets, and reactive triode sputtering with chemically reduced ceramic targets. Since the major part of the study was performed using the second method, and since this method led eventually to the successful preparation of twinfree (010) films, further discussion will be confined to rf sputtering from ceramic targets. It was soon found that films deposited from either stoichiometric or bismuth oxide-rich targets possessed amorphous structures at substrate temperatures lower than 400°C. At higher temperatures, deposits produced from stoichiometric $Bi_4Ti_3O_{12}$ targets were deficient in Bi_2O_3 and comprised two phases, i.e., $Bi_4Ti_3O_{12}$ and the pyrochlore-type $Bi_2Ti_2O_7$ phase. To compensate for the loss of Bi_2O_3 occurring in the film, the target composition was enriched in bismuth oxide, by adding a proportion of the compound $Bi_{12}TiO_{20}$ to the $Bi_4Ti_3O_{12}$ powder used in preparing the final ceramic target. Using a target of composition $0.8Bi_4Ti_3O_{12}+0.2Bi_{12}TiO_{20}$, the required $Bi_4Ti_3O_{12}$ phase alone could be obtained in the grown film at all temperatures in the range of 500 to 720°C. Within this range, described as a "stoichiometric interval" by analogy to similar effects found by Günther (1966) in the deposition of III–V compound films, the $Bi_4Ti_3O_{12}$ phase is thermodynamically stable but excess Bi_2O_3 does not condense. As shown in Fig. 38, at higher substrate temperature similar "stoichiometric intervals" (or stability ranges) occur for film phases progressively more deficient in Bi_2O_3.

Fortunately, the stoichiometric interval for $Bi_4Ti_3O_{12}$ shown in Fig. 38 was found to extend over a temperature range within which epitaxial growth proved to be feasible. In particular, growth at temperatures toward the upper end of this range yielded epitaxial films of excellent structural quality with the orientation relationships as determined from X-ray Weissenberg patterns shown in Table V. Films up to 35 μm in thickness were grown on the (001) epitaxial Pt substrates, and electrical measurements (Takei *et al.*, 1969) showed these to display bulk polarization and permittivity behavior. Growth on (001) MgO yielded an unexpected orientation with the (110) bismuth titanate plane parallel to the substrate surface and with double-positioned azimuthal alignment (Table V). The azimuthal alignment can be understood in terms of the close correspondence between the average oxygen atom (O—O) separation

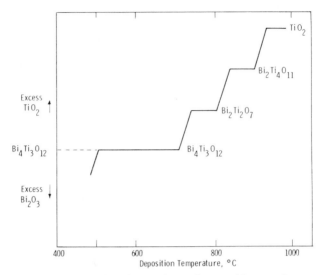

FIG. 38. Schematic representation of dependence of composition on substrate temperature for films rf-sputtered from a target of composition 80% $Bi_4Ti_3O_{12}$: 20% $Bi_{12}TiO_{20}$.

along the c-axis of the titanate and that along the cube axes of MgO. Part of the titanate structure (Dorrian *et al.*, 1971) ia shown in Fig. 39 as viewed along $[1\bar{1}0]$ with the O—O distances along c indicated. Horizontally, the distances are 3.84 Å along [110] and 2.71 Å along the a- and b-axes. Since MgO (see Fig. 39b) has an O—O distance of 4.21 Å along its axes and 2.98 Å along [110], adoption of [001] $Bi_4Ti_3O_{12} \parallel \langle 100 \rangle$ MgO would provide the best match between the structures. Given this alignment, another bismuth titanate direction must align along the perpendicular [010] MgO. The best available fit is with [110] $Bi_4Ti_3O_{12}$, thus resulting in the observed orientation.

The most significant result of the studies on MgO, in relation to the needs for electrooptic display (Takei *et al.*, 1970), was obtained on (110) oriented substrates (Table V). Here, films of mixed orientation with (100) and (010) titanate planes parallel to the substrate surface were obtained. The mode of distribution of these orientations in the film depended upon whether the growth temperature was below or above the Curie temperature of 675°C. For growth below 675°C, the film structure was divided statistically into regions (about 2 to 5 μm in size) possessing (100) or (010) orientations. When grown at higher temperatures (∼725°C), the film developed the high-temperature paraelectric tetragonal structure, but on cooling below the Curie point it transformed to the monoclinic phase and split up into a regular array of (100),(010) stripe-shaped twins (Fig. 40). By removing the epitaxial layer from the substrate and thermally cycling it through the Curie point, the twin

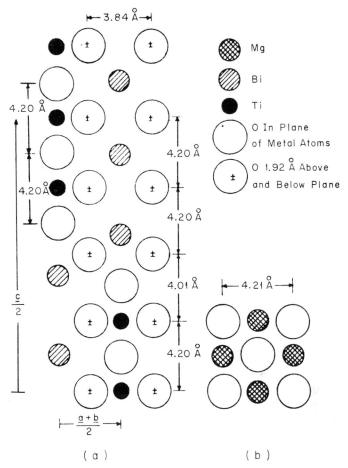

FIG. 39. Structure of $Bi_4Ti_3O_{12}$: (a) portion of structure projected on (110); (b) MgO structure on (001) (after Takei *et al.*, 1970).

structure could be removed and a single (010) orientation produced (Wu *et al.*, 1972). Discounting the multidomain character associated with regions of antiparallel *a*- and *c*-axis orientations, the annealed film now had the orientation required for electrooptic display purposes (Fig. 37).

The generation of the stripe-twin structure in films grown on (110) MgO presumably is associated with relief of stress at the film–substrate interface on cooling from the growth temperature. This stress arises from atomic misfit between the film and substrate lattices, from lattice parameter changes occurring at the tetragonal–monoclinic (Curie point) phase transition (Subbarao,

1961), and from differential thermal expansion between the film and substrate. Optimization of all of these factors so as to favor growth of a twin-free film possessing low residual stress is the desirable aim but clearly presents a complex problem. However, the differential thermal expansion effect alone plays a significant role and can lead to considerable stress in the film, differing in magnitude according to the axial direction involved. The thermal expansion coefficients along a-, b-, and c axes of bismuth titanate are approximately 6×10^{-6}, 11×10^{-6}, and $14 \times 10^{-6}\,°C^{-1}$, respectively (cf. the coefficient for

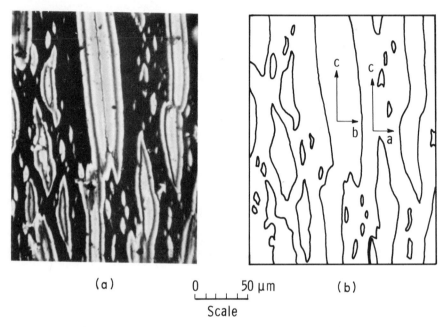

(a)   0 ___ 50 μm   (b)

Scale

Fig. 40. Twin structure of $Bi_4Ti_3O_{12}$ epitaxial film grown on (110) MgO ($b = 12\ \mu m$): (a) optical micrograph taken with crossed polarizers; and (b) diagram showing orientation of twins.

MgO, $14 \times 10^{-6}\,°C^{-1}$). Thus, a thermal expansion mismatch arises on cooling from the growth temperature, along that direction in the film plane lying parallel to the a- or b-axis of the twin structure (see Fig. 40). This generates a strong compressive clamping force along this axis of the film, which presumably also contributes to the adoption of the stripe twin structure on cooling below the Curie point. (Because of the piezoelectric fields associated with this stress, films grown on MgO could not be completely electrically poled to a single-domain state unless they were detached from the substrate and annealed into a single orientation structure.)

In an effort to develop a twin-free structure, other substrate crystals possessing planes structurally similar to the (110) plane of MgO, but with a smaller thermal expansion coefficient than that of MgO, were explored. Among these were spinel ($MgAl_2O_4$), sapphire (Al_2O_3), and rutile (TiO_2). Films grown on (110) $MgAl_2O_4$ possessed excellent optical quality and, moreover, comprised the desired single (010) orientation. The absence of (100),(010) twinning in these films apparently results from the lower thermal-expansion coefficient of spinel ($7.4 \times 10^{-6}\,°C^{-1}$) compared with MgO. A tensile stress (relative to the b-axis of the titanate) generated in the film on cooling from the growth temperature, favors parallelism of the large a-axis ($a = 5.45$ Å), rather than of the b-axis ($b = 5.41$ Å) with the substrate surface. Unfortunately, the relatively large thermal expansion coefficient of the titanate along the c-axis, compared with the coefficient for spinel, leads to cracking and peeling of films thicker than 5 μm. Adherent films thicker than 10 μm were successfully prepared by superimposing a compensating bending stress on the substrate during cooling. After electrically poling these single-orientation (010) films, it was possible to perform optical birefringence and electrooptic switching studies. Birefringence data thus obtained agreed very closely with those for bulk crystals of the titanate (Cummins and Cross, 1968), while electrooptic switching between crossed polarizers yielded contrast ratios higher than 7:1 (Wu et al., 1973). These studies represent the first successful attempt to duplicate bulk-type polarization and electrooptic behavior in thin epitaxial films of a ferroelectric material.

VII. Conclusions and Recommendations

In this survey of the growth of epitaxial films by means of sputtering we have demonstrated that epitaxy, often leading to films of excellent single-crystal quality, has been achieved in a wide variety of metals, alloys, semiconductors, and oxides. It is clear that the cases studied to date represent only a small sampling of the available materials for exploration. The sputtering technique is immediately applicable to the epitaxial growth of any metal or alloy, providing the requisite conditions of cleanliness and target cooling are met. The problem of producing epitaxial semiconductors that are technologically useful may be somewhat more complex, and requires a more complete understanding of the role of electron and ion bombardment in influencing the crystallographic perfection and carrier concentration of the grown film. The studies cited above, in particular those of epitaxial films of Ge and Si, suggest that electrical properties can be optimized by means of growth in a plasma-free region or by careful adjustment of bias-sputtering conditions.

One of the most challenging segments of this field involves the further exploration of semiconductor and insulator compounds of potential use in solid-state components. Examples include III–V compound films (e.g., GaAs, InP, InSb, and InAs) useful in various applications for microwave transistors, Gunn effect devices, and infrared detectors. The presently used chemical-vapor deposition and liquid-phase epitaxy techniques possess significant short-comings, especially in relation to furnishing submicron semiconductor layers of uniform thickness, homogeneous doping, and high crystal quality. Together with newer methods such as molecular-beam epitaxy, sputtering, with its large area capability, offers an attractive alternative to those growth methods.

With the exception of the limited studies made on magnetic oxide films and on ferroelectric bismuth titanate, the field of epitaxial mixed oxides is virtually unexplored. Further work in this area will undoubtedly be undertaken as the available magnetic and ferroelectric oxide films find technological application. The component needs are numerous and varied, ranging from special types of optically, magnetically, and electrically addressed memories and displays, to novel acoustic signal processing devices and pyroelectric infrared vidicons.

References

Adamsky, R. F., Behrndt, K. H., and Brogan, W. T. (1969). *J. Vac. Sci. Technol.* **6**, 542.
Anderson, G. S., Mayer, W. N., and Wehner, G. K. (1962). *J. Appl. Phys.* **33**, 2991.
Arntz, F., and Chernow, F. (1964–1965). *J. Vac. Sci. Technol.* **1–2**, 20.
Bassett, G. A., and Pashley, D. W. (1959). *J. Inst. Metals* **87**, 449.
Bickley, W. P., and Campbell, D. S. (1962). *Vide* **99**, 214.
Brown, A., and Lewis, B. (1962). *J. Phys. Chem. Solids* **23**, 1597.
Bunton, G. V., and Day, S. C. M. (1972). *Thin Solid Films* **10**, 11.
Campbell, D. S., and Stirland, D. J. (1964). *Phil. Mag.* **9**, 703.
Chapman, B. N., and Campbell, D. S. (1969). *J. Phys. C. Solid State Phys.* **2**, 200.
Chopra, K. L., and Randlett, M. R. (1967). *Rev. Sci. Instrum.* **38**, 1147.
Chopra, K. L., Bobb, L. C., and Francombe, M. H. (1963). *J. Appl. Phys.* **34**, 1699.
Chopra, K. L., Randlett, M. R., and Duff, R. H. (1967). *Phil. Mag.* **16**, 261.
Clark, A. H., and Alibozek, R. G. (1968). *J. Appl. Phys.* **39**, 2156.
Cobine, J. D. (1958). "Gaseous Conductors." Dover, New York.
Cohen-Solal, G., Del Valle, J., and Sella, C. (1970). *Congr. Int. Couches Minces, Cannes, Le Vide* **147**, 409.
Comas, J., and Cooper, C. B. (1967). *J. Appl. Phys.* **38**, 2956.
Cummins, S. E., and Cross, L. E. (1968). *J. Appl. Phys.* **39**, 2268.
Cuomo, J. J., Sadagopan, V., DeLuca, J., Chaudhari, P., and Rosenberg, R. (1972). *Appl. Phys. Lett.* **21**, 581.
Davidse, P. D., and Maissel, L. I. (1965). *Trans. Int. Vac. Congr., 3rd, Stuttgart.*
Denoux, M., and Trillat, J. J. (1964). *C. R. Acad. Sci. Paris* **258**, 468.3.
Dorrian, J. F., Newnham, R. E., Smith, D. K., and Kay, M. I. (1971). *Ferroelectrics* **3**, 17.
Evans, T., and Noreika, A. J. (1966). *Phil. Mag.* **13**, 717.
Foster, N. F., and Rozgonyi, G. A. (1966). *Appl. Phys. Lett.* **8**, 221.
Foster, N. F., Coquin, G., Rozgonyi, G. A., and Vannatta, F. (1968). *IEEE Trans. Sonics and Ultrasonics* **SU-15**, 28.
Francombe, M. H. (1963). *In Trans. Nat. Vac. Symp. 10th* (G. H. Bancroft, ed.), p. 316. Macmillan, New York.

Francombe, M. H. (1964). *Phil. Mag.* **10**, 989.

Francombe, M. H. (1966a). *In* "Basic Problems in Thin Film Physics" (R. Niedermayer and H. Mayer, eds.), p. 52. Vandenhoeck and Ruprecht, Göttingen.

Francombe, M. H. (1966b). *In* "The Use of Thin Films in Physical Investigations" (J. C. Anderson, ed.), p. 29. Academic Press, New York.

Francombe, M. H., and Johnson, J. E. (1969). *In* "Physics of Thin Films" (G. Hass and R. E. Thun, eds.), Vol. 5, p. 143. Academic Press, New York.

Francombe, M. H., and Noreika, A. J. (1961a). *Proc. Symp. Elect. Magnetic Properties Thin Metallic Layers* p. 264.

Francombe, M. H., and Noreika, A. J. (1961b). *J. Appl. Phys.* **32**, 975.

Francombe, M. H., and Schlacter, M. (1964). Growth Effects in Epitaxial Films of Gold and Indium Antimonide Prepared by Sputtering [see Francombe, M. H. (1966b)].

Francombe, M. H., Flood, J. J., and Turner, G. L. E. (1962). *Int. Congr. Electron Microsc. 5th, Philadelphia* Vol. I, DD-8. Academic Press, New York.

Francombe, M. H., Noreika, A. J., and Zeitman, S. A. (1967). The Oriented Growth of Thin Film Oxide Dielectrics, AFAL Rep. Contract AF 33 (615)-3814.

Francombe, M. H., Takei, W. J., and Wu, S. Y. (1972). Growth and Properties of Epitaxial Films of Ferroelectric Bismuth Titanate, Technical Rep. AFAL-TR-72-108.

Frerichs, R. (1962). *J. Appl. Phys.* **33**, 1898.

Fukano, Y. (1961). *J. Phys. Soc. Japan* **16**, 1195.

Gillam, E. (1959). *Phys. Chem. Solids* **11**, 55.

Grantham, D. H., Paradis, E. L., and Quinn, D. J. (1970). *J. Vac. Sci. Technol.* **7**, 343.

Griest, A. J., and Flur, B. L. (1967). *J. Appl. Phys.* **38**, 1431.

Günther, K. G. (1966). *In* "The Use of Thin Films in Physical Investigations" (J. C. Andersen, ed.), p. 213. Academic Press, New York.

Hakki, B. W. (1967). *Appl. Phys. Lett.* **11**, 153.

Hansen, M. (1958a). "Constitution of Binary Alloys," p. 2. McGraw-Hill, New York.

Hansen, M. (1958b). "Constitution of Binary Alloys," p. 220. McGraw-Hill, New York.

Hansen, M. (1958c). "Constitution of Binary Alloys," p. 340. McGraw-Hill, New York.

Hansen, M. (1958d). "Constitution of Binary Alloys," p. 678. McGraw-Hill, New York.

Haq, K. E. (1965). *J. Electrochem. Soc.* **112**, 500.

Heavens, O. S. (1964). *In* "Single-Crystal Films" (M. H. Francombe and H. Sato, eds.), p. 381. Pergamon, Oxford.

Herzog, R. F. X., Poschenrieder, W. P., Ruedanaure, F. G., and Satkiewicz, F. G. (1967). *Annu. Conf. Mass Spectrometry, 15th, Denver, Colorado.*

Holland, L. (1960). "Vacuum Deposition of Thin Films." Wiley, New York.

Hutchinson, T. E. (1965). *J. Appl. Phys.* **36**, 270.

Ivanov, R. D., Spivak, G. V., and Kislova, G. K. (1961). *Izv. Akad. Nauk SSSR, Ser. Fir.* **25**, 1524 (*Transl. Bull.* **25**, No. 12).

Jordan, M. R., and Stirland, D. J., unpublished work [see Chapman and Campbell (1969)].

Kay, E. (1963). *J. Appl. Phys.* **34**, 760.

Khan, I. H. (1968). *Surface Sci.* **9**, 306.

Khan, I. H. (1970). *In* "Handbook of Thin Film Technology" (L. I. Maissel and R. Glang, eds.), p. 10–11. McGraw-Hill, New York.

Khan, I. H. (1973). *J. Appl. Phys.* **44**, 14.

Khan, I. H., and Francombe, M. H. (1965). *J. Appl. Phys.* **36**, 1699.

Kloss, F., and Herte, L. (1967). *SCP Solid State Technol.* 45.

Koenig, H. R., and Maissel, L. I. (1970). *IBM J. Res. Develop.* **14**, 168.

Krikorian, E. (1964). *In* "Single-Crystal Films" (M. H. Francombe and H. Sato, eds.), p. 113. Pergamon, Oxford.

Krikorian, E., and Sneed, R. J. (1966). *J. Appl. Phys.* **37**, 3665.

Krikorian, E., and Sneed, R. J. (1967). The Deposition of Single-Crystal Oxide Films by Evaporation and Sputtering, Tech. Rep. No. AFAL-TR-67-139.

Krikorian, E., Longo, R., and Crisp, M. (1972). $Pb_{1-x}Sn_xTe$ Variable Band Gap Alloy System Study. 1st Interim Rep., Contract No. F33615-72-C-1042.

Lau, S. S., and Mills, R. H. (1972). *J. Vac. Sci. Technol.* **9**, 1196.

Layton, C. K., and Cross, K. B. (1967). *Thin Solid Films* **1**, 169 (L).

Lewis, B., and Campbell, D. S. (9967). *J. Vac. Sci. Technol.* **4**, 209.

Lichtensteiger, M., Lagnado, I., and Gatos, H. C. (1969). *Appl. Phys. Lett.* **15**, 418.

Logan, J. S. (1970.) *IBM J. Res. Develop.* **14**, 172.

Maissel, L. I. (1966). *In* "Physics of Thin Films" (G. Hass and R. E. Thun, eds.), Vol. 3, p. 61. Academic Press, New York.

Maissel, L. I. (1970). *In* "Handbook of Thin Film Technology" (L. I. Maissel and R. Glang, eds.), pp. 4–1. McGraw-Hill, New York.

Maissel, L. I., and Schaible, P. M. (1965). *J. Appl. Phys.* **36**, 237.

Maissel, L. I., Jones, R. E., and Standley, C. L. (1970). *IBM J. Res. Develop.* **14**, 176.

Mattox, R. M., and McDonald, J. E. (1963). *J. Appl. Phys.* **34**, 2493.

Molnar, B., Flood, J. J., and Francombe, M. H. (1965). *J. Appl. Phys.* **35**, 3554.

Nickerson, J. W., and Moseson, R. (1964). Semicond. Prods., *Solid State Tech.*

Noreika, A. J. (to be published). Homoepitaxial Layers of Silicon by rf Sputtering.

Noreika, A. J., Francombe, M. H., and Zeitman, S. A. (1969). Research and Development Study of Wide Band-Gap Semiconductor Films. NASA Rep. CR-86395, Contract NAS 12-568.

Pashley, D. W. (1956). *Phil. Mag. Suppl.* **5**, 173.

Pease, R. S. (1960). *Nuovo-Cimento Suppl.* 13.

Perny, G. (1966). *Vide* p. 106.

Reizman, F., and Basseches, H. (1962). *In* "Metallurgy of Semiconductor Materials" (J. B. Schroeder, ed.), p. 169. Wiley (Interscience), New York.

Rozgonyi, G. A., and Hensler, D. H. (1968). *J. Vac. Sci. Technol.* **5**, 194.

Rozgonyi, G. A., and Polito, W. J. (1969). *J. Vac. Sci. Technol.* **6**, 115.

Sato, H., Toth, R. S., and Astrue, R. W. (1964). *In* "Single-Crystal Films" (M. H. Francombe and H. Sato, eds.), p. 393. Pergamon, Oxford.

Sawatzky, E., and Kay, E. (1968). *J. Appl. Phys.* **39**, 4700.

Sawatzky, E., and Kay, E. (1969a). *IBM J. Res. Develop.* **13**, 696.

Sawatzky, E., and Kay, E. (1969b). *J. Appl. Phys.* **40**, 1460.

Schwoebel, R. L. (1963). *J. Appl. Phys.* **34**, 2784.

Sloope, B. W., and Tiller, C. O. (1962). *J. Appl. Phys.* **33**, 3458.

Sloope, B. W., and Tiller, C. O. (1963). *Trans. Nat. Vac. Symp., 10th* (G. H. Bancroft, ed.), p. 339. Macmillan, New York.

Sosniak, J. (1971). *J. Appl. Phys.* **4**, 1802.

Spencer, E. G., and Schmidt, P. H. (1971). *J. Vac. Sci. Technol.* **8**, 552.

Stuart, R. V., and Wehner, G. K. (1962). *Trans. Nat. Vac. Symp., 9th* (G. Bancroft, ed.), p. 160. Macmillan, New York.

Subbarao, E. C. (1961). *Phys. Rev.* **122**, 804.

Takei, W. J., Formigoni, N. P., and Francombe, M. H. (1969). *Appl. Phys. Lett.* **15**, 256.

Takei, W. J., Formigoni, N. P., and Francombe, M. H. (1970). *J. Vac. Sci. Technol.* **7**, 442.

Theurer, H. C., and Hauser, J. J. (1964). *J. Appl. Phys.* **35**, 554.

Thomas, R. N., and Francombe, M. H. (1971). *Surface Sci.* **25**, 357.

Thomson, M. W. (1968). *Phil. Mag.* **18**, 361.

Unvala, B. A., and Pearmain, K. (1970). *J. Mater. Sci.* **5**, 1014.

Unvala, B. A., Woodcock, J. M., and Holt, D. B. (1968). *Brit. J. Appl. Phys. J. Phys. D Ser. 2* **1**, 11.

Valletta, R. M., Perri, J. A., and Riseman, J. (1966). *Electrochem. Technol.* **4**, 402.

von Ardenne, M. (1956). "Tabellen der Elektronenphysik, Ionenphysik and Übermikros-kopie," Vol. I, p. 653. VEB Deutscher Verlag der Wissenschaften, Berlin.

Wallis, G., Wolsky, S. P., and Pittelli, E. (1965). Investigation of Ultra-High Vacuum Sputtered Thin Films. Contract No. A.F. 19(628)-3840.

Wehner, G. K. (1955). *Advan. Electron. Electron Phys.* **7**, 239.

Wehner, G. K. (1962). U.S. Patent 3,021,271.

Wehner, G. K., and Anderson, G. S. (1970). *In* "Handbook of Thin Film Technology" (L. I. Maissel and R. Glang, eds.), pp. 3–1. McGraw-Hill, New York.

Weissmantel, C., Fiedler, O. Hecht, G., and Reisse, G. (1972). *Thin Solid Films* **13**, 359.

Willardson, R. K., and Beer, A. C. (eds.) (1970). "Semiconductors and Semimetals," Vol. 5, Infrared Detectors, pp. 111–253. Academic Press, New York.

Winters, H. F., and Kay, E. (1972). *J. Appl. Phys.* **43**, 794.

Wolsky, S. P., Shooter, D., and Zdanuk, E. J. (1962). *Trans. Nat. Vac. Symp.*, *9th* p. 164. Macmillan, New York.

Wolsky, S. P., Pivkowski, T. R., and Wallis, G. (1965). *J. Vac. Sci. Technol.* **2**, 97.

Woodyard, J. R., and Cooper, C. B. (1964). *J. Appl. Phys.* **35**, 1107.

Wu. S. Y., Takei, W. J., Francombe, M. H., and Cummins, S. E. (1972). *Ferroelectrics* **3**, 217.

Wu, S. Y., Takei, W. J., and Francombe, M. H. (1973). *Appl. Phys. Lett.* **22**, 1.

Zaslavskii, A. I., Zvinchuk, R. A., and Tutov, A. G. (1955). *Dok. Akad. Nauk. SSSR* **104**, 409.

Zozime, A., Sella, C., and Cohen-Solal, G. (1972). *Thin Solid Films* **13**, 373.

2.6   Liquid-Phase Epitaxy

E. A. Giess and R. Ghez

IBM Thomas J. Watson Research Center
Yorktown Heights, New York

I. Introduction

Semiconductor, magnetic, and optical device requirements demand fabrication processes for multilayered solid-state structures. The physical properties of these structures invariably depend on the thickness and uniformity of their individual layers as well as on their chemical composition and crystalline perfection. Most devices require such thin layers (of the order of 1 μm) that polished-crystal platelets, sliced from bulk-grown crystals, cannot meet the specifications. However epitaxial growth, in particular liquid-phase epitaxy

(LPE), can easily produce multilayered structures that fulfil device requirements.

Liquid-phase epitaxy is the process by which a thin single-crystal film grows from a very dilute molten solution on a flat, oriented, single-crystal substrate (seed). Although growth from concentrated solutions is possible, dilute solutions: (i) provide slower growth rates, hence better thickness control, (ii) allow lower growth temperatures, hence better films both in physical perfection and stoichiometry, and (iii) are less susceptible to unwanted crystallization. If the lattice parameter and the thermal expansion coefficient of the film closely match those of the substrate, then high-quality pseudomorphic epitaxy can result. Successful epitaxy also relies on uniform, clean, damage-free substrates with low dislocation counts. Silicon semiconductor technology provides a good foundation for substrate preparation techniques.

LPE technology has advanced rapidly because of its similarity to bulk-crystal growth from seeded solutions, the subject of an extensive literature (see, for example, Laudise, 1963; White, 1965; Elwell and Neate, 1971; Wilcox, 1971; Mullin, 1972; Brice, 1973). In both cases the seed–crystal interface is important because it determines a particular crystallographic orientation, and because dislocations and other imperfections can propagate from a defective seed into the growing crystal. However, because deposited material adjacent to the interface is a significant fraction of the total volume of films grown by LPE, we expect substrate melt-back, initial transient growth during the establishment of a solute boundary layer, and substrate-surface perfection and preparation to be more critical factors in LPE than in bulk-crystal growth. Nevertheless, LPE films can be grown on an hourly basis, whereas bulk-crystal growth may require weeks, a decided technological advantage of the former over the latter.

This chapter draws heavily on the authors' experience with magnetic-garnet films. The next section touches only briefly on the LPE growth of semiconductor compounds and device requirements because these topics have been reviewed very adequately by Casey and Trumbore (1970), Dawson (1972), and Elwell and Scheel (1974). The third section considers garnet systems in some detail, from the points of view of both film preparation and growth-rate measurements. Finally, the fourth section deals with crystallization models, their mathematical analysis, experimental test, and justification.

II. Compound Semiconductor and Magnetic-Garnet LPE Technology

A. MATERIALS

Although LPE could be applied to a wide variety of systems, two classes of technologically important materials have been studied extensively: first the

III–V compound semiconductors, and later the magnetic rare-earth iron garnets. Interestingly, the literature concerned with these two classes of materials is clearly divided; papers in one class rarely cross-reference the other. While the physical chemistry of semiconductor LPE processes differs from that of garnets, the physics and phenomenology do not. It is therefore surprising that more comparisons have not been made. For instance, films of both material classes are produced from molten solutions: LPE processes for compound semiconductors often use a group III solvent such as Ga. Garnets are grown from fluxed melts, usually with $PbO-B_2O_3$ or $BaO-B_2O_3$ solvents. On the other hand, the wetting properties of the solvents, their reactivity, and viscosity dictate particular growth techniques. Compound semiconductors are usually grown from graphite or SiO_2 crucibles in a neutral or reducing atmosphere. Their molten solutions do not wet the crucible materials. Garnets are grown from Pt crucibles in an oxidizing atmosphere (air). Their fluxed melts do wet Pt.

B. TECHNIQUES AND DEVICES

Most of the LPE growth techniques were originally developed for semiconductors. We do not dwell on methods that rely on establishing thermal gradients, such as Deitch's (1970) cooler substrate mode. We describe briefly the "tipping," "sliding-boat," and "dipping" techniques, which differ principally in their geometric arrangement. All three techniques can also be designed so that the seed and its nutrient medium remain in a spatially homogeneous temperature region, although the (almost constant) temperature can be programmed to change in the course of time.

1. *Nonisothermal Techniques*

In the tipping technique, a substrate and a saturated molten solution lie on opposite ends of a boat-shaped crucible initially tilted with the solution at its lower end (see Fig. 1). When the growth temperature is reached, the crucible is tipped, causing the solution to flow over and immerse the substrate. Growth terminates when the crucible is tipped back to its original position. This method was first used by Nelson (1963) to fabricate Ge tunnel diodes and GaAs laser diodes. In the sliding-boat technique (Hayashi *et al.*, 1969; Blum and Shih, 1971), the substrate is positioned in a machined graphite holder that can slide to contact, in sequence, several wells containing saturated solutions. This method can give rise to a form of "volume-limited" growth, because the wells can contain only a very small amount of solute, and spontaneous nucleation is thereby avoided as shown by Blum and Shih. Donahue and Minden (1970) were first to report a volume-limited-type growth technique: a horizontal cylindrically shaped boat rotates about its axis to bring the melt into contact with the substrate. Dipping was first introduced by Linares *et al.* (1965) to grow garnets, and by Rupprecht (1967) to grow GaAs. A preheated substrate, held either in a vertical or horizontal plane, is plunged into a solution for a prescribed period of time. In their original form, all three techniques rely on a programmed temperature drop to establish a supersaturation at the substrate–solution interface. Dipping is simpler and faster than tipping. The latter, however, permits easy atmosphere control; the former does not. The sliding-boat technique has not been used to grow garnet films because Pt does not slide on Pt.

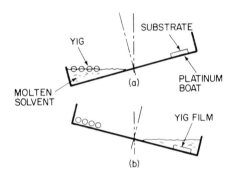

FIG. 1. Tipping apparatus (Linares, 1968); position of boat and molten solution during (a) solution equilibrium, and (b) film growth.

We refer to Dawson's (1972) review for applications to semiconductor devices, e.g., LEDs, lasers, and microwave-device structures. Shaw (1973) has discussed the principles of semiconductor material selection and the fabrication of several microwave devices. He has also compared LPE to other forms of epitaxial growth such as chemical-vapor deposition (CVD).

By the dipping and tipping techniques, Linares *et al.* (1965; Linares, 1968) first produced rare-earth iron-garnet films, $Y_3Fe_5O_{12}$ (YIG) on $Gd_3Ga_5O_{12}$ (GGG) substrates, for microwave applications. Bobeck *et al.* (1970) demonstrated that "bubbles" (cylindrical magnetic domains) in flux-grown garnet platelets could be used for magnetic devices. Shortly thereafter, Shick *et al.* (1971) produced garnet films by the tipping method and operated shift registers fabricated with these films. Other devices, such as optical waveguides (Tien *et al.*, 1972), take advantage of the transparency of Al and Ga garnet substrates and of the low optical scattering losses in high-perfection iron-garnet LPE films. Varnerin (1971) discussed the requirements for bubble-domain materials, e.g., garnet properties, substrate–film matching criteria, epitaxial methods of growth, and defects in LPE films.

2. *Isothermal Techniques*

Garnet films have been grown isothermally from undercooled fluxed melts (Levinstein *et al.*, 1971) on substrates rotating axially in a horizontal plane (Giess *et al.*, 1972b; White and Wood, 1972). The growth process was studied extensively (cf. Section III) and is reasonably well understood from a theoretical point of view (cf. Section IV). This mode of growth is geometrically similar to the Czochralski method but is not primarily dependent on heat flow to achieve supersaturation. At the present time, the bubble-domain-device technology is a major motivation for further development of the LPE process. For example, Hewitt *et al.* (1973) reported pilot line production of over 1000 LPE films. We note that LPE by dipped horizontal rotating substrates was applied independently to semiconductors (Sudlow *et al.*, 1972). Woodall (1971) grew (GaAl) As films by an isothermal technique using a sliding-boat technique. Unlike the usual techniques for compound semiconductor growth, his method is nearly independent of the thermal properties of the growth apparatus. It involves neither temperature programming controls nor heat flow to achieve crystallization. Supersaturation is controlled by mixing known volumes of two isothermal solutions having different compositions.

III. Garnet-Film Processing

A. Substrate-Crystal Growth and Characterization

Flux-grown rare-earth gallium garnet crystals (Nielsen, 1960) are sufficiently large and perfect to qualify for use as LPE substrates. More economical to produce, however, are crystals grown by the Czochralski method first used by Linares (1964a) to grow GGG. Czochralski crystals are pulled at 10^{-4} cm/sec, two orders of magnitude faster than flux-grown crystals, including LPE films. Also, pulled crystals with diameters larger than 1 in. are commercially available (Keig, 1973). The conditions for growing GGG boules and the factors influencing physical perfection were investigated by Linares (1964a), Brandle *et al.* (1972), Brandle and Valentino (1972), Heinz *et al.* (1972), Brandle and Barns (1973), Carruthers *et al.* (1973), Cockayne and Roslington (1973), Cockayne *et al.* (1973), Kyle and Zydzik (1973), and O'Kane *et al.* (1973). Boules are pulled from melts contained in rf-heated iridium crucibles. The growth atmosphere is either dry N_2 or Ar, usually containing a few percent of O_2. If the growth temperature is excessive or if the atmosphere is too rich in O_2, small Ir particles form in the melt and are incorporated into the boule. The rapid growth direction of gallium garnets is [111], which is fortunately the preferred orientation of garnet films. Boules rotating at 40 to 80 rpm during pulling induce a planar growth interface; otherwise (211) facets form and produce a centrally located core defect which is mechanically strained. The higher rotation rates, however, require slower pulling rates and enhance the formation of striations (conforming to the growth interface shape) along the boule axis. Striations are less serious defects than are core defects. Extensive studies of substrate defects were made by Chaudhari (1972), Glass (1972, 1973), Miller (1972, 1973), Belt et al. (1973), Belt and Moss (1973), and Matthews et al. (1973).

B. Substrate-Surface Preparation

Film perfection depends crucially on the availability of defect-free, clean substrates. Levinstein et al. (1973) found that GGG substrates must be free of dislocations and strains, as well as free of dust, grease, and pits or scratches due to the polishing process. Although boules are pulled on oriented seeds, they are also usually oriented by the Laue X-ray method before sawing. A diamond saw, cutting normal to the growth direction, slices the boules into wafers. O'Kane et al. (1973) sliced boules on a multiple-blade cutter using a 30-μm SiC abrasive slurry. Wafers are about 0.7 mm thick after sawing. Saw damage can be etched away by inserting a wafer into fresh commercial 85% orthophosphoric acid (H_3PO_4) contained in a Pt crucible at 160°C, and then by raising the bath temperature to 300°C in 10 min (Miller, 1972). To avoid polishing substrates, Robertson et al. (1973a) proposed liquid-phase homo-epitaxy involving wafer melt-back in a PbO–B_2O_3 fluxed melt of GGG and subsequent cooling to deposit a fresh GGG epilayer. Both of these techniques are helpful to evaluate substrates, but they do not produce reliable flat surfaces for LPE.

The best results are obtained by polishing wafers sequentially on laps with a series of graded diamond abrasives, followed by polishing on a softer lap with a chemical–mechanical, alkaline–silica polishing agent (e.g., Syton, registered trademark of Monsanto Co.). Usually about 0.1 mm of material is removed during polishing.

Dislocation etch pits are revealed by a 10-min etch in a 1:1 mixture of H_2SO_4:H_3PO_4 at 140°C (O'Kane et al., 1973). The sulfuric–phosphoric acid mixture is more stable than H_3PO_4 alone, which dehydrates upon heating and becomes viscous in time. Hewitt et al. (1973) and Miller (1973) etched wafers in H_3PO_4 for 30 sec at 160°C. After etching, wafers were quenched in heavy oil at 70°C to reduce thermal shock. Then they were ultrasonically agitated in a heated bath of Alconox (registered trademark of Alconox, Inc.) detergent solution containing 15 wt% KOH. Finally, the wafers were rinsed in deionized water, blown dry, and inspected between crossed polarizers for dislocations and strain. This procedure does not degrade good surfaces; it is a cleaning step to remove polishing compound, dirt, and grease. Furthermore, the process causes any surface damage present to develop features that scatter light from an intense source and are therefore visible to the naked eye.

C. Film Composition and Lattice Match

Both device considerations and the film–substrate lattice parameter match (fit) dictate the film composition. Most garnet films are designed to fit approximately Czochralski grown GGG substrates, which have a lattice parameter $a_0 \cong 12.383$ Å. Mee *et al.* (1969), Van Uitert *et al.* (1970, 1971), Nielsen (1971), Giess *et al.* (1971a,b), and Tolksdorf *et al.* (1972) discussed the design of garnet compositions for bubble devices fabricated both from sliced bulk crystals and from LPE films. Earlier, Bobeck *et al.* (1969), Gianola *et al.* (1969), and Thiele (1970) had defined the general magnetic property requirements for bubble-domain devices.

The lattice parameters of mixed (solid solution) garnets can be calculated from published values (Geller, 1967; Winkler *et al.*, 1972) of the end-member garnets using an additive (Vegard) relationship. Geller *et al.* (1969) measured the linear thermal expansion coefficients α_T for garnets in the range of 20 to 1200°C. They found $\alpha_T = 9.18 \times 10^{-6}/°C$ for GGG and $\alpha_T = 10.35 \times 10^{-6}/°C$ for YIG. Thus films that fit a substrate at the growth temperature shift toward a tensile stress state when cooled. In CVD film studies, Mee *et al.* (1969) and Besser *et al.* (1972) determined that the unstrained lattice mismatch $\Delta a = a_s - a_f$ (the room-temperature lattice parameters of the substrate and film respectively) should be within the limits $+0.01$ Å (tension) and -0.02 Å (compression). Blank and Nielsen (1972) and Tolksdorf *et al.* (1972) came to the same conclusion for LPE films. Furthermore, Blank and Nielsen (1972) showed that films do not nucleate under conditions of excessive mismatch. Matthews and Klokholm (1972) explained the tensile stress limit of garnet films in terms of Griffith's (1920) crack theory that predicts an increasing stress limit with decreasing film thickness. Carruthers (1972) reviewed the literature on the origin, magnitude, and configuration of hereroepitaxial stresses in thin oxide films.

In the next subsection we discuss segregation phenomena, such as film constituent distribution and impurity incorporation. These affect film composition through the choice of flux systems and growth conditions.

D. Fluxed-Melt Compositions

1. *Garnet Solubility*

A complete phase-equilibrium diagram for a given crystal-growth system reveals liquidus temperatures and crystallization paths (saturation versus temperature). Fluxed melts for garnet LPE growth contain at least four components: a rare-earth oxide (RE_2O_3), Fe_2O_3, PbO, and B_2O_3. Although the associated phase diagrams are complex (Holtzberg and Giess, 1970), insight results from the use of a pseudobinary system: the solid-garnet phase (the film composition) and the liquid phase (the remaining constituents that form the molten solution). For dilute (ideal) solutions, the garnet solubility in this pseudobinary system can be expressed as

$$W_g/W \propto \exp(-\Delta H/RT) \qquad (1)$$

where W_g is the garnet weight, W the total melt weight, ΔH the heat of solution, R the gas constant, and T the absolute temperature. For a $EuYb_2Fe_5O_{12}$ system, Ghez and Giess (1974) found $\Delta H \cong 25$ kcal/mole. Given the melt composition, Eq. (1) determines the supersaturation of the melt as a function of temperature.

2. Solvent Volatility

Most film studies used a $PbO-B_2O_3$ flux system, the basis of earlier bulk-crystal growth investigations by Titova (1959). This system suffers less from solvent vaporization than do PbO- or PbF_2-based systems. Solvent vaporization modifies the supersaturation and the crystallization path; it generally obeys an Arrhenius-type relationship (Giess, 1966; Quon and Potvin, 1971). Using vapor weight-loss data from a typical LPE melt, Roland (1972) estimated a heat of vaporization of 45.7 kcal/mole over the range 800–980°C.

Linares (1962, 1964b) used a low-volatility $BaO \cdot 0.6B_2O_3$ fluxed melt to grow bulk YIG crystals by a seeded growth-pulling technique. For garnet LPE, Hiskes et al. (1973) developed a 41 mol% $BaO-41B_2O_3-18BaF_2$ flux system. Although viscous, this system appears promising because it (i) has almost negligible vapor losses, (ii) contains a high proportion of garnet in solution, and (iii) exhibits only a slight tendency toward garnet constituent segregation, with effectively no Ba contamination of films.

3. Segregation Effects

Selective segregation of melt constituents during crystallization results in a crystal product, the composition of which differs from that of the melt. Burton et al. (1953; Burton and Slichter, 1958) studied the impurity redistribution in elemental semiconductors grown under steady-state conditions. Let C_L, C_i, C_s, and C_e be the impurity concentration in the bulk solution, at the crystal–solution interface, in the solid, and at equilibrium, respectively. Burton et al. defined the effective, interfacial, and equilibrium distribution coefficients $k_L = C_s/C_L$, $k_i = C_s/C_i$, $k_e = C_s/C_e$, respectively. They derived the relation

$$(1-k_L)/k_L = [(1-k_i)/k_i] \exp(-fA) \qquad (2)$$

where f is the (steady-state) growth rate, and A is a constant that measures diffusional limitations. We note that C_L, C_s, C_e, k_L, and k_e are measurable quantities, but that C_i and k_i are generally not. The usual assumption that $k_i \cong k_e$, i.e., the phase diagram determines the interfacial distribution coefficient, is valid only if interface reaction rates are not rate limiting. If these rates are comparable to diffusional-rate limitations, then k_i will depend, a priori, on f as well as on the equilibrium properties of the system (Baralis, 1968; Brice, 1971).

Fluxed garnet melts are obviously more complex than binary semiconductor systems. Accordingly, Nielsen et al. (1967) introduced the concept of a normalized effective distribution coefficient α to measure the segregation of Ga relative to Fe during growth of Ga-substituted YIG crystals. During epitaxial growth of $(EuY)_3(FeGa)_5O_{12}$ garnet films, Giess et al. (1972b, 1973b) found that the mole ratio (normalized distribution coefficient)

$$\alpha_{Ga} = \frac{[Ga \div (Ga + Fe)]_{film}}{[Ga \div (Ga + Fe)]_{melt}} \tag{3}$$

ranged from 1.94 to 1.74 when the growth rate was varied from 0.98 to 5.18 cm/10^6 sec. In $(EuY)_3(FeGa)_5O_{12}$, the Fe and Ga ions occupy the same lattice sites, while Eu and Y ions (and Pb impurity) occupy another set of lattice sites. Thus α_{Eu} and α_{Pb}, measures of the Eu and Pb segregation, can be defined as in Eq. (3). Experiments show that $\alpha \to 1$ exponentially as $f \to \infty$. Consequently, although $\alpha_{Pb} \ll 1$, films grown rapidly in PbO-fluxed melts contain more Pb in solid solution than films grown slowly (Giess et al., 1972a; Stacy et al. 1973; Robertson et al., 1973b). The first authors also noted a tendency of films grown at low temperatures to contain more Pb than those grown nearer the liquidus temperature. The Pb impurity increases the film lattice parameter significantly and promotes compressive biaxial stress. Blank et al. (1973) found an activation energy of 27.6 Kcal/mole for Pb incorporation (at $f = 1.3$ cm/10^6 sec) and observed that α_{Ga} increases with growth temperature.

All these observations are consistent with the generalization: $\alpha \to 1$ as $f \to \infty$. Moreover, the greater the disparity in ionic radii between ions competing for a lattice site, the greater the segregation effect. The film also is enriched in the constituent that tends to increase its melting point (Giess et al., 1973a). In conclusion, segregation effects must be considered when calculating a fluxed melt batch to grow a specific garnet film composition.

4. Phase Equilibria

Nielsen and Dearborn (1958) determined the phase-equilibria relationships for the pseudoternary $PbO-Y_2O_3-Fe_2O_3$. They found four primary phases: hematite (Fe_2O_3), magnetoplumbite ($PbFe_{12}O_{19}$), orthoferrite ($YFeO_3$), and

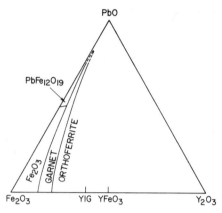

FIG. 2. Pseudoternary PbO–Fe_2O_3–Y_2O_3 phase equilibrium diagram showing magneto-plumbite, hematite, garnet, and orthoferrite primary phase areas (Nielsen and Dearborn, 1958).

garnet (YIG) (see Fig. 2). They also observed the incongruent melting behavior of garnet, i.e., garnet will not crystallize in the pseudobinary system PbO–garnet. By adding Fe_2O_3 in excess of the amount required to form stoichiometric garnet, they formulated fluxed melts that crystallize garnet as the primary solid phase. It is important to recognize that the excess Fe_2O_3 is part of the solvent phase. Without this excess Fe_2O_3 in solution (in the pseudobinary), orthoferrite precipitates as the primary phase under an air atmosphere. If the excess Fe_2O_3 is too great, however, either magnetoplumbite or hematite is the primary phase.

Blank and Nielsen (1972) studied phase equilibrium relationships in fluxed garnet systems based on PbO–B_2O_3. They related the primary phase and liquidus (saturation) temperatures of these melts to certain molar ratios R of the constituents. Blank and Nielsen (1972) defined

$$R_1 \equiv \sum Fe_2O_3, Ga_2O_3 / \sum RE_2O_3 \tag{4}$$

which is $9.0 \div 0.72 = 12.5$, for the molar fluxed melt of Giess et al. (1972b) shown in Table I;

$$R_2 \equiv Fe_2O_3 / Ga_2O_3 \tag{5}$$

i.e., 7.25;

$$R_3 \equiv PbO / B_2O_3 \tag{6}$$

i.e., 15.6; and

$$R_4 \equiv \frac{\sum \text{garnet constituent oxides}}{\text{total melt}} \tag{7}$$

TABLE I

MOLAR FLUXED MELT FOR GROWING
$(EuY)_3(FeGa)_5O_{12}$ FILMS

Constituent	Mole %
Eu_2O_3	0.156
Y_2O_3	0.564
Fe_2O_3	7.91
Ga_2O_3	1.09
PbO	84.83
B_2O_3	5.45

i.e., 0.0972. Orthoferrite is the primary phase unless the ratio $R_1 > 12$. For stoichiometric garnet, $R_1 = 5/3$. Magnetoplumbite or hematite forms when $R_1 > 29$. Generally, R_1 should be as small as possible, otherwise the yield of garnet is reduced. This consideration is important when a large number of films must be grown. However, large R_1 stabilizes supercooled melts against spontaneous nucleation.

Stein and Josephs (1973) observed that the garnet phase exists from $R_1 = 12$ to at least $R_1 = 66$ in the $(EuY)_3(FeGa)_5O_{12}$ system. They found the liquidus temperature T_L to be linear in concentration, and to increase by 320°C/mol% RE_2O_3 added. They also observed a T_L increase of about 20°C/mol% Fe_2O_3 added. Blank and Nielsen (1972) calculated the changes $\Delta T_L/\Delta R_1 \simeq -12.44$°C/unit; $\Delta T_L/\Delta R_4 \simeq +45.3$°C/unit; and $\Delta T_L/\Delta R_3 \simeq +8.49$°C/unit. The R_3 effect tends to keep T_L constant as PbO evaporates.

5. Melt-Composition Calculation and Preparation

The molar composition in Table I is based on a melt composition used by Levinstein et al. (1971) viz., 0.72 mol% RE_2O_3, 9.0$(FeGa)_2O_3$, 84.83PbO, and 5.45B_2O_3. The Table I melt composition produces a garnet film composition of approximately $Eu_{0.6}Y_{2.4}Fe_{3.9}Ga_{1.1}O_{12}$ chosen by estimating the combination of constituents that (i) gives a saturation magnetization $4\pi M \simeq 180$ G (knowing the magnetic effect of Ga substitution for Fe), and (ii) fits GGG substrates. Taking $\alpha_{Eu} = 0.925$ and $\alpha_{Ga} = 1.82$ (Giess et al., 1972b) and using Eq. (3), the mol% $Eu_2O_3 = (0.6 \div 3)(0.72 \div 0.925) = 0.156$ and the mol% $Ga_2O_3 = (1.1 \div 5)(9.0 \div 1.82) = 1.09$.

The fluxed melt must be completely dissolved before film growth. The RE_2O_3 constituents are very refractory and therefore require special attention in the melt-dissolution procedure. The initial melting in a covered Pt crucible should be at ~ 1200°C for 4 hr or more, depending on the degree of comminution and mixing of the reactants. Crucibles and substrate holders must be pure Pt because the melt corrodes Rh or Ir alloys of Pt.

E. FILM-GROWTH PROCEDURE

Originally garnet LPE films were produced by tipping with cooling (Linares, 1968; Shick *et al.*, 1971; Plaskett *et al.*, 1973). However, Levinstein *et al.* (1971) demonstrated the possibility of dipping under isothermal conditions because fluxed-garnet melts tend to supercool. Giess *et al.* (1972b) extended the isothermal dipping technique to include axial rotation of substrates held in a horizontal plane. Rotation improves control of the growth process by inducing a steady state in the diffusion boundary layer at the film–melt interface.

Figure 3 shows the apparatus used by Ghez and Giess (1974). Tolksdorf *et al.* (1972) developed a substrate holder consisting of a set of Pt wire fingers that have low heat capacity and that are easily cleaned. Substrates can be preheated prior to dipping by being lowered to a position just above the melt (see Fig. 3) or by being held in a separate furnace located above the growth chamber (Ghez and Giess, 1973). PbO vapor can condense on the surface of a substrate during preheating above the melt and produces lens shaped interfacial defects in the film (Chaudhari, 1972). Preheating in a separate furnace avoids this

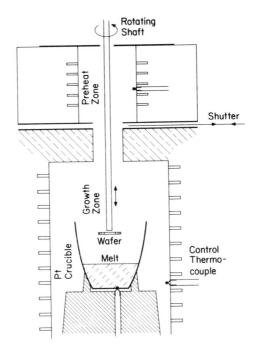

FIG. 3. LPE apparatus, with preheater chamber, for isothermal dipping (Ghez and Giess, (1974).

difficulty. The growth chamber must be free of dirt and as isothermal as possible. The furnace cover and roof can be lined with a thin Pt sheet, which functions as a heat reflector and prevents dirt from contaminating and spuriously nucleating the melt. Blank and Nielsen (1972) position a Pt reflector above the crucible. This reflector also functions as a baffle to dampen atmospheric convection and to lessen circulation of dust.

The growth temperature is usually the most critical growth parameter. The degree of undercooling $\Delta T = T_L - T$ (liquidus minus growth temperature) is typically 5–15°C (Hewitt et al., 1973). The growth rate is approximately linear with ΔT in this range, hence 1% control of f requires temperature control within 0.05 to 0.15°C. The substrate rotation rate ω must be uniform and controlled to within 2 rpm for 1% control of f at $\omega \simeq 36$ rpm (Giess et al., 1972b). In general, f increases approximately as $\omega^{1/2}$. The wafer must not wobble lest the hydrodynamic flow pattern of the melt become unstable. The authors use a substrate immersion mechanism controlled by digital timers that can be preset to an accuracy of 0.1 sec. They find immersion times generally reproducible to better than 0.5 sec.

A growth experiment begins by preheating a cleaned substrate to T in a period of about 3 min to avoid cracking from thermal shock. Additional preheating time is allowed to ensure that the substrate and holder are equilibrated at T. Then the substrate is immersed into the melt while undergoing axial rotation. After withdrawal from the melt, the substrate is held immediately above the melt and spun at $\omega > 600$ rpm to remove any flux droplets adhering to the film and holder. The substrate then is slowly withdrawn (>2 min) from the furnace to avoid cracking from thermal shock. While still mounted on the holder, substrates are cleaned sequentially in warm dilute acetic acid, detergent solution, and deionized water rinse.

F. Measurement Techniques

Film thicknesses can be determined nondestructively from reflectance interference fringes. By varying the wavelength λ (Wemple and Tabor, 1973), the thickness is

$$h = \frac{N}{2} \left[\frac{n_1}{\lambda_1} - \frac{n_2}{\lambda_2} \right]^{-1} \tag{8}$$

where N is the number of fringes between wavelengths λ_1 and λ_2 (typically 5500 and 6330 Å), and n_1 and n_2 are the corresponding film indices of refraction. The film indices of refraction vary significantly with composition and with λ; pure iron garnets have $n > 2.4$ at 6330 Å while Ga-substituted compositions have $n \simeq 2.35$. This method of thickness measurement is very precise and reproducible (1%); however, accuracy is always doubtful unless the dispersion relation $n(\lambda)$ is known. Ghez and Giess (1973) find evidence for a systematic error in interferometric thickness measurements, which they attribute to a film-composition gradient at the substrate interface. In the initially deposited garnet, this gradient presumably results from rapid growth during the transient regime prior to steady-state growth.

Film compositions have been measured by the electron microprobe technique (Giess et al., 1972a, b; Blank et al., 1973; Hiskes and Burmeister, 1973; Robertson et al., 1973b). Bulk-grown crystals of simple end-member garnets are frequently used standards in electron-microprobe analyses. Measurement precision (at a 95% confidence level) of major constituents is about ±3%, while the accuracy is about ±5%. This nondestructive analytical method is therefore especially useful to compare films. Measurements of the Pb impurity content, which ranges from <1 to 10%, are less precise and less accurate. Radioactive tracers grown into films (Janssen et al., 1973) help to measure ions common to both the film and the substrate. Despite the complexity of most garnet films, analytical results from different laboratories and by different techniques are in general agreement.

G. KINETICS OF FILM GROWTH

In dipping experiments with axial substrate rotation, the growth rate and the film thickness depend on the rotation rate, the growth temperature, and the melt composition C_L. Two distinct sets of experiments on the LPE growth of $EuYb_2Fe_5O_{12}$ have been reported.

In the first set (Ghez and Giess, 1973), T and C_L were constant ($T = 880°C$, $\Delta T \simeq 50°C$), but ω was varied in the range 0–169 rpm. Figure 4 summarizes the data. For $\omega > 0$, an initial transient decays rapidly (in less than 4 sec) into a steady state, where h is a linear function of time t. The positive extrapolated intercept h_0 ($\simeq 0.1 \ \mu m$) at $t = 0$ is statistically significant; it demonstrates the existence of the initial transient. The growth rate increases with ω and can be analyzed as proposed by Brice (1967). As observed earlier in the discussion of segregation effects, large growth rates ensure more uniform composition throughout the film thickness. When $\omega = 0$, the steady state is never achieved. The thickness h increases linearly with $t^{1/2}$, clearly suggesting a time-dependent diffusive process.

In the second set of experiments (Ghez and Giess, 1974) ω and t were constant (100 rpm and 122 sec), but C_L and T were varied. For small values of the undercooling ΔT, the (steady-state) growth rate f is linear with ΔT. However, f becomes sublinear and eventually constant for increasing values of ΔT. This behavior suggests an activated rate-limiting process at the interface.

Section IV,B summarizes the theory developed to explain the above experiments. We refer the reader to the original papers for details. Present

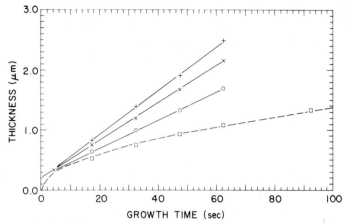

FIG. 4. Isothermally grown film thickness versus growth time for various rotation rates (Ghez and Giess, 1973). The points plotted are: □, 0 rpm; ○, 36 rpm; ×, 100 rpm; and +, 169 rpm.

measurements of LPE growth kinetics are sufficiently precise to warrant comparisons with low-temperature solution growth, a field recently reviewed by Bennema *et al.* (1973) and by Ohara and Reid (1973).

IV. Models of Epitaxial Growth

In the following section we discuss crystallization models applicable to LPE. To this end, we first describe the physical processes and their mathematical characterization. Then we summarize several analytical and numerical solutions for the dipping and tipping procedures. We emphasize that the establishment of a solute boundary layer at the growth interface initially induces transients in the growth rate, even in the case of forced convection by substrate rotation.

A. Governing Equations

Contrary to the growth of elements from the melt where heat flow mainly determines the growth rate, solution growth often proceeds by combined mass transport and subsequent incorporation reaction at the crystal–solution interface. Thermal gradients due to temperature differences in the growth apparatus and to the dissipation of the crystallization heat are generally neglected. Strictly speaking, the following description is only valid for one-component systems. However, in multicomponent systems, growth is often limited by the low rate constants of one of the solute constituents (or complexes) called the growth unit. In this circumstance, a one-component description applies to the growth unit. In Fig. 5, adapted from White and Wood (1972), we give a one-dimensional representation of the concentration profile $C(x,t)$ of growth units in solution. Far from the interface, this concentration must approach the input concentration C_L, i.e.,

$$C(\infty, t) = C_L \qquad (9)$$

If the solution has been carefully homogenized prior to the introduction of the seed, then the initial condition must be

$$C(x, 0) = C_L \qquad (10)$$

The crystallization (dissolution) process depletes (feeds) this source (sink) of material, and the interfacial concentration $C_i = C(0, t)$ lies below (above) C_L. The relative values of C_L and C_i depend on those of C_L and the equilibrium concentration C_e. This latter is determined by the phase diagram and is a function of the growth temperature as well as of the concentration of the other components in solution. The situation in Fig. 5 corresponds to a supersaturated solution $C_L > C_e$. Viewed as a problem of supply and demand, the supply of growth units by diffusion and convection through the solution must balance their possible rate of incorporation into the crystal. Departures of the interfacial concentration from the equilibrium concentration can thus be expected. The simplest mathematical expression of this two-step process, transport and reaction, is

$$D(\partial C/\partial x)_{x=0} = K(C_i - C_e)^n \qquad (11)$$

where n is the order of the reaction. The diffusion coefficient D and the reaction

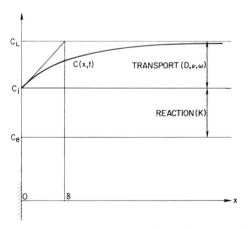

FIG. 5. Solute concentration profile showing driving force for diffusion through the boundary layer and for kinetic processes at the interface (adapted from White and Wood, 1972).

constant K are, a priori, functions of temperature and concentration. This K bears no relation to the segregation coefficient discussed in Section III, D, 3. It is a lumped parameter that hopefully describes the processes occurring at the interface: adsorption and desorption, dissociation and chemical reaction, nucleation, surface migration, and capture at growth sites. We note that fast kinetics, $K \to \infty$, implies that C_i is anchored to its equilibrium value C_e; the process is then transport limited. On the other hand, slow kinetics, $K \to 0$, implies $C_i \simeq C_L$; the process is then reaction limited. Concentration profiles are solutions of the diffusion equation

$$\partial C/\partial t = D(\partial^2 C/\partial x^2) + v(\partial C/\partial x) \tag{12}$$

where v is a velocity resulting from convection (free and forced) and from the growth rate itself. We emphasize that correct solutions of Eq. (12) under the conditions (9)–(11) have been obtained *only* in steady state when v is either a constant, or a known function of position. However, the convective term $v(\partial C/\partial x)$ can be neglected in many cases of time-dependent growth ($\partial C/\partial t \neq 0$). Once the concentration profile is known, mass conservation at the interface yields the growth rate

$$f = (C_s - C_i)^{-1} D(\partial C/\partial x)_{x=0} \tag{13}$$

where C_s is the concentration of growth units in the crystal. Then the thickness of the deposited film results by integrating the growth rate, i.e.,

$$h(t) = \int_0^t f(t') \, dt' \tag{14}$$

From the exact concentration profile of Fig. 5, Levich (1942, 1962) and Burton *et al.* (1953) define a solute boundary layer of thickness δ that correctly describes the interfacial gradient, i.e.,

$$(C_L - C_i)/\delta = (\partial C/\partial x)_{x=0} \tag{15}$$

Such a representation is useful in many transport problems and was first introduced by Prandtl (1904, cf. Schlichting, 1968, for a modern account) in hydrodynamics. There, at high Reynolds numbers, the region of fluid flow can be divided into two parts: a large region of nonviscous flow, and a thin layer of fluid adjacent to surfaces where viscosity is essential and where normal velocity gradients are large enough to satisfy the nonslip conditions. Introducing a solute boundary layer affords considerable mathematical simplification. A priori, δ depends on the physical parameters of the system: D, K, and the kinematic viscosity v. It also depends on the rate of convective flow (Burton *et al.*, 1953; Burton and Slichter, 1958; Wilcox, 1969, 1971), and generally changes with time in cases of time-dependent growth.

We now turn to methods of solution for the dipping and tipping procedures. Well-documented accounts of the diffusion equation and some of its solutions can be found in the standard textbooks of Crank (1956), Carslaw and Jaeger (1959), and Luikov (1968).

B. Mathematical Solutions for Dipping

The dipping procedure is merely the introduction of a seed crystal or substrate into a supersaturated solution. Crystallization can be obtained with or without (i) imposed temperature lowering, and (ii) convective motion of the fluid. The following model (Ghez and Giess, 1973, 1974) assumes that the growth process is isothermal and that the horizontally supported substrate rotates axially. Rotating discs are widely used to investigate electrode processes and the transport properties of electrolytes (Levich, 1962). Our solution explicitly accounts for an initial transient. This latter decays into a well-known steady state (Bircumshaw and Riddiford, 1952), analyzed by Brice (1967) and Garside (1971) in detail.

We assume that the substrate, rotating at an angular frequency ω, sets up a steady-state forced convective flow. By dimensional analysis it can be shown that such a steady state is established in a time less than δ^2/D ($\propto \omega^{-1}$). We solve the time-dependent Eq. (12) without the convective term (v would contain f, which is time dependent), but we account for forced convection by replacing the boundary condition (9) with

$$C(\delta, t) = C_L \tag{16}$$

Here δ is derived from the steady-state solution of Eqs. (9)–(13) (Levich, 1942; Burton et al., 1953), i.e.,

$$\delta = 1.6 D^{1/3} v^{1/6} \omega^{-1/2} \tag{17}$$

a formula valid only for large values of the Schmidt number v/D. We note that δ is independent of f (when $f\delta/D \ll 1$) and of K. The assumption embodied in Eq. (16) is certainly valid for short deposition times because then the concentration profile is insensitive to the boundary condition far from the interface. The assumption is also valid for large values of time since the profile approaches the known steady state. For dilute solutions the condition $C_L \ll C_s$ is often satisfied, and the integral (14) can be evaluated in closed form. Because nonlinear time-dependent solutions are unavailable, we assume a first-order reaction, i.e., $n = 1$ in Eq. (11). We then find the following expressions for the growth rate and for the thickness of the deposited film:

$$f = \frac{D(C_L - C_e)}{\delta C_s} \left[\frac{1}{1+r} + 2 \sum_{n=1}^{\infty} \frac{\exp(-\alpha_n^2 Dt/\delta^2)}{1+r+r^2\alpha_n^2} \right] \tag{18}$$

$$h = \frac{\delta(C_L - C_e)}{C_s} \left[\frac{Dt/\delta^2}{1+r} + 2 \sum_{n=1}^{\infty} \frac{1-\exp(-\alpha_n^2 Dt/\delta^2)}{\alpha_n^2(1+r+r^2\alpha_n^2)} \right] \tag{19}$$

where $r = D/K\delta$, and where the α_n's are the positive nonzero roots of $\tan\alpha + r\alpha = 0$. The dimensionless parameter r represents the relative importance of transport to kinetics. For example, $r \to 0$ implies that interface kinetics are very fast; transport then limits the rate of growth.

Equations (18) and (19) lead to the following observations. First, we note that both f and h depend on time. Second, after an initial exponentially decaying transient, the growth rate approaches the well-known steady-state expression

$$f_\infty = C_s^{-1}(C_L - C_e)(D/\delta)(1+r)^{-1} \qquad (20)$$

Third, we note that f, f_∞, and h depend (i) on temperature T through K and C_e, and to a lesser degree through D and v, (ii) on stirring rate ω through δ [cf. Eq. (17)], and (iii) on substrate orientation through K. Fourth, at steady state, Eq. (19) can be written as

$$h \simeq h_0 + f_\infty t \qquad (21)$$

where h_0 is a *positive* quantity due to the initial transient. The initial offset h_0 and the relaxation time required to reach steady state are analyzed elsewhere (Ghez and Giess, 1973, 1974). Fifth, both f and h are linear in the supersaturation $(C_L - C_e)/C_s$ in accordance with our assumption of first-order kinetics. Defining the liquidus temperature T_L as the solution of

$$C_L = C_e(T_L) \qquad (22)$$

we expand f_∞ to first order in the undercooling $\Delta T = T_L - T$ to obtain

$$f_\infty \simeq C_s^{-1}(\Delta T/T_L)(T_L \partial C_e/\partial T)(D/\delta)(1+r)^{-1} \qquad (23)$$

all quantities being evaluated at $T = T_L$. For small deviations from equilibrium ($\Delta T/T_L \ll 1$), the growth rate is thus linear with undercooling. For large values of ΔT, we expect the growth rate to fall off for two reasons: (i) The equilibrium concentration C_e is generally not linear with temperature; the liquidus curve can often be represented by an Arrhenius expression [cf. Eq. (1)]

$$C_e \propto \exp(-\Delta H/RT) \qquad (24)$$

where ΔH is the heat of solution and R the gas constant. The supersaturation therefore goes to the limiting value C_L/C_s at low temperatures. (ii) The reaction constant explicit in $r = D/K\delta$ decreases rapidly, and the curve $f_\infty(T)$ must flatten out. The interfacial reaction is then dominant. Ghez and Giess (1974) have explained magnetic-garnet growth kinetics on the basis of the ideal behavior Eq. (24), and of a simple activation process with energy $\Delta E \simeq 37$ kcal/mole

$$K \propto \exp(-\Delta E/RT) \qquad (25)$$

Sixth, it is known that growth and dissolution rates are often asymmetric under equal absolute values of supersaturation (Mullin and Garside, 1968; Klein Haneveld, 1971), although Eqs. (18) and (19) are not. Thus the reaction constant K, which we have seen must account for all surface processes, is generally different for growth and dissolution. Seventh, we emphasize that data must be analyzed according to the *full* Eq. (20) (Brice, 1967; Garside, 1971). Otherwise (Blank *et al.*, 1973; Rode, 1973) writing Eq. (20) as

$$f_\infty = C_s^{-1}(C_L - C_e)(D/\delta)_{\text{eff}} \qquad (26)$$

one obtains values of the diffusion coefficient that are a factor $1 + r$ smaller than the actual ones, and that appear to have unusually high activation energies. Finally, our model holds in the limiting case of zero rotation rate ($\omega = 0$) which implies $\delta \to \infty$. We refer to Ghez and Giess (1973) for the analogs of Eqs. (18) and (19), and give the asymptotic expression for the thickness, which is borne out by experiment:

$$h \simeq C_s^{-1}(C_L - C_e)[2(Dt/\pi)^{1/2} - (D/K)] \qquad (27)$$

In this case steady state is never achieved. The thickness scales as $t^{1/2}$ at a rate proportional to the supersaturation and D. The *negative* offset at $t = 0$, due to the initial transient, is inversely proportional to K. We expect it, therefore, to become more important at the lower growth temperatures. If an independent, accurate determination of $D(T)$ were known, the functions $C_e(T)$ and $K(T)$ could be measured according to Eq. (27).

Further analysis, possibly numerical, is necessary to assess those factors neglected in the above treatment: free convection, dissipation of the crystallization heat, initial undercooling due to substrates being at a different temperature than the solution, nonlinear kinetics, i.e., $n \neq 1$ in Eq. (11), and the influence of the rate of immersion, to name but a few.

C. MATHEMATICAL SOLUTIONS FOR TIPPING

The tipping procedure relies on a programmed homogeneous temperature drop that lowers C_e. There are several causes for solute transport. First, any initial supersaturation must be relieved. Second, even if the initial concentration is at equilibrium with the seed, the imposed drop of C_e with T (and thus with time) causes C_i to drop below its initial equilibrium value, creating a diffusive flux toward the interface. We emphasize that cooling necessarily implies time-dependent growth, i.e., a steady-state growth rate can exist only under very special conditions. Third, free convection causes a material flow toward the interface which results in well-known nonuniform growth morphologies (see, for example, Crossley and Small, 1973). In the following, we do not consider forced convection since few experimental designs allow for it. We present some analytic solutions first; then we discuss numerical methods.

1. *Analytical Methods*

At time $t = 0$, the solution at temperature T_0 flows over the substrate. The initial values of the equilibrium concentration and reaction constant are

$$C_0 = C_e(T_0), \qquad K_0 = K(T_0) \tag{28}$$

However, the initial concentration C_L can differ from C_0, i.e., the seed can be subject to an initial supersaturation σ

$$C(x,0) = C_L = C_0(1+\sigma) \tag{29}$$

Next, following Tiller and Kang (1968), the concentration can be viewed as the sum $C = C_1 + C_2$, both functions C_1 and C_2 satisfying the diffusion equation (12), such that C_1 satisfies the boundary condition (11) (with $n = 1$ only) and the initial condition $C_1(x,0) = C_0$, whereas C_2 satisfies

$$C_2(0,t) = 0, \qquad C_2(x,0) = \sigma C_0 \tag{30}$$

The boundary conditions far from the interface are

$$C_1(\infty,t) = C_0, \qquad C_2(\infty,t) = \sigma C_0 \tag{31}$$

for a semi-infinite system, and

$$(\partial C_1/\partial x)_{x=l} = (\partial C_2/\partial x)_{x=l} = 0 \tag{32}$$

for a bounded layer of solution of thickness l. Physically, C_2 accounts only for the initial supersaturation σ; C_1 depends only on the kinetics at the interface. Once profiles for C_1 and C_2 are determined, the corresponding growth rates, f_1 and f_2, and thicknesses, h_1 and h_2, follow from Eqs. (13) and (14). The total growth rate and thickness of deposited film are then

$$f = f_1 + f_2, \qquad h = h_1 + h_2 \tag{33}$$

Solutions for C_2 are well known (Crank, 1956, p. 30). In fact, the full Stefan problem has been solved for an unbounded medium (Crank, 1956; Carslaw and Jaeger, 1959). For low supersaturation, i.e., $\sigma C_0/C_s \ll 1$, the concentration, growth rate, and thickness are

$$C_2(x,t) = \sigma C_0 \, \mathrm{erf} \, x/2(Dt)^{1/2}$$
$$f_2(t) = (\sigma C_0/C_s)(D/\pi t)^{1/2} \tag{34}$$
$$h_2(t) = (2\sigma C_0/C_s)(Dt/\pi)^{1/2}$$

The thickness h_2 scales with $t^{1/2}$; it is linear in supersaturation but does *not* depend on the cooling rate.

To date, all analytical solutions for C_1 depend on the assumption of a *constant* cooling rate b. Thus, the temperature lowering is described by

$$T(t) = T_0 - bt \tag{35}$$

There are two classes of solutions: those that assume the reaction constant K to be so large as to anchor the interfacial constant C_i to its equilibrium value C_e, and those that assume a finite value for K. In the first class, we list in order of complexity the solutions by Mitsuhata (1970), Small and Barnes (1969), Tiller and Kang (1968), Minden (1970), and Ghez (1973). Mitsuhata (1970) assumed a uniform concentration of growth units in solution but his results on etching suggest a diffusive process. Small and Barnes (1969) solved the diffusion equation for very short deposition times when the liquidus curve can be approximated by a linear segment. Introducing the (inverse) slope

$$m^{-1} = (\partial C_e / \partial T)_{T = T_0} \tag{36}$$

they found a thickness–time dependence

$$h_1 = (4b/3C_s m)(D/\pi)^{1/2} t^{3/2} \tag{37}$$

Thus the thickness scales as $t^{3/2}$; it is linear in the cooling rate and is inversely proportional to the liquidus slope at the initial temperature. Tiller and Kang (1968) considered a bounded medium of thickness l and applied Austin's (1932) solution of the diffusion equation. The boundary condition at the interface can be written as

$$C_i = C_e(T(t)) = C_0 \exp(-t/\tau) \tag{38}$$

where the Arrhenius expression (24) for C_e is expanded for small fractional temperature changes $bt/T_0 \ll 1$ [cf. Eq. (35)], and where the relaxation time

$$\tau = mC_0/b = RT_0^2/b\,\Delta H \tag{39}$$

describes departures of C_i from its initial value C_0. Introducing the diffusion length $\lambda = (D\tau)^{1/2}$, the characteristic ratio

$$a = l/(D\tau)^{1/2} = l/\lambda \tag{40}$$

and the approximation $C_s \gg C_i$, we derive the thickness from their growth rate:

$$h_1 = (lC_0/C_s)\left[1 - a^{-1}\tan a \exp(-t/\tau) + 2a^{-2}\sum_{n=1}^{\infty}\frac{\exp(-\beta_n^2 t/\tau)}{\beta_n^2(\beta_n^2 - 1)}\right] \tag{41}$$

where $\beta_n = (\pi/2a)(2n-1)$ label the decaying exponentials in the sum. It can be shown that the behavior (37) is obeyed for $t/\tau \ll 1$, but we see that the final thickness cannot exceed lC_0/C_s, which is related to the total amount of material in solution. In Fig. 6 we represent Eq. (41) for various values of a. For large values of a, growth proceeds according to Eq. (37) for long periods of time. However, when $l \lesssim \lambda\,(a \lesssim 1)$, we see that the thickness approaches its maximum value more quickly. Minden (1970) solved the same problem for an infinite

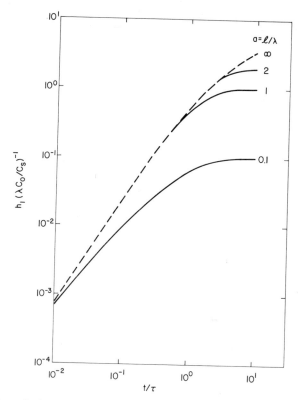

FIG. 6. Dimensionless thickness versus time derived from Tiller and Kang's (1968) model for various normalized thicknesses of bounded melt. The dashed line is derived from Minden's (1970) analysis for an unbounded melt.

medium $(l \to \infty)$. Integrating his expression for the growth rate, we obtain

$$h_1 = (\lambda C_0/C_s)\left[2(t/\tau\pi)^{1/2} - 2\pi^{-1/2}\exp(-t/\tau)\int_0^{(t/\tau)^{1/2}} \exp(u^2)\,du\right] \quad (42)$$

Equation (42) is represented as a dashed line in Fig. 6. We see that it coincides with the thickness (41) derived from Tiller and Kang's model when $a = l/\lambda \gg 1$ and when t/τ is not too large. Ghez (1973) extended Minden's calculations to the full liquidus (24) and to large fractional temperature changes. Minden's calculations are accurate except when the heat of solution ΔH is small in comparison with the initial thermal energy RT_0, and when the temperature drop is considerable.

The second class of analytic solutions results from the work of Ghez and Lew (1973, 1975). These authors considered the effect of the reaction constant

K on the tipping procedure, an effect that was emphasized above concerning the dipping procedure. Previously, Tiller and Kang (1968) attempted to account for interfacial kinetics, but their model relied on solutions of Laplace's equation, an inadmissible procedure for essentially time-dependent situations. In accordance with Eqs. (25), (28), and (35), we write the reaction constant as

$$K = K_0 \exp(-\gamma t/\tau) \tag{43}$$

where $\gamma = \Delta E/\Delta H$ is the ratio of the activation energy for reaction to the heat of solution. Closed-form solutions in terms of elementary functions are unavailable, but Ghez and Lew (1973) set up an integral equation which the flux to the interface must satisfy. They solved this equation numerically and analyzed its asymptotic behavior. The following conclusions can be drawn: (i) For small values of time ($t \ll \tau$), the thickness behaves as

$$h_1 = (K_0 b/2C_s m) t^2 \tag{44}$$

in contrast to Eq. (37), i.e., the thickness increases with t^2 instead of $t^{3/2}$, and it depends on the initial rate constant K_0 instead of D. If this K-process is operative, then a graph of h_1 vs t^2 would be linear and its slope strongly temperature dependent. (ii) For long deposition times

$$f_1 = (C_0 K_0/C_s) \exp(-\gamma t/\tau) \tag{45}$$

i.e., the growth rate decays exponentially, not as $t^{-1/2}$ predicted by Minden. It follows that even if the initial rate constant K_0 is large, the activation energy implicit in $\gamma = \Delta E/\Delta H$ lowers the reaction constant, resulting in an interface limited process. Then, even though we are dealing with an unbounded medium, the thickness achieves a *limiting value*. (iii) Since now $C_i \neq C_e$, an effective supersaturation develops at the interface and causes instabilities even in the absence of constitutional supercooling. (iv) The combined effect of temperature-activated reaction *and* diffusion constants can also be calculated (Ghez and Lew, 1975). The effect of the activation energy for diffusion is small unless it is comparable to the heat of solution. We refer the reader to the original paper for details. (v) One must always bear in mind that if the solution is even slightly supersaturated, the supersaturation effect, Eq. (34), will *initially* dominate all other effects, e.g., Eqs. (37) or (44).

2. Numerical Methods

Analytical methods are well suited to experimental situations involving simple cooling programs and simple phase diagrams [cf. Eqs. (35) and (24)]. The effect of the various physical parameters (time, initial temperature, cooling rate, etc.) on the solutions thus derived is easy to discuss. Unfortunately analytical methods cannot always handle complex situations that arise in multicomponent systems. In a series of noteworthy papers, Crossley and Small (1971, 1972a, b) have applied finite difference methods to the growth of GaAs and (GaAl)As. In the first place they easily account for multi-component diffusion and use experimentally determined liquidus data. Further, their method allows for any cooling program and includes the effect of growth rate in the convective term of Eq. (12). Last, they simulate homogeneous nucleation events in solution. Although the III–V compounds seem to crystallize at equilibrium ($K \to \infty$, and $C_i = C_e$), numerical methods are suited to any form of boundary conditions at the interface. Numerical methods, however, require a separate evaluation for each set of values of the physical parameters. In their first paper, Crossley and Small (1971) calculated the As concentration profiles and the thickness of deposited film for three situations: bulk diffusion in the liquid, with and without nucleation, and boundary-layer diffusion with nucleation. They simulated a nucleation event by specifying a constant concentration beyond the point in the solution where the actual concentration minus its equilibrium value (at that temperature) corresponds to an effective undercooling of 20°K. Calculations for boundary-layer thicknesses of 1 and 2 mm simulate the effect of liquid convection. The region of the solution affected by nucleation increases with the cooling rate. At a rate $b = 10°K/min$ with nucleation, the bulk-diffusion profile is very similar to diffusion through a 1-mm boundary layer. These models were compared to experimental data. Crossley and Small concluded that the bulk-diffusion model describes horizontal dipping adequately, whereas the boundary-layer model accounts for the convection present in vertical dipping systems. Later Crossley and Small (1972a) applied the bulk-diffusion model to the (GaAl)As system. The calculated thickness begins approximately as a $t^{3/2}$ law [cf. Eq. (37)]; it is calculated as a function of initial mole fraction of Al in solution and of the cooling rate. They also calculated the solid composition as a function of film thickness and of temperature. Changing the initial Al content of the solution has a drastic influence on these solid composition curves due to the relative values of the segregation coefficients of Al and As. In all cases their results are at variance with Ilegems's (1970) thermodynamic estimates based on phase equilibrium diagrams, showing the important of nonequilibrium processes. Their last paper (Crossley and Small, 1972b) demonstrates the importance of an initially undercooled solution to obtain smooth epilayers.

V. Areas for Further Investigation

The mass production of reproducible, uniform, large area, defect-free epilayers having precisely controlled composition depends on further understanding of LPE growth processes. We believe that careful experiments would answer important questions in the following areas.

Garnet LPE growth by the dipping technique relies on the stability of undercooled fluxed melts against spontaneous nucleation. The nature and lifetime of this metastable state are of special concern if films are to be grown sequentially without periodic interruptions for remelting the solution. Additional thermodynamic analyses, such as Berkes and White's (1969) on alkali borate flux liquids, and models analogous to those evolved in glass studies (Hench and Freiman, 1971), would help indicate the structure and degree of association of the melt.

From the viewpoint of kinetics, the diffusing species should be determined and their diffusion coefficients measured by means other than growth rates. The experiments of Tolksdorf et al. (1972) on GGG spheres clearly show that the growth rate is orientation dependent. Yet the nature of the interfacial reaction(s) remains to be elucidated. For example, it is not known which of the many surface phenomena are rate-limiting. Segregation of both impurities and film constituents cause compositional changes that affect the physical properties of the epilayers. Segregation effects should be thoroughly investigated as a function of growth temperature, growth rate, and melt composition. A complete theory of nonequilibrium distribution coefficients and their dependence on substrate orientation is still lacking. Aleksandrov and Loginova (1973) have recently addressed this problem.

Finally, LPE films show a remarkable morphological stability, yet they are grown under conditions where the usual constitutional supercooling criterion (Tiller et al., 1953) predicts instability. It is well known (Parker, 1970; Sekerka, 1973) that surface diffusion, surface energy, and interface kinetics are all stabilizing factors. These should be assessed for LPE growth.

ACKNOWLEDGMENTS

We wish to thank F. Holtzberg, M. I. Nathan, A. Reisman, and T. O. Sedgwick for their advice and support. D. Elwell and H. J. Scheel kindly sent us a prepublication copy of the LPE chapter from their forthcoming book. Our colleagues J. M. Blum, T. S. Plaskett, K. K. Shih, and J. M. Woodall supplied us with useful information drawn from their experience with LPE film growth techniques. Our secretary, Linda Callahan, deserves special thanks for her help and patience during the preparation of the manuscript. The second author's wife, Ann Ghez, provided invaluable editorial skills and encouragement when "The sedge has withered from the lake/And no birds sing."

Symbols

$a = l/\lambda$	characteristic ratio in Tiller and Kang's model
a_0	lattice parameter
a_s, a_f	room-temperature substrate and film lattice parameters
b	constant cooling rate
$C(x,t)$	concentration profile of growth units (or impurity) in solution

C_e, C_i, C_L	equilibrium, interfacial, and bulk melt concentrations
C_s	concentration in the crystalline film
C_0	equilibrium concentration at the initial temperature T_0
D	diffusion coefficient of growth units (or impurity) in solution
$f(t)$	growth rate
f_∞	steady-state growth rate
$h(t)$	thickness of deposited film
h_0	extrapolated thickness at $t = 0$
k_e, k_i, k_L	equilibrium, interfacial, and effective distribution coefficients
K	interfacial reaction constant
K_0	interfacial reaction constant at the initial temperature T_0
l	thickness of a bounded solution
$m = (\partial T/\partial C_e)_{C_e = C_0}$	liquidus slope
n	order of the interfacial reaction; index of refraction; index of terms in series expansions
$r = D/K\delta$	characteristic ratio in Ghez and Giess's model
R	gas constant
R_1-R_4	molar ratios defined by Blank and Nielsen
t	time
T	growth temperature
T_L	liquidus temperature
T_0	initial temperature
v	total convective velocity
W_g, W	garnet and total melt weights
x	distance normal to growth interface
α_T	thermal-expansion coefficient
$\alpha_{RE}, \alpha_{Ga}, \alpha_{Pb}$	normalized segregation coefficients
α_n	roots of $\tan \alpha + r\alpha = 0$, label terms in Ghez and Giess's series solution
$\beta_n = (\pi/2a)(2n-1)$	label terms in Austin's series solution
$\gamma = \Delta E/\Delta H$	energy ratio
δ	solute boundary-layer thickness
$\Delta a = a_s - a_f$	room-temperature lattice mismatch
ΔE	activation energy for interfacial reaction
ΔH	heat of solution
$\Delta T = T_L - T$	undercooling
λ	diffusion length $(D\tau)^{1/2}$ in tipping models; optical wavelength
v	kinematic viscosity (viscosity per unit density)
σ	relative supersaturation ratio
$\tau = mC_0/b$	relaxation time in tipping models
ω	angular frequency of a rotating substrate

References

Aleksandrov, L. N., and Loginova, R. V. (1973). *Sov. Phys-Crystallogr.* **17**, 905.

Austin, J. B. (1932). *J. Appl. Phys.* **3**, 179.

Baralis, G. (1968). *J. Crystal Growth* **3, 4**, 627.

Belt, R. F., and Moss, J. P. (1973). *Mater. Res. Bull.* **8**, 1197.

Belt, R. F., Moss, J. P., and Latore, J. R. (1973). *Mater. Res. Bull.* **8**, 357.

Bennema, P., Boon, J., Van Leeuwen, C., and Gilmer, G. H. (1973). *Krist. Tech.* **8**, 659.

Berkes, J. S., and White, W. B. (1969). *J. Crystal Growth* **6**, 29.

Besser, P. J., Mee, J. E., Glass, H. L., Heinz, D. M., Austerman, S. B., Elkins, P. E., Hamilton, T. N., and Whitcomb, E. C. (1972). *In* "Magnetism and Magnetic Materials" (C. D. Graham, Jr. and J. J. Rhyne, eds.), *A.I.P. Conf. Proc. No. 5*, pp. 125–129. A.I.P., New York.

Bircumshaw, L. L., and Riddiford, M. A. (1952). *Quart. Rev.* **6**, 157.

Blank, S. L., and Nielsen, J. W. (1972). *J. Crystal Growth* **17**, 302.

Blank, S. L., Hewitt, B. S., Shick, L. K., and Nielsen, J. W. (1973). *In* "Magnetism and Magnetic Materials" (C. D. Graham, Jr. and J. J. Rhyne, eds.), *A.I.P. Conf. Proc. No. 10*, pp. 256–70. A.I.P., New York.

Blum, J. M., and Shih, K. K. (1971). *Proc. IEEE* **59**, 1498.

Bobeck, A. H., Fischer, R. F., Perneski, A. J., Remeika, J. P., and Van Uitert, L. G. (1969). *IEEE Trans. Magn.* **MAG-5**, 544.

Bobeck, A. H. *et al.* (1970). *Appl. Phys. Lett.* **17**, 131.

Brandle, C. D., and Barns, R. L. (1973). *J. Crystal Growth* **20**, 1.

Brandle, C. D., and Valentino, A. J. (1972). *J. Crystal Growth* **12**, 3.

Brandle, C. D., Miller, D. C., and Nielsen, J. W. (1972). *J. Crystal Growth* **12**, 195.

Brice, J. C. (1967). *J. Crystal Growth* **1**, 161.

Brice, J. C. (1971). *J. Crystal Growth* **10**, 205.

Brice, J. C. (1973). "The Growth of Crystals from Liquids." North-Holland Publ., Amsterdam.

Burton, J. A., and Slichter, W. P. (1958). *In* "Transistor Technology" (H. E. Bridges, J. H. Scaff, and J. N. Shive, eds.), Vol. 1, Chapter 5. Van Nostrand-Reinhold, Princeton, New Jersey.

Burton, J. A., Prim, R. C., and Slichter, W. P. (1953). *J. Chem. Phys.* **21**, 1987.

Carruthers, J. R. (1972). *J. Crystal Growth* **16**, 45.

Carruthers, J. R., Kokta, M., Barns, R. L., and Grasso, M. (1973). *J. Crystal Growth* **19**, 204.

Carslaw, H. S., and Jaeger, J. C. (1959). "Conduction of Heat in Solids," 2nd ed. Oxford Univ. Press (Clarendon), London and New York.

Casey, H. C. Jr., and Trumbore, F. A. (1970). *Mater. Sci. Eng.* **6**, 69.

Chaudhari, P. (1972). *IEEE Trans. Magn.* **MAG-8**, 333.

Cockayne, B., and Roslington, J. M. (1973). *J. Mater. Sci.* **8**, 601.

Cockayne, B., Roslington, J. M., and Vere, A. W. (1973). *J. Mater. Sci.* **8**, 382.

Crank, J. (1956). "The Mathematics of Diffusion." Oxford Univ. Press (Clarendon), London and New York.

Crossley, I., and Small, M. B. (1971). *J. Crystal Growth* **11**, 157.

Crossley, I., and Small, M. B. (1972a). *J. Crystal Growth* **15**, 268.

Crossley, I., and Small, M. B. (1972b). *J. Crystal Growth* **15**, 275.

Crossley, I., and Small, M. B. (1973). *J. Crystal Growth* **19**, 160.

Dawson, L. R. (1972). *Progr. Solid State Chem.* **7**, 117–38.

Deitch, R. H. (1970). *J. Crystal Growth* **7**, 69.

Donahue, J. A., and Minden, H. T. (1970). *J. Crystal Growth* **7**, 221.

Elwell, D., and Neate, B. W. (1971). *J. Mater. Sci.* **6**, 1499.

Elwell, D., and Scheel, H. J. (1974). "Crystal Growth from High Temperature Solutions." Academic Press, New York.

Garside, J. (1971). *Chem. Eng. Sci.* **26**, 1425.

Geller, S. (1967). *Z. Kristallogr.* **125**, 1.

Geller, S., Espinosa, G. P., and Crandall, P. B. (1969). *J. Appl. Crystallogr.* **2**, 86.

Ghez, R. (1973). *J. Crystal Growth* **19**, 153.

Ghez, R., and Giess, E. A. (1973). *Mater. Res. Bull.* **8**, 31.

Ghez, R., and Giess, E. A. (1974). *J. Crystal Growth* **27**, 221.

Ghez, R., and Lew, J. S. (1973). *J. Crystal Growth* **20**, 273.

Ghez, R., and Lew, J. S. (1975). *J. Crystal Growth* (to be published).

Gianola, U. F., Smith, D. H., Thiele, A. A., and Van Uitert, L. G. (1969). *IEEE Trans. Magn.* **MAG-5**, 558.

Giess, E. A. (1966). *J. Amer. Ceram. Soc.* **49**, 104.

Giess, E. A., Calhoun, B. A., Klokholm, E., McGuire, T. R., and Rosier, L. L. (1971a). *Mater. Res. Bull.* **6**, 317.

Giess, E. A., Argyle, B. E., Calhoun, B. A., Cronemeyer, D. C., Klokholm, E., McGuire T. R., and Plaskett, T. S. (1971b). *Mater. Res. Bull.* **6**, 1141

Giess, E. A., Argyle, B. E., Cronemeyer, D. C., Klokholm, E., McGuire, T. R., O'Kane, D. F., Plaskett, T. S., and Sadagopan, V. (1972a). *In* "Magnetism and Magnetic Materials" (C. D. Graham, Jr. and J. J. Rhyne, eds.), *A.I.P. Conf. Proc. No. 5*, pp. 110–14. A.I.P., New York.

Giess, E. A., Kuptsis, J. D., and White, E. A. D. (1972b). *J. Crystal Growth* **16**, 36.

Giess, E. A., Guerci, C. F., Kuptsis, J. D., and Hu, H. L. (1973a). *Mater. Res. Bull.* **8**, 1061.

Giess, E. A., Cronemeyer, D. C., Ghez, R., Klokholm, E., and Kuptsis, J. D. (1973b) *J. Amer. Ceram. Soc.* **56**, 593.

Glass, H. L. (1972). *Mater. Res. Bull.* **7**, 385.

Glass, H. L. (1973). *Mater. Res. Bull.* **8**, 43.

Griffith, A. A. (1920). *Phil. Trans. Roy. Soc.* **A221**, 163.

Hayashi, I., Panish, M. B., and Foy, P. W. (1969). *IEEE J. Quant. Electron.* **5**, 211.

Heinz, D. M., Moudy, L. A., Elkins, P. E., and Klein, D. J. (1972). *J. Electron. Mater.* **1**, 310.

Hench, L. L., and Freiman, S. W. (1971). "Advances in Nucleation and Crystallization of Glasses." Amer. Ceram. Soc., Ohio.

Hewitt, B. S., Pierce, R. D., Blank, S. L., and Knight, S. (1973). *IEEE Trans. Magn.* **MAG-9**, 366.

Hiskes, R., and Burmeister, R. A. (1973). *In* "Magnetism and Magnetic Materials" (C. D. Graham, Jr. and J. J. Rhyne, eds.), *A.I.P. Conf. Proc. No. 10*, pp. 304–308. A.I.P., New York.

Hiskes, R., Felmlee, T. L., and Burmeister, R. A. (1973). *J. Electron. Mater.* **1**, 458.

Holtzberg, F., and Giess, E. A. (1970). *In* "Ferrites" (Y. Hoshino, S. Iida, and M. Sugimoto, eds.), *Proc. Int. Conf.* pp. 296–302. Univ. Tokyo Press, Tokyo.

Ilegems, M. (1970). Ph.D. Thesis, Stanford Univ., California.

Janssen, G. A. M., Robertson, J. M., and Verheijke, M. L. (1973). *Mater. Res. Bull.* **8**, 59.

Keig, G. A. (1973). *In* "Magnetism and Magnetic Materials" (C. D. Graham, Jr. and J. J. Rhyne, eds.), *A.I.P. Conf. Proc. No. 10*, pp. 237–55. A.I.P., New York.

Klein Haneveld, H. B. (1971). *J. Crystal Growth* **10**, 111.

Kyle, T. R., and Zydzik, G. (1973). *Mater. Res. Bull.* **8**, 443.

Laudise, R. A. (1963). *In* "The Art and Science of Growing Crystals" (J. J. Gilman, ed.), pp. 252–273. Wiley, New York.

Levich, V. G. (1942). *Acta Physicochem. URSS* **17**, 257.

Levich, V. G. (1962). "Physicochemical Hydrodynamics." Prentice-Hall, Englewood Cliffs, New Jersey.

Levinstein, H. J., Licht, S., Landorf, R. W., and Blank, S. L. (1971). *Appl. Phys. Lett.* **19**, 486.

Levinstein, H. J., Nielsen, J. W., and Varnerin, L. J. (1973). *Bell Lab. Record* **July/Aug.** 209.

Linares, R. C. (1962). *J. Amer. Ceram. Soc.* **45**, 307.

Linares, R. C. (1964a). *Solid State Commun.* **2**, 229.

Linares, R. C. (1964b). *J. Appl. Phys.* **35**, 433.

Linares, R. C., (1968). *J. Crystal Growth* **3**, **4**, 443.

Linares, R. C., McGraw, R. B., and Schroeder, J. B., (1965). *J. Appl. Phys.* **36**, 2884.

Luikov, A. V. (1968). "Analytical Heat Diffusion Theory." Academic Press, New York.

Matthews, J. W., and Klokholm, E. (1972). *Mater. Res. Bull.* **7**, 213.

Matthews, J. W., Klokholm, E., Sadagopan, V., Plaskett, T. S., and Mendel, E. (1973). *Acta Met.* **21**, 203.

Mee, J. E., Pulliam, G. R., Archer, J. L., and Besser, P. J. (1969). *IEEE Trans. Magn.* **MAG-5**, 717.

Miller, D. C. (1972). *J. Electron. Mater.* **1**, 499.

Miller, D. C. (1973). *J. Electrochem. Soc.* **120**, 678.

Minden, H. (1970). *J. Crystal Growth* **6**, 228.

Mitsuhata, T. (1970). *Jap. J. Appl. Phys.* **9**, 90.

Mullin, J. W. (1972). "Crystallization," 2nd ed. Butterworth, London and Washington, D.C.

Mullin, J. W., and Garside, J. (1968). *Trans. Inst. Chem. Eng.* **46**, T11.

Nelson, H. (1963). *RCA Rev.* **24**, 603.

Nielsen, J. W. (1960). *J. Appl. Phys.* **31**, 51S.

Nielsen, J. W. (1971). *Met. Trans.* **2**, 625.

Nielsen, J. W., and Dearborn, E. F. (1958). *J. Phys. Chem. Solids* **5**, 202.

Nielsen, J. W., Lepore, D. A., and Leo, D. C. (1967). *In* "Crystal Growth" (H. S. Peiser, ed.), pp. 457–61. Pergamon, Oxford.

Ohara, M., and Reid, R. C. (1973). "Modelling Crystal Growth Rates from Solutions." Prentice-Hall, Englewood Cliffs, New Jersey.

O'Kane, D. F., Sadagopan, V., Giess, E. A., and Mendel, E. (1973). *J. Electrochem. Soc.* **120**, 1272.

Parker, R. L. (1970). *Solid State Phys.* **25**, 151–299.

Plaskett, T. S., Klokholm, E., Hu, H. L., and O'Kane, D. F. (1973). *In* "Magnetism and Magnetic Materials" (C. D. Graham and J. J. Rhyne, eds.), *A.I.P. Conf. Proc. No. 10*, pp. 319–23. A.I.P., New York.

Prandtl, L. (1904). *Proc. Int. Math. Congr., 3rd, Heidelberg* pp. 484–91.

Quon, H. H., and Potvin, A. J. (1971). *J. Crystal Growth* **10**, 124.

Robertson, J. M., Van Hout, M. J. G., Janssen, M. M., and Stacy, W. T. (1973a). *J. Crystal Growth* **18**, 294.

Robertson, J. M., Wittekoek, S., Popma, Th. J. A., and Bongers, P. F. (1973b). *Appl. Phys.* **2**, 219.

Rode, D. L. (1973). *J. Crystal Growth* **20**, 13.

Roland, G. W. (1972). *Mater. Res. Bull.* **7**, 983.

Rupprecht, H. (1967). *In Proc. Int. Symp. GaAs, 1st, Reading, 1966* p. 57. Inst. Phys. Soc., London.

Schlichting, H. (1968). "Boundary Layer Theory," 6th ed. McGraw-Hill, New York.

Sekerka, R. F. (1973). *In* "Crystal Growth: An Introduction" (P. Hartman, ed.), pp. 403–443. North-Holland Publ., Amsterdam.

Shaw, D. W. (1973). *J. Electron. Mater.* **2**, 255.

Shick, L. K., Nielsen, J. W., Bobeck, A. H., Kurtzig, A. J., Michaelis, P. C., and Reekstin, J. P. (1971). *Appl. Phys. Lett.* **18**, 89.

Small, M. B., and Barnes, J. F. (1969). *J. Crystal Growth* **5**, 9.

Stacy, W. T., Janssen, M. M., Robertson, J. M., and Van Hout, M. J. G. (1973). *In* "Magnetism and Magnetic Materials" (C. D. Graham, Jr. and J. J. Rhyne, eds.), *A.I.P. Conf. Proc. No. 10*, pp. 314–318. A.I.P., New York.

Stein, B. F., and Josephs, R. M. (1973). *In* "Magnetism and Magnetic Materials" (C. D. Graham, Jr. and J. J. Rhyne, eds.), *A.I.P. Conf. Proc. No. 10*, pp. 329–32. A.I.P., New York.

Sudlow, P. D., Mottram, A., and Peaker, A. R. (1972). *J. Mater. Sci.* **7**, 168.

Thiele, A. A. (1970). *J. Appl. Phys.* **41**, 1139.

Tien, P. K., Martin, R. J., Blank, S. L., Wemple, S. H., and Varnerin, L. J. (1972). *Appl. Phys. Lett.* **21**, 207.

Tiller, W. A., and Kang, C. (1968). *J. Crystal Growth* **2**, 345.

Tiller, W. A., Jackson, K. A., Rutter, J. W., and Chalmers, B. (1953). *Acta Met.* **1**, 428.

Titova, A. G. (1959). *Sov. Phys.-Solid State* **1**, 1714.

Tolksdorf, W., Bartels, G., Espinosa, G. P., Holst, P., Mateika, D., and Welz, F. (1972). *J. Crystal Growth* **17**, 322.

Van Uitert, L. G., Bonner, W. A., Grodkiewicz, W. H., Pictroski, L., and Zydzik, G. J. (1970). *Mater. Res. Bull.* **5**, 825.

Van Uitert, L. G., Gyorgy, E. M., Bonner, W. A., Grodkiewicz, W. H., Heilner, E. J., and Zydzik, G. J. (1971). *Mater. Res. Bull.* **6**, 1185.

Varnerin, L. J. (1971). *IEEE Trans. Magn.* **MAG-7**, 404.

Wemple, S. H., and Tabor, W. J. (1973). *J. Appl. Phys.* **44**, 1395.

White, E. A. D. (1965). *In* "Technique of Inorganic Chemistry" (H. B. Jonassen and A. Weissberger, eds.), Vol. IV, pp. 31–64. Wiley (Interscience), New York.

White, E. A. D., and Wood, J. D. C. (1972). *J. Crystal Growth* **17**, 315.

Wilcox, W. (1969). *Mater. Res. Bull.* **4**, 265.

Wilcox, W. (1971). *In* "Preparation and Properties of Solid Materials" (R. A. Lefever, ed.), Vol. 1, pp. 37–182. Dekker, New York.

Winkler, G., Hansen, P., and Holst, P. (1972). *Philips Res. Rep.* **27**, 151.

Woodall, J. M. (1971). *J. Electrochem. Soc.* **118**, 150.

CHAPTER 3

EXAMINATION OF THIN FILMS

3.1   Studies of Thin-Film Nucleation and Growth by Transmission Electron Microscopy

Helmut Poppa

Ames Research Center, NASA
Moffett Field, California

I. Introduction

Different forms of electron microscopy (EM) in general and transmission electron microscopy (TEM) in particular have long been employed as powerful research tools for the study of thin films and their conditions and mechanisms of formation. It is not our intention in this chapter to discuss the conventional applications of TEM, since a number of excellent reviews exist already that deal with general TEM techniques (Reimer, 1967; Hirsch et al., 1965; Thomas, 1962) or with their application to specific thin-film problems (Anderson, 1966; Pashley, 1965, 1970; Hass and Thun, 1964–1967; Khan, 1970; Chopra, 1969). Instead, a progress report is presented that concentrates its attention on some of the more recent advances in TEM methods that have resulted in the ability to study various stages of formation of thin condensed layers in more detail than previously possible. Furthermore, because of the very nature of TEM, the emphasis in this discussion will be placed on methods to study the very early stages of thin-film nucleation and growth, and on techniques favoring the investigation of processes responsible for the introduction of epitaxial order.

The main impetus for progress in the application of TEM techniques to problems of thin-film growth is directly tied to recent advances in related fields of interest. Here, one has to mention specifically the technology of achieving and characterizing UHV environments, modern methods of surface analysis such as low-energy electron diffraction (LEED), high-energy electron diffraction (HEED), Auger electron spectroscopy (AES), and, lately, secondary ion-mass spectroscopy (SIMS) and low-energy ion backscattering, the advances made in very high-resolution phase contrast electron microscopy, and the combination of EM with thin-film deposition from the vapor phase by UHV in situ electron microscopy in UHV.

The two experimental approaches that will be specifically discussed in this chapter (see also Poppa et al. 1972) resulted from an effective combination of some of the new analytical techniques. They are basically different in their methodological character: postdeposition TEM methods are based on step kinetic studies of low-rate vapor depositions performed under extremely well-controlled substrate surface and evaporation conditions. The depositions are performed on bulk substrate materials in conventional UHV systems, and the final results are obtained by subsequent TEM inspection of suitably prepared and hopefully representative test samples. In situ TEM techniques on the other hand are primarily designed to supply information about the deposition kinetics in the most direct way and on a real time basis. In situ results are always influenced, however, by complicating effects of the imaging electron beam. In recent years, significant improvements have been reported for both experimental approaches but the progress made is also accompanied by a more knowledgeable awareness of specific weaknesses and restrictions of each technique. It is our main goal in this chapter to point out the known and the potential problem areas of these advanced analytical methods that are suitable for in-depth studies of thin-film nucleation and growth. It will also be demonstrated that these new techniques can be used to reach reliable conclusions regarding some of the most fundamental and interesting surface physical processes related to the formation of thin films when the different analytical approaches are combined in a judicious and appropriate way.

II. Postnucleation Transmission Electron Microscopy

A. TREATABLE SUBSTRATE–OVERGROWTH COMBINATIONS AND SPECIMEN PREPARATION

1 Control of Deposition and Substrate Surface Conditions and Selection of Deposition Materials

The most important basic prerequisites for meaningful thin-film-growth studies by postdeposition TEM techniques are extremely well-controlled deposition and substrate-surface conditions during condensation and during the subsequent TEM specimen preparation phase. The major deposition parameters include substrate temperature, impinging flux, and residual gas environment, and they can all be controlled to a high degree of accuracy by making proper use of modern UHV condensate flux and residual gas-monitoring techniques. One of the significant advantages of postdeposition film-growth studies is the possibility of depositing onto bulk substrates in a conventional UHV system that also contains the necessary facilities for the application of any one of the recently developed powerful surface analytical techniques such as LEED, HEED, AES, low-energy ion back scattering, and SIMS.

The substrate materials for nucleation studies are usually nonmetallic. Sometimes the substrate materials are semiconducting, but usually they will be insulating materials due to the low free energy of substrate surfaces required for three-dimensional nucleation and growth behavior (Bauer, 1958). The careful selection of substrate–overgrowth material combinations that will lead to the so-called Vollmer–Weber type of three-dimensional film growth can often determine a priori the overall success of the entire research project. One should, for instance, take into account not only the availability of sufficiently pure and UHV compatible substrate materials but also consider carefully such material properties as the secondary electron-emission coefficient for low- and medium-energy electrons of surface-probing primary electron beams to avoid electrical charging when analyzing the crystallography and chemical composition of substrate surfaces.

The specific substrate material must also lend itself to the application of one of the methods found suitable for reproducibly preparing well-defined, "clean" surfaces prior to vapor deposition. Which of these cleaning procedures is successful for a specific substrate material has to be most carefully checked and verified under existing vacuum conditions. In general, one can attempt cleaning either by thermal treatment in vacuum or reactive gas environments, by ion bombardment, by vapor deposition of fresh surface layers, or by mechanical cleavage in vacuum. For many insulating single-crystal materials, cleavage is the most suitable method of preparing clean substrate surfaces in situ because the structural damage caused by ion bombardment can normally not be annealed out afterwards. Medium-high free surface energy semiconductor materials, on the other hand, can often be very satisfactorily cleaned either by cleavage or by repeated ion sputtering and subsequent annealing (Farnsworth *et al.*, 1959).

2. *Inspection of Overgrowth Only*

The most powerful asset of TEM is its high spatial-resolution capability, important consequences of which are the necessity for very thin electron transparent microscope specimens and high specimen electron irradiation loads of the order of several ampere seconds per square centimeter. Both requirements are most easily fulfilled in film-deposition studies if only the thin overgrowth layer is inspected, which is suspended in or strengthened by a more or less structureless support film of excellent stability under intense electron irradiation. Evaporated low-density support films of carbon or SiO have long been used in conventional TEM with good success and are often found satisfactory for the type of film-growth investigations discussed here if their limitations are kept in mind. These amorphous support films do actually exhibit appreciable phase contrast eigenstructures at the highest magnifications and sometimes do not absolutely secure the original fine structure of the deposit (Jaeger *et al.*, 1969; Bachmann and Hilbrand, 1966). Preparation artifacts of this type deserve particular attention when considering the possibility of detaching an overgrowth–support film composite from its original bulk substrate.

Single-crystal support films offer the advantage of lower phase contrast background structures although they are more difficult to prepare. Thin graphite (Moran, 1960), mica, and MoS_2 substrates can be routinely prepared by cleavage techniques (Palatnik *et al.*, 1970). The use of high-temperature recrystallized amorphous carbon films, exhibiting relatively large single-crystal areas of up to 1000 Å diam, has been reported (White *et al.*, 1971). Highly single-crystalline films of BeO have also been prepared by electron-beam evaporation techniques (Mihama *et al.*, 1974).

The possibility of separating substrate and overgrowth is basically determined by the different chemical activities of substrate and overgrowth materials, which, therefore, have to be known when selecting substrate–overgrowth combinations suitable for TEM studies. Great caution is however indicated when relying solely on tabulated macroscopic chemical solubility data for this purpose. One has to keep in mind that very small amounts of deposit material are usually involved, especially when studying very early stages of film growth, and chemical reaction rates can be quite different with very small single-crystalline particles. Such small deposit particles can also be lost during the dissolution of the substrate material by the mechanical stripping action during the detachment process. For these reasons, it is definitely advisable to compare experimentally different methods of preparing representative overgrowth specimens for TEM, particularly if quantitative nucleation measurements are involved and if substrate and overgrowth cannot be separated very easily by chemical means. As an example, it is known that Au films deposited on UHV cleaved mica can be chemically stripped from the substrate by dissolving the mica in HF. However, when comparing particle-density measurements with those of similar specimens prepared by thinning the mica from the back and not detaching the gold deposit, it was found (Elliot, 1972) that many particles were lost during the chemical detachment procedure.

Conditions for chemically stripping representative deposit layers are more favorable for easily dissolved substrate materials. Unquestionably, the popularity of alkali halide substrate materials for studying epitaxial film-growth mechanisms (Green *et al.*, 1969, 1970) is due to a large extent to their water solubility (and easy cleavability). Extensions of this approach to other ionic substrates that cleave easily along {111} planes, rather than the cubic {100} planes of the alkali halides, have also been reported (Koch and Vook, 1971). Thin epitaxial layers of alkali halides are evaporated at room temperature onto air-cleaved mica and subsequently serve as substrates for the metal deposition. The epitaxial intermediate layers then permit easy detachment of the epitaxial metal layer by floating them off in water. Au, Ni, Cu, and Ag continuous epitaxial films of good quality and in {111} orientation have been prepared successfully in this way. In general, however, there is no doubt that the necessity for significant differences in chemical activity of substrate and overgrowth materials required for substrate–overgrowth separation severely limits the applicability of postdeposition TEM methods to film-growth studies.

3. *Simultaneous Inspection of Substrate and Overgrowth*

Some of the major disadvantages of investigating the overgrowth only can be eliminated by not detaching the deposited film that is to be investigated from its substrate. When it is possible to examine substrate and overgrowth simultaneously, such important specimen preparation considerations as chemical reactivities and detachment artifacts do not limit the selection of substrate and overgrowth materials. Although the simultaneous inspection technique imposes new substrate material requirements of electron transparency and good stability under the imaging electron beam (electrical charge-up and decomposition), one also gains the ability to assess, most directly, epitaxial relationships between substrate and overgrowth. The highly sensitive and well-known TEM technique of measuring this substrate–overgrowth relationship through moiré effects also becomes feasible.

The requirement of simultaneous TEM inspection of substrate and overgrowth defines the class of substrate materials suitable for this approach. It is necessary, of course, to be able to thin the materials sufficiently either before or after depositing the overgrowth film. Three techniques are known at present that accomplish the thinning of the substrate in different ways or at different stages of the specimen preparation process.

The first technique is based on conventional TEM specimen preparation methods. Thin flakes of such layer materials as graphite (Darby and Wayman, 1970), molybdenite (Honma and Wayman, 1965), or mica (Allpress and Sanders, 1967; Jaeger et al., 1967; Poppa et al., 1971) are prepared by multiple cleavage (Hirsch et al., 1965) and mounted on regular EM microscope specimen grids. The overgrowth film is then deposited under controlled conditions and after removal from the deposition vacuum system the composite specimens are ready for TEM inspection. The main disadvantage of this approach is the virtual impossibility of preparing clean substrate surfaces in this way because heat treatment and gaseous etching alone are usually insufficient for cleaning these tiny flakes of crystalline substrate materials. Sputtering and subsequent annealing (Farnsworth et al., 1958) can, in principle, be used for cleaning semiconducting materials before deposition, but no systematic studies of this kind have been reported in literature.

Green (Green et al., 1970) has developed a different technique that is based upon the ease of good epitaxial growth of the lead chacogenides on alkali halide substrates. In this approach an electron transparent layer of high crystalline perfection (of, for instance, lead selenide) is prepared on a KCl bulk crystal immediately preceding the deposition of the metal deposit, the growth of which is to be studied. After removal from the deposition system the substrate film combined with the overgrowth film is easily detached by floating off in water. The detached film is then prepared on regular specimen grids.

The third technique of studying substrate and overgrowth simultaneously relies upon bulk substrates of easy cleaving layer materials that are thinned after the overgrowth layer has been deposited (Poppa et al., 1971). In the case of mica, two different approaches have been successfully employed for post-deposition thinning. One can combine multiple mechanical cleavage with final thinning by ion-beam milling (Elliot, 1972), or the gelatin method (Palatnik et al., 1970) can be employed. This latter method consists of putting a small drop of gelatin on top of the metal deposit and letting it dry. When the gelatin's viscosity and the drying conditions are adjusted carefully (Heinemann and Poppa, 1972; Lee, 1974), the drying gelatin cleaves off a slice of the mica surface that is suitable for TEM inspection (after dissolving the gelatin in water). The main advantages of the gelatin technique are its simplicity and its efficiency in preparing many specimens of large electron transparent area. In contrast, ion-bombardment thinning is often very tedious, although extremely thin, small specimen areas can eventually be obtained that contain valuable epitaxial information when viewed under high resolution TEM conditions (Fig. 1).

4. *Reproducibility of Deposition Results and TEM Specimen Preparation Artifacts*

The two most important factors that determine the reliability of film-deposition studies at present are the reproducible preparation of well-defined substrate surfaces and the elimination of artifacts during the subsequent preparation of thin specimens suitable for high-resolution TEM analysis.

The crystallography and chemical composition of the substrate surface, averaged over an area of approximately 0.3 mm^2, can be determined fairly accurately by probing the surface with such techniques as LEED, HEED, AES, and ESCA immediately prior to deposition of the overgrowth (Poppa *et al.*, 1971). The crystallographic and morphological microstructure of the substrate surface, however, escape detection by these averaging techniques, and there is ample proof that surface roughness, cleavage, or localized defect structures on a microscopic scale can influence considerably the microstructure and orientation of vapor deposits (Green *et al.*, 1969, 1970; Sato and Shinozaki, 1970; Bethge, 1969; Green, 1973). It would be highly desirable if more attention were paid to the morphological and defect microstructure of substrate surfaces in connection with epitaxial deposition studies. There exist quite sensitive and relatively easy methods of assessing these surface properties by high-resolution surface replication and surface decoration.

This characterization of surface morphologies by replica methods is probably one of the presently most underrated surface-analysis techniques. This is undoubtedly due to misconceptions that can be traced back to the usage of replicas during early phases of transmission electron microscopy. Seldom is it realized that surface details of the order of 10 to 30 Å can presently be resolved routinely by significantly improved replication techniques. The basic requirements for achieving high resolution in a surface replica are high directionality and high supersaturation of the shadow casting vapor under deposition conditions that minimize "grain growth" by coalescence, growth, or recrystallization processes. Although no satisfactory systematic studies are available at present, extensive empirical evidence shows that refractory metals deposited at cryogenic substrate temperatures, possibly combined with recrystallization-impeding additives such as carbon or other refractories, can lead to extremely fine-grained shadow-casting deposits (Hayeck *et al.*, 1969, Hayeck and Schwabe. 1971). As might be expected, the quality of these deposits depends a great deal upon the nature of the substrate materials because of the known influence of various surface parameters on heterogeneous nucleation processes. Specifically, adsorption and surface-diffusion energies and preferred sites for

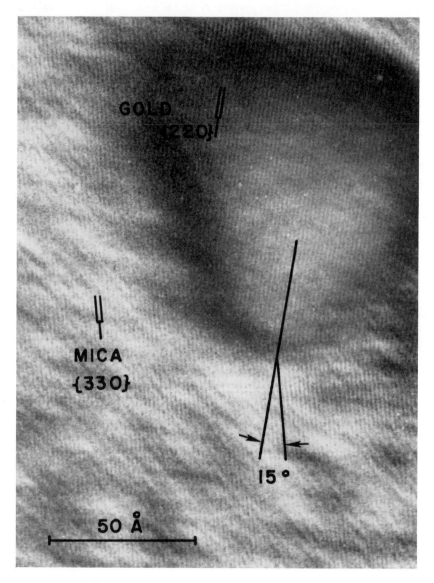

FIG. 1. High-resolution lattice image of misaligned gold crystallite on mica substrate taken with tilted illumination. The substrate $(d_{(330)} = 1.49$ Å$)$ and the overgrowth $(d_{(220)} = 1.42$ Å, compressed$)$ lattices are simultaneously imaged (with permission, North Holland Publ. Co.).

nucleation are of importance. On mica, for example, lateral resolutions of artifical step structures of 20 to 30 Å separation have been achieved (Abermann, 1972) and step heights of 10 to 20 Å have been distinguished with Ta–W shadow-casting layers. Even simple Pt deposits have been shown to resolve cleavage steps of less than 10 Å in height if the depositions are performed at low temperatures on silicon single-crystal surfaces (Abbink et al., 1968).

The results obtained with high-resolution surface replicas are often influenced considerably by decoration effects, which, in pure form, comprise the second technique useful for surface-structure characterization. In principle, the decoration of surface features by vapor depositing extremely thin metal layers (Bassett, 1958) can be understood as a special case of heterogeneous nucleation from the vapor phase at preferred nucleation sites. The preferred surface sites can characterize either morphological or atomistic defect features of the substrate surface, and a theoretical explanation of the phenomenon of decoration has been proposed that is based on classical nucleation principles (Chakraverty and Pound, 1963). Preferred nucleation in atomistic terms has been discussed in detail by several investigators recently (Venables, 1973; Stowell and Hutchinson, 1971; Markov and Kashchiev, 1973). The decoration technique has been developed to a high degree of sophistication by Bethge and co-workers (Bethge, 1962a, b), who applied it specifically to the study of the atomistic defect structure of alkali halide surfaces. It has become apparent, however, through recent nucleation studies on alkali halides, that the orientation of films deposited on single-crystal substrates can also be significantly influenced by the action of preferred substrate surface sites (Poppa et al., 1971; Green, 1973). Sato and Shinozaki (1970), for example, presented some evidence that epitaxially well-aligned deposit particles of Au on NaCl were preferentially located along cleavage steps. Although these particular results could not be verified by other investigators using advanced TEM techniques for particle-orientation determination (Green, 1973; Heinemann and Poppa, 1972), much additional evidence for the influence of surface defects on epitaxial film growth exists. Here one has to mention the large body of work regarding the well-documented effects of electron or X-ray irradiation of substrates during epitaxial growth (Green et al., 1969; Stirland, 1966) or the similar effects of impurities in doped substrate materials (Toth and Cicotte, 1968; Birjiga et al., 1972).

Another major area of concern for the reliability of postdeposition TEM studies is described by the term "preparation artifacts." This terminology is usually used in conventional EM to define undesirable changes of the original microstructure of specimens during preparation for TEM inspection. In the case of thin-film-growth studies, preparation artifacts determine the probability that the deposit structures that are finally observed by high-resolution TEM resemble exactly the as-deposited structure of the film. The opportunities for such undesirable changes of film structure to occur are plentiful in post-deposition studies at various stages of specimen preparation: postdeposition annealing, cooling to room temperature, application of additional fixing layers, exposure to the atmosphere, removal from the substrate and thinning, and even exposure to the imaging electron beam. Examples for all these effects have been observed repeatedly by many workers in the field, but insufficient attention is usually paid to them because they are difficult to assess and often complex in nature (Poppa et al., 1972).

The only way to accomplish the much needed check on TEM preparation procedures accurately and efficiently is by employing in situ TEM techniques. For Au on mica (Elliot, 1972; Poppa et al., 1971), for instance, it was possible to simulate in the in situ microscope many postdeposition preparation steps including prolonged annealing at deposition temperature, cooling to room temperature, exposure to atmosphere, reevacuation, and deposition of a conducting carbon-fixing layer. The main advantage of this approach is, of course, the ability to monitor changes in microstructure of exactly the same specimen area or even individual deposit crystallites. However, only selected substrate–overgrowth systems are suitable for in situ microscopy studies (see Section III).

Considering the many possible sources of error in postdeposition film-growth studies, it is obvious that proper caution should always be exercised when interpreting the results of poorly documented deposition experiments. The additional data that would assure reliability are usually difficult to obtain, however, because of the time-consuming nature of modern UHV procedures, and similar considerations apply to efforts of providing for deposition results of sufficient statistical reliability. Since UHV procedures are absolutely essential for meaningful film-growth studies, a major requirement for efficient, systematic research in this field is the incorporation of some sort of multiple-specimen capability into any experimental approach to the problem. Various possibilities of achieving this goal will be discussed in the following sections of this chapter.

B. MEASUREMENT OF NUCLEATION AND GROWTH KINETICS

1. Step-Kinetic Approach

In postdeposition studies the measurement of film nucleation and growth-kinetic processes can only be accomplished by a step-kinetic technique where substrate specimens are exposed to the condensing vapor flux for successively longer times. As pointed out previously, one major requirement for valid kinetic measurements is the evaluation of a fairly large number of deposition steps or equivalent deposition times. Also, the depositions usually have to cover a range of substrate temperatures T and a range of vapor impingement

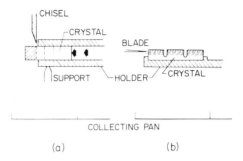

FIG. 2. Two suitable techniques of preparing clean multiple-substrate specimens by in situ vacuum cleavage of brittle (a), or layer structure (b) ionic bulk single crystals.

fluxes R (Robinson and Robins, 1974), and one has to explore the effects of other influential deposition parameters. These parameters include substrate predeposition treatment, residual gas environment during deposition, substrate exposure to particle or electromagnetic irradiation before or during deposition, etc.

Three different UHV compatible approaches have been successfully used to provide a multiple-specimen capability. The first two techniques are applicable to substrate materials that cleave easily. Figure 2a demonstrates schematically the "salami technique" applied to brittle bulk-substrate crystals (e.g., alkali halides, MgO, CaF$_2$, etc.) that have to be cleaved with the help of a chisel (Green *et al.*, 1969). This cleavage device was further refined so that three bulk crystals could be simultaneoulsy prepared. For layer materials like mica that can easily be cleaved by splitting with a sharp blade, the arrangement of Fig. 2b has proven feasible, and up to 40 specimens have been prepared by repetitively splitting a 0.6-mm thick and subdivided block of mica. A specimen marking and collecting system has to be combined with each particular cleavage arrangement.

A third multiple-specimen approach consists of using a specimen holder that can accommodate a large number of individual small substrate crystals that can successively be brought into the deposition position. Often a carousel-type holder design is in use, and the individual substrate crystals are usually of a material the surface of which can be reproducibly cleaned by some means other than cleavage. Repetitive ion bombardment and subsequent annealing are useful cleaning methods for metal or semiconductor materials that could be used here.

2. Nucleation Rate and Saturation Density

Consider the important physical properties that typify the film-growth process and that can be measured by TEM. For deposition systems that are characterized by a three-dimensional nucleation and growth mode, the general situation can be schematically described as in Fig. 3.

The nucleation phase is characterized by the successive appearance of stable nuclei that grow from critical size nuclei until they are large enough to be detected by TEM. The time t and number density $n(t)$ scales vary significantly with the degree of supersaturation

$$S = p/p_e \qquad \text{where} \quad p = R/(2\pi mkT)^{1/2}$$

where p_e is the equilibrium vapor pressure, R the impinging vapor flux, m the mass of impinging particles, k the Boltzmann constant, and T the substrate temperature. Under conventional deposition conditions the kinetic processes are quite fast and are often difficult to measure. In principle, however, one can determine by measurements of this kind the nucleation rate I and the maximum number density of deposit particles n_{max} as a function of such influential deposition parameters as T, R, substrate-surface state, and residual gas environment (Robinson and Robins, 1974). Theoretical expressions exist for the induction time t_0, I, and n_{max}. They are based upon various nucleation theories (Bauer, 1958; Frankl and Venables, 1970; Venables, 1973; Lewis, 1970a,b; Routledge and Stowell, 1970; Stowell, 1972; Stowell and Hutchinson, 1971a,b; Walton, 1962; Zinsmeister, 1966, 1968, 1969, 1971) and relate the measured properties to important nucleation physical parameters.

It is obvious that for measurements of this kind it is most desirable to be able to detect stable nuclei of the smallest possible size. The main problem is not one of EM resolution but of how to distinguish the small deposit particles by contrast differences from background structures of the supporting film.

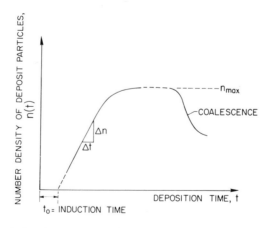

FIG. 3. Schematic dependence on deposition time of density of heterogeneously nucleated overgrowth particles. The nucleation rate is approximated by $I = \Delta n/\Delta t$.

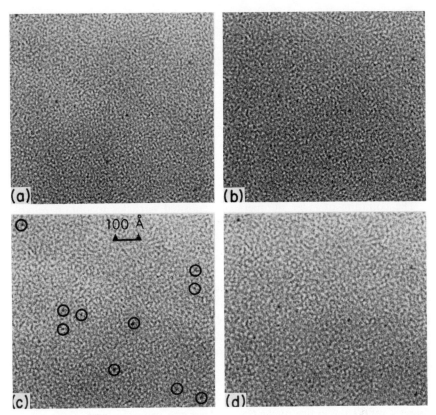

FIG. 4. Detection of very small stable clusters of deposit atoms by high-resolution EM. A through-focus series distinguishes gold deposit particles (diameter < 10 Å) from phase-contrast background features of the support film.

Stable nuclei smaller than 10 Å in diameter can be identified on low-background structure substrate materials like mica (see Fig. 4) if conventional high-performance electron microscopes are used and if the phase contrast structure of the support film is properly subtracted by examining through focus series of micrographs.

In most cases of more practical importance, the initial stages of film growth are very short and almost never resolved. The nucleation rate I is very high and the stable deposit particles are very small. With high microscope resolution one can usually determine n_{max}, however, although caution is indicated to make sure that the n_{max} values obtained are valid. This means that serious consideration has to be given to the possible occurrence of coalescence, and this can often be extremely difficult to decide.

If reliable n_{max} data can be compiled as a function of substrate temperature and impingement flux, they can be used to elucidate film-growth mechanisms. As an example, it was concluded from the independence of n_{max} on R and T in the case of Au on MgO (Robins and Rhodin, 1964) that the nucleation process was not random but controlled by substrate impurities. Later experiments with Au on higher-purity MgO showed (Moorhead and Poppa, 1972) that a conventional n_{max} behavior could be restored. This is not, however, a sufficient condition for the absence of preferred surface sites for it is at least conceivable (Elliot, 1972; Markov and Kashchiev, 1973) that a spectrum of sites that are thermally deactivated can exist on the substrate surface.

3. Particle Growth, Size Distribution, and Condensation Coefficient

The counting of large numbers of deposited particles to obtain reliable statistical results can be very tedious and can constitute a significant source of error in nucleation and growth measurements. The introduction of automated image computers has improved the situation considerably by replacing hand-operated counters.[†]

The use of automated image analyzers (Schmeisser, 1974) becomes almost a necessity for the determination of (a) accumulated particle areas needed to measure surface coverages, which have been used instead of $n(t)$ expressions in some nucleation models (Stowell, 1972; Stowell and Hutchinson, 1971a,b) to simplify the theoretical interpretation of results; (b) average particle sizes (particle diameter D) as a function of time, $D = D(t)$, which can eventually lead to the nucleation independent determination of the adsorption energy (Sigsbee, 1971); and (c) particle size distributions, $dn/dD = f(t)$, which indirectly reflect the entire nucleation and growth processes and which can be compared with computer simulations of film-growth experiments (Robertson, 1971).

The improved statistics and higher accuracies of particle measurements obtainable with automated image-analysis systems will also affect the possibility of making valid condensation-coefficient determinations. The measurement of the total condensed mass still depends, also, on the exact three-dimensional shape of the deposit particles in addition to their number and lateral size. However, the height of small deposit particles is very difficult to determine by EM means, and for very small particles it is practically impossible to do this with any reasonable degree of accuracy. Uncertain geometrical scaling factors have to be used, therefore, for condensation-coefficient determinations from electron micrographs, and more reliable methods of measuring the total deposited mass are very desirable indeed. One

---

[†] Commercial instruments available from: IMANCO (Quantimet 720), Bausch and Lomb (QMS), Zeiss (Videomat), Leitz (Classimat).

approach to this problem makes use of X-ray fluorescence techniques that can easily be combined in most modern high-performance TEMs with conventional EM and ED measurements. Either wavelength or energy dispersive attachments are in use to detect the characteristic X rays of elements down to an atomic number of about 11 with good efficiency. Fuchs (1966) has shown that with this type of attachment the average thickness of nickel films can be determined with the respectable accuracy of 1% for 50-Å thick layers and with 10% accuracy for a 10-Å thick film. In these cases, the integration time for data collection was only 3 min. It should be possible, therefore, to improve the sensitivity of these measurements appreciably by employing longer integration times in conjunction with advanced methods of data collection and computer processing (Beaman and Isasi, 1972).

C. HIGH-RESOLUTION TRANSMISSION ELECTRON MICROSCOPY

1. Usefulness for Film-Growth Studies

The main impetus for the renewed interest in high-resolution electron microscopy as applied to film-growth studies is due to the progress made recently in obtaining and interpreting lattice images of thin metal crystals. A capability of reproducibly resolving lattice spacings that correspond to the three innermost Debye–Scherrer diffraction rings of a representative fcc metal like gold ($a_0 = 4.08$ Å) with

$$d_{111} = 2.35 \text{ Å}, \qquad d_{200} = 2.04 \text{ Å}, \qquad d_{220} = 1.44 \text{ Å}$$

finds obvious applications in the following areas related to basic thin-film properties:

(a) Detailed examination of crystallographic structure and composition of small deposit particles including the phenomenon of multiple twinning.

(b) Detection of lattice-parameter changes in very small crystallites and, therefore, the most direct determination of average stresses in overgrowth islands.

(c) Direct determination of the orientation of individual deposit crystallites either in relationship to other crystallites in detached overgrowth layers or in relationship to the local substrate lattice in composite substrate–overgrowth EM specimens.

(d) Detection of very small stable nuclei.

The systematic correlation of these truly microstructural deposit properties with changing experimental deposition parameters can be expected to lead to valuable insights into fundamental film-growth mechanisms on an atomistic scale. Particularly in the field of epitaxy, this kind of information should help to distinguish the importance of different stages of deposition for the development of film order. Basic questions concerning the influence of deposition conditions and/or deposition stages—nucleation, growth, or the coalescence phase—upon the final orientation of an epitaxial film can in principle be answered most accurately by lattice-imaging techniques.

Current limitations in the practical application of lattice-imaging TEM are mainly due to high instrumental-performance requirements, to unavoidable structural damage in electron-radiation-sensitive substrate materials, and to a lack of understanding of recently emerging phase-contrast imaging principles. The radiation damage in compound EM specimens like mica is a direct consequence of the high specimen-current densities required by the extremely high electron-optical magnifications needed for lattice imaging. Figure 5 shows this directly in a lattice image of a mica specimen that was exposed to the imaging electron beam for several minutes before the micrograph was taken. The mica lattice structure has already been destroyed in many areas of the high-magnification micrograph. The beam action is also displayed by the low-magnification insert that shows the entire beam-illuminated specimen area. Use of an efficient EM image intensifier can appreciably extend the time available to the microscopist before the lattice structure of such a specimen is destroyed. The reduction in specimen exposure when using image intensification as compared to conventional EM internal photography has been estimated recently to be of the order of $50 \times$ (Heinemann et al., 1972).

2. Resolution and Image Constrast

Both phase contrast and amplitude contrast EM image-forming mechanisms are of importance to the interpretation of film-growth studies by TEM, and a very abbreviated review of the imaging principles involved will be given. Since it is impossible, in the context of this review, to discuss the field of phase-contrast image theory in more depth, the interested reader is referred to the extensive original literature on this subject (Hanszen et al., 1963; Hanszen and Morgenstern, 1965, 1967; Lenz, 1971; Thon, 1971).

Since phase contrast is basically tied to the coherency of the illuminating electron beam, it is convenient to define the application of phase- or amplitude-contrast principles to the separation δ of predominantly coherently illuminated specimen details given by Lenz (1965):

$$\delta = \lambda/\pi\alpha_{ill} \qquad (1)$$

where λ is the electron wavelength that is 0037 Å for 100 kV electrons. For an illumination aperture $\alpha_{ill} = 5 \times 10^{-4}$ rad, $\delta \approx 24$ Å, and one can, therefore conclude that amplitude-contrast phenomena will dominate when imaging specimen details with $\delta \gtrsim 24$ Å. Thus, the measurement of some characteristic features of small deposit particles such as their shape, number density, and

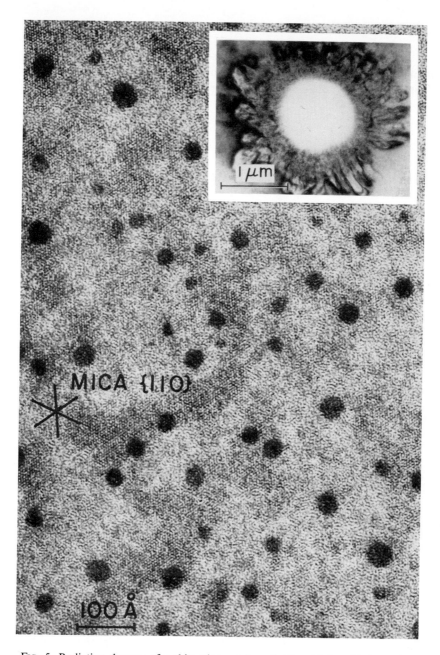

FIG. 5. Radiation damage of a thin mica specimen (with gold deposit) by prolonged exposure to the imaging electron beam. Localized destruction of regular mica lattice structure (high magnification), and general radiation damage appearance of beam area (low-magnification insert) (with permission, North Holland Publishing Co.).

size distribution will be determined by resolution values that are applicable to amplitude objects. The respective well-known theoretical point-to-point resolution limit δ_{pp} that minimizes spherical and diffraction error is (Haine and Mulvey, 1954)

$$\delta_{pp} \approx 0.43(C_s \lambda^3)^{1/4} \tag{2}$$

for an optimum objective lens angular aperture of

$$\alpha_{opt} \approx 1.4(\lambda/C_s)^{1/4} \tag{3}$$

where C_s is the spherical-aberration constant. One notices that the $C_s^{1/4}$-dependence of the resolution on the spherical-aberration constant C_s makes it very difficult to improve the resolution by designing objective lenses with smaller C_s. Table I shows theoretical point resolutions according to Eq. (2)

TABLE I

CALCULATED POINT-TO-POINT RESOLUTIONS δ_{pp} FOR THREE DIFFERENT COMMERCIAL OBJECTIVE LENSES OPERATING AT 100 kV ACCELERATING VOLTAGE[a]

	Siemens regular model 101 objective	Siemens model 101 objective with shortened focal length	Siemens model 101 special objective
C_F	2.1 mm	1.6 mm	1.6 mm
C_s	2.9 mm	2.25 mm	1.35 mm
δ_{pp}	2.7 Å	2.5 Å	2.2 Å
α_{opt}	$>4.4 \times 10^{-3}$ rad	4.7×10^{-3} rad	5.4×10^{-3} rad

[a] $\lambda = 0.037$ Å.

that can be achieved with three different objective lenses of varying C_s, a lens parameter usually given by the microscope manufacturer or easily measured using thin crystalline island films (Heinemann, 1971). In Table I, C_F is the chromatic-aberration constant, which will be discussed later. (According to Table I, it would be impossible with presently available commercial objective lenses to resolve structures of much less than approximately 2-Å separation with amplitude contrast only; and the necessary illumination aperture angles to image such separation predominantly with amplitude contrast would have to be quite high, about one order of magnitude higher than normal.)

If the structural specimen details to be resolved are smaller than approximately δ of Eq. (1), phase contrast dominates the image formation processes. As will be proven later, a different resolution formula applies for resolving phase structures of periodic separation Λ with an image contrast of 30% (Heinemann, 1972)[†]:

$$\delta_\Lambda = 0.85(\lambda C_F \, \Delta W/W)^{1/2} \tag{4}$$

Here, $\Delta W/W$ denotes the sum of all relative chromatic instabilities

$$\frac{\Delta W}{W} = \frac{\Delta U}{U} + \frac{\Delta V_0}{U} + \frac{\Delta E}{eU} + 2k\frac{\Delta I}{I} \tag{5}$$

where $\Delta U/U$ and $\Delta I/I$ are the relative stabilities of the accelerating voltage U and the objective lens current I, respectively, ΔE the energy loss of the imaging electrons through inelastic collisions in the specimen, and ΔV_0 the natural energy width of the electron beam leaving the electron gun of the microscope.[‡]

Equation (4) applies for the regular mode of imaging with paraxial specimen illuminating [as does Eq. (2) for amplitude contrast considerations], but it is immediately clear by comparing δ_{pp} with δ_Λ that the resolution of phase structures is quite sensitive to C_F and $\Delta W/W$ ($\frac{1}{2}$ power only) and does not depend on C_s at all. Obviously, these results will have to be kept in mind when considering the effectiveness of possible instrumental improvements later.

In Fig. 6, δ_Λ is plotted against $\Delta W/W$ according to (4) for the two values of the chromatic aberration constant also used in Table I and for 100-keV acceleration voltage.

The resolutions obtainable according to Fig. 6 for phase structures are somewhat lower (for a realistic value of $\Delta W/W \approx 1 \times 10^{-5}$) than those given for the classical amplitude structures in Table I, and further instrumental improvements resulting in smaller $\Delta W/W$ values are within reach of present-day TEM technology. Resolution curves of the type shown in Fig. 6 should always be consulted when dealing with lattice image applications under paraxial illumination conditions.

3. Specific Instrumental Improvements

A number of specific improvements of high performance electron microscopes for lattice imaging purposes follow directly when accepting the validity of the resolution equation, Eq. (4).

---

† δ_Λ was calculated by assuming a reduction of fringe-image contrast to 30% of what the contrast would have been without chromatic changes of objective-lens defocus $\delta(\Delta z)$ caused by a chromatic instability. $\Delta W/W$ causes changes in defocus Δz according to $\delta(\Delta z) = C_F \, \Delta W/W$ (Vorobev and Vyazigin, 1967).

‡In some modern electron microscopes, the objective lens is sometimes operated partly in magnetic saturation and, therefore, $\Delta I/I > \Delta B/B = k \, \Delta I/I$, where B represents the magnetic field on the optical axis, i.e., $k \leqslant 1$. For the normal objective lens of the Siemens 101, operated with increased lens current for shorter focal length, $k \approx 0.65$.

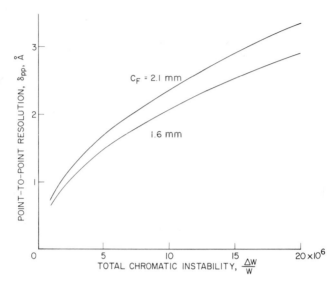

Fɪɢ. 6. Point-to-point resolution for phase structures in the axial illumination mode as a function of total chromatic instability. The resolution curves are shown for two objective lenses with different chromatic aberration constants C_F.

First, it is usually rather easy to shorten the focal length of the commercial objectives by increasing the lens current. In the case of a Siemens 101 microscope, the lens current was boosted from 460 to 580 mA by adding one F2A tube (parallel) in the existing regulated lens current supply and adding a 700 Ω wirewound resistor (parallel) in the lens-current control module. This results in a shorter focal length ($2.7 \rightarrow 2.3$ mm) and correspondingly smaller C_F ($2.1 \rightarrow 1.6$ mm) and C_s ($2.9 \rightarrow 2.25$ mm) lens-aberration constants.

Second, it is imperative to lower $\Delta W/W$. The objective lens current and the accelerating voltage are usually sufficiently stabilized to about 2×10^{-6} each in modern high-performance instruments. The other contributions to $\Delta W/W$ in Eq. (5) are, however, more problematic. The energy losses ΔE of the imaging electrons in the specimens are unavoidable, in principle, but the number of inelastically scattered electrons can be kept low by working with sufficiently thin specimens. The last contribution to $\Delta W/W$ is ΔV_0 which can be the most influential and most troublesome because of the so-called "Boersch effect" (Boersch, 1954). This effect causes the natural energy width ΔV_0 of the emitted electrons to become a sensitive function of the total beam current. For high current densities in the beam, the resulting energy width can easily increase tenfold over its normal value of $\Delta V_0 \approx 0.3$–0.5 eV for thermionically emitted electrons. Therefore, $\Delta V_0/U$ can assume values of the order of 3–5×10^{-5} for $U = 100$ keV, which then becomes the dominant term in Eq. (5). For this reason alone it is absolutely essential to use pointed filaments for high-resolution work because the high directionality ("Richtstrahlwert" often misleadingly termed "brightness") of these filaments permits the use of much smaller beam currents for the same image brightness. [It was found that the "Boersch effect" becomes noticeable with a pointed filament in the Siemens 101 microscope if the beam currents exceeds 2 μA (Heinemann, 1972).]

Improvements in the electron gun should be always combined with facilities for dc heating of the filament. Although the ac character of the conventional filament-heating current is too small to influence ΔV_0 directly, it has been shown experimentally (Heinemann, 1972) that resolution improvements are actually achieved by dc heating, possibly due to the elimination of fluctuating electrical-charging effects in the microscope specimen.

Another critical factor that significantly affects the performance of high-resolution instruments concerns the sensitivity and accuracy of conventional microscope-alignment techniques. Most critical are the alignment procedures used for voltage center and astigmatism correction. Both can be efficiently corrected by employing Bragg reflection images of small vapor deposited crystallites (Heinemann, 1971). The frequent use of this procedure can only be highly recommended.

New imaging techniques that are based on different approaches for eliminating chromatic image aberrations should be discussed at this point because they include instrumental improvements such as special condensor and objective lens apertures. The following paragraphs, however, will deal with this general subject area in detail and the discussion of the use of specialized lens aperture shall, therefore, be deferred.

4. *Different TEM Methods for Imaging Phase Structures*

(a) Axial bright-field illuminating EM (three-beam case): According to Abbe's theory of image formation by a lens, one needs at least two imaging waves originating at the object to form an image. A plane electron wave diffracted by the specimen into the direction $\pm\alpha_0$ (see Fig. 7) will be concentrated into two diffraction spots, each located at a distance $D/2$ from the optical axis in the back focal plane of the objective, and the zero order and at least one of the two diffracted beams have to be admitted through the lens aperture to form an electron image with good resolution.

The contrast with which a periodic phase specimen of periodicity $\lambda/\alpha_0 = 2\lambda f/D$ is imaged when the zero- and both first-order diffracted ($\pm\alpha_0$) electron waves interfere in the image plane behind the objective lens depends on the phase shift suffered by the diffracted waves when passing through the objective lens. This phase angle γ is given by the Scherzer equation (Scherzer, 1949)

$$\gamma = (\pi/2\lambda)(C_s\alpha_0^4 - 2\,\Delta z\,\alpha_0^2) \tag{6}$$

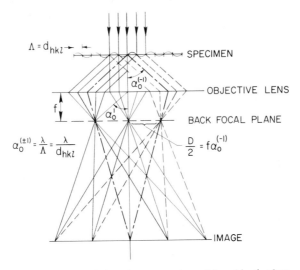

FIG. 7. Geometric imaging parameters of the objective lens.

for an objective with the spherical aberration constant C_s, which is defocused by Δz from the Gaussian focus setting.

If (6) is plotted as function of α_0 with the defocus $\Delta z = \text{const}$ as a parameter one obtains a family of curves shown in Fig. 8. These curves are called *aberration characteristics* and are plotted in Fig. 8 for $C_s = 2.25$ mm.

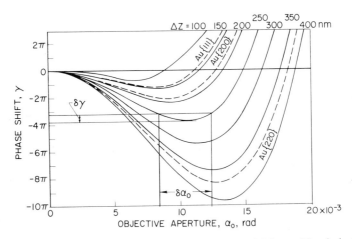

FIG. 8. Aberration curves for an objective lens with $C_s = 2.25$ mm. The dashed curves show the phase shifts pertaining to the conical illumination imaging ($\alpha_0 = \hat{\theta}_B$) of the three lowest index sets of gold lattice planes. The phase angle is given by Eq. (6) and λ (100 kV) = 0.037 Å.

Equation (6) can be rearranged so as to express α_0 (or $\Lambda = \lambda/\alpha_0$) as a function of C_s, Δz, and γ. If now only those values of γ are considered that result in maximum phase contrast (Hanszen and Morgenstern, 1965),[†] namely,

$$\gamma = (2n-1)(\pi/2) \qquad (n = \text{integer}) \tag{7}$$

one obtains

$$\alpha_0(n, \Delta z) = \left[\frac{\Delta z}{C_s} \pm \left(\frac{\Delta z^2}{C_s^2} + \frac{(2n-1)\lambda}{C_s}\right)^{1/2}\right]^{1/2} \tag{8}$$

Equation (8) is plotted in Fig. 9, again for $C_s = 2.25$ mm. The resulting curves are the phase transfer characteristics of the objective lens and can best be used to describe the selective transfer of preferred specimen periodicities. This feature of any objective lens can be clearly demonstrated when considering the electron optical imaging of an amorphous specimen—like an evaporated carbon film—that can be thought of as containing a continuous Fourier spectrum of specimen periodicities Λ_n. Which of these Λ's are imaged with best (phase) contrast for a present defocus setting of Δz can be immediately seen from Fig. 9 and directly demonstrated with light optical-diffraction patterns of high-resolution carbon-specimen micrographs (Thon, 1965).

It is clear, therefore, that in this case one can never obtain a "true" electron optical image of objects built up by a continuous spectrum of space frequencies $1/\Lambda$ because any objective lens, even one without spherical aberration, will act as a filter for some of these space frequencies. It is also obvious, however, from the phase-transfer characteristic that the imaging conditions for specimens with only one specific periodicity $\Lambda = d_{hkl}$ are always well fulfilled for one particular defocus setting Δz_{opt} (see Figs. 9 and 10). For crystalline specimens of the kind important in epitaxial film-growth studies, the selective transfer properties of electron lenses are, therefore, not harmful. On the contrary, it will be shown soon that high-phase contrast imaging conditions for selected d_{hkl} specimen periodicities can very easily be determined.

The phase contrast imaging features of Figs. 8 and 9 that are most important for our applications can be quantitatively evaluated most easily by considering the expression for the sensitivity of the phase angle γ to changes in the diffraction angle α_0 and changes in the defocus Δz. From the basic equation (6)

---

† Equation (7) is only applicable in the case of "three-beam interference," i.e., in the case where zero order and *both* diffracted first orders of diffraction are superimposed in the image plane. This is the standard, bright-field mode of operation. In all other cases like single-sideband holography with circular objective apertures (Downing, 1972), central dark-field microscopy (Heinemann *et al.*, 1972), regular dark-field microscopy, or tilted illumination (Dowell, 1963), as well as conical illumination (Heinemann and Poppa, 1970), no space frequency selectivity due to Eq. (7) takes place and image interpretation is facilitated.

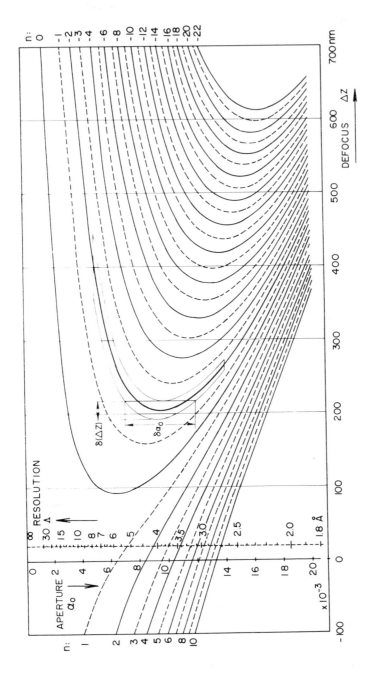

Fig. 9. Phase-contrast transfer characteristic for an objective lens with $C_s = 2.25$ mm and $\lambda(100 \text{ kV}) = 0.037$ Å. The apexes define the values of optimum defocus $\Delta z_{opt} = C_s \alpha_0^2$ for given specimen periodicites $d_{hkl} = \lambda/\alpha_0$. The effect of chromatic defocus changes $\delta(\Delta z)$ and the range of specimen periodicities $\delta(\Lambda)$, transferable with sufficient phase contrast, can be directly assessed.

we obtain

$$\delta\gamma = (2\pi\alpha_0/\lambda)(C_s\alpha_0^2 - \Delta z)\Big|_{\Delta z = \text{const}} \delta(\alpha_0) - (\pi\alpha_0^2/\lambda)\Big|_{\alpha_0 = \text{const}} \delta(\Delta z) \qquad (9)$$

For $\alpha_0 = \text{const}$, we find

$$|\delta\gamma| = (\pi\alpha_0^2/\lambda)\,\delta(\Delta z) \qquad (10)$$

In other words, the larger $\alpha_0 = \lambda/d$ or the smaller d, the more sensitive the phase contrast will be to changes in Δz. Furthermore, the horizontal distances of the curves of the phase-contrast transfer characteristic of Fig. 9 are constant for a given α_0, and for the special case of $\delta\gamma = \pi/2$ we obtain

$$|\delta(\Delta z)| = \lambda/2\alpha_0^2 = d^2/2\lambda \qquad (11)$$

If the Δz-sensitivity is to be assessed quantitatively, we first have to recall that all chromatic instabilities $\Delta W/W$ can in principle be related to corresponding changes in defocus (Vorobev and Vyazigin, 1967) through

$$\delta(\Delta z) = C_F\,\Delta W/W \qquad (12)$$

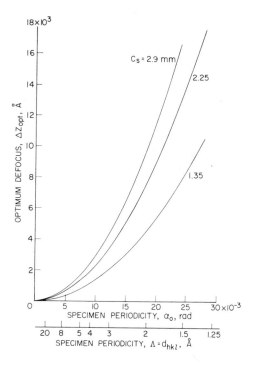

FIG. 10. Calculated optimum defocus values ($\Delta z_{\text{opt}} = C_s\alpha_0^2$) for high phase-contrast imaging of different specimen periodicities $\Lambda = d_{hkl}$.

In addition, use has to be made of the fact that a phase shift γ of the diffracted wave causes a lateral shift Δx of the image of a periodic d_{hkl} specimen (Hanszen 1965). In the simple two-beam case this image shift is given by

$$\Delta x = (d_{hkl}/2\pi)\, \delta\gamma \tag{13}$$

and random changes in γ will, therefore, cause a blurring of the specimen image according to

$$\delta(\Delta x) = \frac{d_{hkl}}{2\pi}\, \delta\gamma = \frac{d_{hkl}}{2}\frac{\alpha_0^{\,2}}{\lambda}\, \delta(\Delta z) = \frac{\lambda}{2\,d_{hkl}}\, C_F\left(\frac{\Delta W}{W}\right) \tag{14}$$

Since it can be shown that a variation of $\delta\gamma = \pm 0.7\pi$ can be allowed if a decrease in image contrast to 30% of its original value is acceptable (Heinemann, 1972), we now obtain the previously used [See Eq. (4)] resolution formula for phase structures. In terms of $\Delta W/W$ this equation defines the the total chromatic instabilities allowable if lattice spacings d_{hkl} are to be resolved:

$$\Delta W/W \approx 1.4 d_{hkl}^2/\lambda C_F \tag{15}$$

For $C_F = 1.6$ mm and $d_{hkl} = 1.44$ Å, this formula demands $\Delta W/W \approx 5 \times 10^{-6}$ for 100 keV electrons. This kind of stability is beyond the performance capability of most high-resolution instruments, and other ways of lowering the influence of chromatic instabilities have to be sought if this kind of resolution is to be obtained in lattice images (see Section II, C, 4, b).

Returning now to Eq. (9), it can be seen that the imaging of periodic lattice-phase structures can also be influenced by changes in α_0 even if all chromatic instabilities are neglected:

$$\delta\gamma = (2\pi\alpha_0/\lambda)(C_s\alpha_0^{\,2} - \Delta z)\, \delta\alpha_0 \tag{16}$$

Changes in the diffraction angle α_0 are due either to a finite angular aperture of the specimen illumination α_{i11} or to a range of periodicities Λ existing in the specimen. As discussed earlier, $\delta\Lambda$ can either cover a small continuous band of space frequencies in amorphous specimens or a range of discrete lattice spacings $\delta(\lambda/d_{hkl})$ in crystalline specimens. Therefore,

$$\delta\alpha_0 = \alpha_{i11} + \delta(\lambda/d_{hkl}) \tag{17}$$

and from (16) one can deduce the optimum defocus setting for which the least sensitivity of γ to changes in α_0 is obtained:

$$\Delta z_{\text{opt}} = C_s\alpha_0^{\,2} = C_s\,(\lambda/d_{hkl})^2 \tag{18}$$

The optimum defocus for best imaging of a given set of lattice planes corresponds to the apex of the curves in the phase transfer characteristic in Fig. 9, and is explicitly plotted in Fig. 10 for a $C_s = 2.25$ mm. Figure 9 demonstrates the large defocus values necessary for imaging small lattice spacings and also shows that setting the optimum defocus is less critical for larger lattice spacings or correspondingly smaller diffraction angle α_0. This latter feature and the generally much decreased contrast sensitivity of lattice images to changes in α_0 for small values of α_0 can be seen most directly from the aberration curves of Fig. 8. The improved imaging properties for small diffraction angles are an important result and contribute to the success of the special imaging technique that follows.

(b) Conical bright-field illumination EM: The image deteriorating effect, caused by phase differences between electron beams passing through different zones of an objective lens, can be minimized by special conditions of specimen illumination in two ways. First, if the angular width of illuminating beam α_{ill}, as well as the diffraction angle α_0, are kept small, then, as shown in the previous section, the changes in phase angles $\delta\gamma$ between neighboring rays through the same lens zone are small. Secondly, one can make use of only those imaging electron beams that pass through the same lens zone extending in rotational symmetry around the optical axis of the lens.

In practice these imaging requirements can be fulfilled by illuminating the specimen not axially as usual but under the Bragg angle

$$\theta_B = \lambda/2d_{hkl} \tag{19}$$

of the corresponding specimen periodicity d_{hkl} that is to be imaged. The zero-order beam will then pass through the lens under the angle $\alpha_0^{(0)} = \theta_B$ whereas the diffracted first-order beam will pass into the lens under $\alpha_0^{(+1)} = -\theta_B$ (the other first-order diffracted beam is not used for imaging). This kind of high-resolution EM is known as the "tilted illumination method" (Dowell, 1963). The good resolutions in lattice images obtained with this technique are due to the elimination of phase shifts between imaging beams and, therefore, to the elimination of image contrast reducing lateral image shifts [Eq. (13)]. In this way one eliminates in effect the chromatic aberration of the objective lens; however, the resulting high resolutions are obtained for only one set of lattice planes running perpendicular to the direction of tilt. Komoda (1966) later generalized the technique so that more than one set of lattice planes in a single-crystal specimen could be imaged simultaneously with high resolution. For this purpose the illumination had to be tilted so that the zero-order reflection and a small number of other neighboring reflections are arranged on a circle around the optical axis in the back focal plane of the objective.

In spite of the most impressive results obtained with Komoda's approach, it is clear that its application is limited to single-crystal films and the imaging of a few sets of lattice planes only. This serious deficiency for the study of early stages of thin-film growth, which is usually characterized by many small crystallites of often strongly varying azimuthal orientation, was eliminated by introducing a new form of specimen illumination—rotation symmetric tilted illumination, also known as "conical specimen illumination" (Heinemann and Poppa, 1970). The various aspects of different techniques of tilted illumination are pictorially summarized in Fig. 11. An example of the application of this technique will be discussed later.

Another very worthwhile feature of the conical specimen-illumination technique is the possibility of not only imaging lattice planes of any random azimuthal orientation but also of simultaneously imaging all lattice planes with spacings

$$d_{hkl} \geqslant \lambda/2\theta_B \qquad (20)$$

This can easily be seen when considering the "diffraction pattern" that applies to the conical illumination case. Each spot of a regular spot pattern is now represented by an annular ring (Fig. 12b), the opening of which is determined by the dimensions of the annular condensor aperture, its distance above the specimen, and the additional condensor effect of the objective lens. If in Fig. 12b it is assumed only one set of lattice planes d_{hkl} causes diffraction as in Fig. 12c, then the original zero-order ring with center at 0_0 is displaced to the second solid-line circle with center 0_1. The displacement for this first set of lattice planes is, of course, $f\lambda/d_{hkl}$ if f is the objective focal length. The corresponding coherent and symmetric imaging beams are designated A and B.

A second parallel set of lattice planes of spacing d'_{hkl} with $d'_{hkl} > d_{hkl}$ results in another dashed diffraction circle. It can be seen that again two coherent imaging beams A' and B' can be found that although not symmetric to 0_0 will still lead to high resolution imaging of the d'_{hkl} lattice planes because they both pass through the same lens zone.

For lattice planes with different azimuthal orientations, these considerations can easily be generalized to prove that with conical illumination it is possible to obtain improved high-resolution lattice images of a wide range of respective lattice spacings.

(c)   Selected-zone dark-field EM: The principle of practically eliminating the influence on EM resolution of the chromatic instabilities $\Delta W/W$ by using only such electron beams for imaging that have passed through the same lens zone is also applicable to axial specimen illumination. However, in contrast to tilted or conical illumination, it follows for axial illumination that the undiffracted zero-order beam has to be excluded from the image-formation process. As a consequence, we deal with a high-resolution dark-field technique in this case. In order to restrict the diffracted beams contributing to image formation only to those that quite exactly pass through the same small lens zone, an accurate selection of image forming beams has to be performed. This is accomplished (Heinemann et al., 1972) by placing a micromachined and precisely dimensioned annular objective lens aperture diaphragm in the back focal

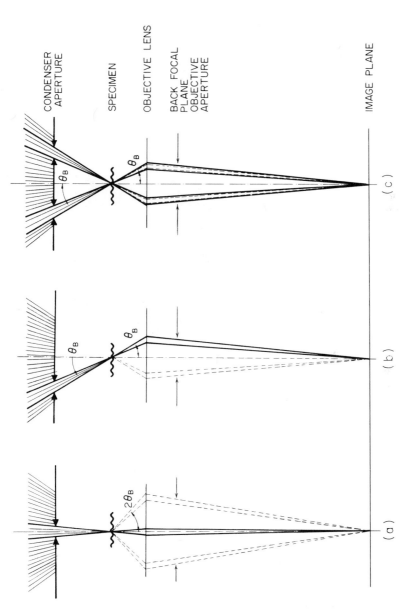

FIG. 11. Comparison of three different modes of specimen illumination used in high-resolution electron microscopy. (a) Axial illumination (centered illumination aperture); (b) tilted illumination (off-center aperture); (c) conical illumination (annular aperture).

plane of the EM objective which selects the lens zone appropriate to the specimen used. The width of the annular opening determines the maximum $\delta(\alpha_0)$ of the imaging beams. In this way $\delta(\alpha_0)$ can be kept small, which in combination with the drastic reduction of chromatic aberration effects leads to a considerable improvement in resolution over conventional EM with axial illumination. The imaging quality obtained with this technique in terms of resolution power is not as high as that with conical specimen illumination because $\alpha_0 = 2\theta_B$ as opposed to $\alpha_0 = \theta_B$ for the conical illumination case. However, the very much improved image contrast of the selected-zone darkfield (SZDF) technique is a definite advantage over the conical bright-field

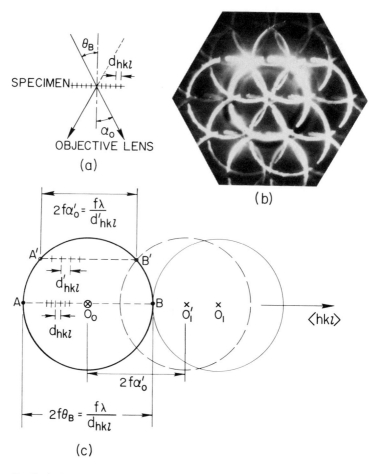

FIG. 12. Conical specimen illumination. (a) Illumination geometry; (b) diffraction pattern of mica specimen; (c) geometrical relationships in the back focal plane.

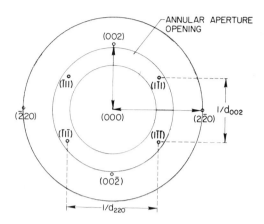

FIG. 13. Selection of imaging Bragg reflections in the back focal plane of the objective lens by an annular aperture as used in selected-zone dark-field electron microscopy (SZDM).

method. The necessity of having to use different objective-lens annular apertures for different specimens and for selecting different zones or Bragg reflections is not as bothersome as it may appear. The needed apertures are very small, and micromachining techniques have been developed to the point (Heinemann *et al.*, 1972) where it is easily possible to place a relatively large number of annular apertures of varying dimensions on one conventional objective-lens aperture.

The main applications of SZDF microscopy are in the areas of high-resolution lattice imaging and in determining spatial orientation distributions in thin-film specimens by collective dark-field microscopy for all azimuthal directions. Orientation measurements will be discussed later, but some of the peculiarities of SZDF microscopy for lattice-imaging purposes shall be briefly discussed.

Let us suppose an annular aperture is being used that selects only the {111} reflections of an epitaxial film specimen (Heinemann *et al.*, 1972). Normally, one would expect that only the two pairs of {111} reflections symmetric to the zero-order beam in Fig. 13 would lead to the imaging of lattice fringes with the well-known (Hanszen and Morgenstern, 1965) $\frac{1}{2}d_{\langle 111 \rangle}$ spacing.[†] Experimentally, however, one finds {200} and {220} lattice fringes in addition to the {111} half-spacing fringes. The unexpected fringe images have been tentatively

---

[†] The lattice images can be interpreted as formed by interference of electron waves originating at diffraction spots in the back focal plane, and the diffraction pattern can simply be treated as a diffraction grating. It is obvious, then, why the elimination of the zero–order beam leaves two symmetric {111} reflections with twice the "grating periodicity." The corresponding lattice image will show half the regular spacing.

termed "pseudolattice images" (Heinemann *et al.*, 1972), and the following simple explanation to these images is given.

The interpretation[†] (see Fig. 13) is based on the interference for image formation of unsymmetrical coherent diffracted beams, an assumption also made previously when explaining the conical illumination imaging of lattice spacings larger than $\lambda/2\theta_B$. Referring to Fig. 13, this means that the interference of, for instance, the $(\bar{1}1\bar{1})$ and the $(1\bar{1}\bar{1})$ beams will cause lattice-fringe images equivalent to {220}-type fringes that would result from the interference of the—here excluded—(220) beam and the zero-order (000) beam. The same explanation would apply to the unsymmetrical $(1\bar{1}\bar{1})$ and $(1\bar{1}1)$ beams and the equivalent (000) and (002) beams in the case of {200}-type lattice fringes.

In spite of some open questions at the present time concerning the exact interpretation of these pseudolattice images, it is clear that they have been observed experimentally and that they can be utilized to determine the local orientation of the diffracting specimen. This interesting topic of application will be discussed in detail in the next section.

5. *Some Specific Applications of Improved Lattice-Imaging Techniques to Epitaxial Thin-Film Nucleation and Growth Problems*

Some of the improved lattice-imaging methods discussed in the preceding pages were developed with the express purpose of shedding more light on problem areas of fundamental importance for the understanding of the complex phenomenon of epitaxy. A few specific examples will be discussed to demonstrate when phase-contrast lattice images of thin films have been used with good success. These examples are also intended to serve as proof that lattice images of very high resolution can be obtained, almost routinely, with present-day high-performance microscopes.

The microstructure of small overgrowth particles contains information about their detailed crystallographic character and their orientation that can be revealed by lattice images only. Therefore, it is possible to study simple and highly complex, multiply twinned particles (MTPs), as shown in Fig. 14, that do not exhibit the well-known gross contrast features of regular MTPs (Ino *et al.*, 1964).

The crystallographic orientation of such a small crystallite is also an important film-growth feature. However, the dimensions of deposit particles in early stages are so small that conventional selected-area diffraction (SAD) techniques are hopelessly inadequate. Evaluation of the lattice image of a

---

† An alternate interpretation of the pseudolattice fringes considers the interference of two sets of {111}-type lattice planes in the specimen, resulting in the formation of a moiré pattern with two sets of moiré fringes. These moiré fringes would also be perpendicular to each other and display fringe spacings of exactly d_{200} and d_{220}.

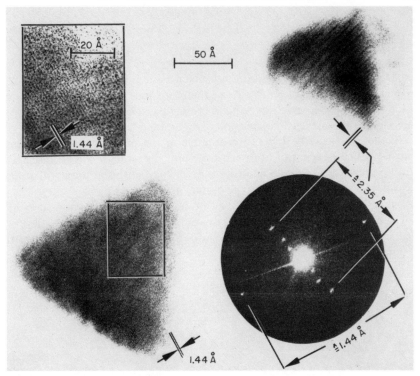

FIG. 14. High-resolution imaging of lattice planes in small gold-deposit crystallites and corresponding light optical diffraction pattern. "SAD patterns" of specimen areas as small as about 50 Å in diameter can be obtained (with permission, Claitor's Publ. Div.).

particle not only by visual inspection of the EM photo plate but by further light optical analysis represents a SAD equivalent usable for these small overgrowth particles. The EM photo plate with lattice image fringes can be treated as a light optical-diffraction grating and a light optical-diffraction pattern obtained from this object displays all directional and spatial fringe information in an integrated form (see Fig. 14). Lattice-fringe spacings can be measured with high precision, and confusing multitudes of several sets of low-contrast lattice fringes can be unraveled. The orientation of specimen areas—or deposit crystallites—as small as about 50 Å has been measured in this way.

The simultaneous resolution of substrate and overgrowth lattices was briefly mentioned before (Fig. 1). It is evident that in every epitaxial system the orientational relationship of substrate to overgrowth is of paramount interest. Of course, if one can resolve simultaneously both the substrate *and* the overgrowth lattices, then the crystallographic relations can be studied in great detail. This has been achieved for gold deposits on mica substrates and is demonstrated in Fig. 1. In this context it is interesting to note that one determines the respective (220) lattice spacing of the epitaxial small Au crystallite to

$$d_{220} = 1.42 \text{ Å}$$

if spacing of the mica lattice in Fig. 1 is taken as standard ($d_{330} = 1.49$ Å). An appreciable shrinkage of $\Delta d/d \approx 2\%$ is, therefore, indicated. This compression of the gold lattice is presently being interpreted as a consequence of the large compressive surface tension forces on a crystallite of such small size.

Another particularly illustrative example of the practical value of simultaneous lattice imaging of both substrate and overgrowth will be discussed because there seems to exist no other experimental means at present that has the capability of providing this type of information.

Figure 15 shows gold particles growing on an epitaxial thin-film substrate of lead sulfide in (100) orientation. The parallel moiré fringe spacing in the low-magnification part of Fig. 15 is 9 Å. Moiré fringes are usually assumed to

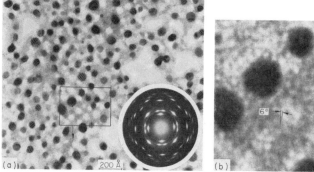
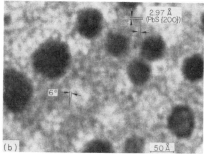

FIG. 15. (a) Low- and (b) high-magnification electron micrographs of gold particles on a lead sulfide substrate film. The gold crystallites are in perfect local alignment with the PbS substrate as proven by the directly resolved substrate lattice. The rotational misalignment of moiré fringes in the lower magnification image would have led to an erroneous interpretation of epitaxial results.

be well suited for studying epitaxial relationships and Fig. 15a clearly demonstrates the existence of small rotational misalignments of the individual gold crystallites. However, the high-magnification lattice image of the same specimen in Fig. 15b proves that the observed orientational misalignments are not real since each gold crystallite is in perfect epitaxial-orientation relationship with the local substrate lattice. The small local variations in the substrate-lattice structure, a consequence of the preparation of this substrate by vapor deposition, can be detected only with the help of direct lattice images.

D. ORIENTATION DETERMINATION IN THIN ISLAND FILMS

1. *Conventional TED and TEM Approaches*

Of the conventional methods for determining the orientation in thin island films, the discussion will be limited to those techniques that can be easily and directly combined with high-resolution electron microscopy. The information on film growth that follows from the evaluation of genuine diffraction techniques only has the main disadvantage of being obtained by integration over a very large specimen area, and thus microstructural details are a priori excluded. Layers that grow predominantly in two-dimensional form can usually be analyzed best with such highly sensitive techniques as RHEED or LEED, and excellent reviews on this subject are available (Bauer, 1969; Prutton, 1971; Sickafus and Bonzel, 1971). Often the combination of diffraction data from microscopically small specimen areas and microstructural morphology information of high spatial resolution is necessary, however, to correlate substrate surface and deposition conditions with film properties. Also, studies of different mechanisms that may be responsible for the introduction of epitaxial order in three-dimensional island films usually necessitate high-resolution microscopy information. Suitable TED–TEM approaches include selected area diffraction (SAD), on- and off-axis dark-field EM, and moiré techniques, all of which suffer from (a) a lack of sensitivity for analyzing the orientation of very thin films, (b) a lack of accuracy for quantitative measurements of the degree of orientation of textured films, and (c) the inability to obtain diffraction data from very small specimen areas.

Lack of sensitivity for orientation measurements on very thin films is a particular disadvantage when studying very early stages of film growth. The main problem of all transmission diffraction techniques considered in this context results from the very large ratio of substrate to overgrowth diffraction intensities. The reduction of substrate thickness will, of course, improve the situation. However, there are obvious limits imposed by mechanical specimen stability and specimen-preparation considerations.

The high substrate-background intensities can be significantly reduced when using imaging rather than simple diffraction techniques. This is shown schematically in Fig. 16 where the ratio of background to deposit intensity is given by the ratio of the dotted to the shaded areas. The simplest way of obtaining orientation information from the image of a deposit island would be its characteristic crystallographic shape, but other more sophisticated techniques that are more generally applicable are available and will be mentioned later. The extremely high average mass sensitivity of any imaging method for analyzing three-dimensional island films is due to the fact that the properties of a single small deposit particle can in principle be determined. In other words, "films" of extremely small average "thickness" can be investigated.

A further disadvantage of conventional ED techniques is the result of difficulties encountered when quantitatively assessing the degree of orientation in partially ordered deposits even if sufficient diffraction information is provided by a thicker film. There have been attempts to characterize the state of order in deposits of this kind by somehow evaluating the degree of "partial polycrystallinity," i.e., by measuring the transformation of a single-crystal spot pattern into a polycrystalline Debye–Scherrer ring pattern. The length of partial rings is compared to the length of complete rings, but the results obtained are usually unsatisfactory and not suitable for valid quantitative studies.

The same is true for the application of conventional dark-field methods, which are based upon the selection of a small portion of the entire diffraction pattern. The selection is accomplished by a small circular objective lens aperture and the corresponding TEM image displays appreciable intensity only in those parts of the specimen that diffract into the selected portion of the diffraction pattern. This technique is practically useless when applied to partially ordered layers showing diffraction rings, since it would be very cumbersome to probe along a Debye–Scherrer ring in all azimuthal directions with a small circular dark-field aperture.

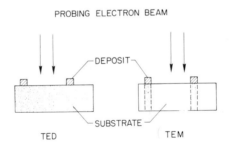

FIG. 16. Simplified schematic comparison of transmission electron diffraction (TED) and transmission electron microscopy (TEM) approaches used to determine the orientation of thin island deposits.

Speaking in terms of high-resolution, high-magnification TEM work, the relatively large selected area of the specimen that can be analyzed by the usual SAD methods is also a significant shortcoming. Conventional SAD leads to the simultaneous orientation analysis of a large number of deposit particles in most thin-film specimens. By using special illumination conditions, the spot size of the illuminating electron beam can certainly be reduced considerably, to about 100-Å diameter in good scanning electron microscopes, but this has so far not been made generally available in conjunction with conventional TEM instruments.[†]

Crystallographic information of individual small deposit particles can be obtained in the most simple way in cases where the deposit particles display well recognizable habits. Although this kind of orientation determination is certainly limited to particular substrate–overgrowth combinations, and usually applicable only to particles that are several hundred angstroms in diameter, it can sometimes be quite helpful in identifying interfacial and other habit planes of overgrowth crystallites. The relative values of the interfacial and surface free energies of the deposit crystallite and the substrate material will determine the equilibrium shapes of simple deposit crystallites (Green *et al.*, 1970). However, composite deposit particles of more complicated crystallographic structure, as for instance multiply twinned particles that are so often encountered (Ino *et al.*, 1964) in vapor deposits, will certainly not be amenable to an orientation analysis of this type. An example of well-shaped and irregular-shaped island deposits of the same deposition material (Ag) on two different, low free-surface-energy substrate materials is presented in Fig. 17, which is included to demonstrate the strong specificity that can be encountered in epitaxial systems.

The study of moiré effects between overlapping thin crystals of different relative orientation and/or different lattice parameter can probably be classified as the most powerful conventional microscopic imaging method that has unique capabilities of providing quantitative orientation information (Hirsch *et al.*, 1965; Matthews, 1960; Jacobs *et al.*, 1966; Jesser and Kuhlmann–Wilsdorf, 1968). It is advisable, however, to keep in mind the rather long list of specific drawbacks exhibited by this approach when judging its general applicability to thin-film nucleation and growth problems:

(a) Substrate and deposit must be simultaneously examined by TEM because double diffraction is necessary for moiré fringes to occur.

(b) The lattice spacings of lattice planes of substrate and overgrowth crystals that are parallel to the imaging beam must have a suitable ratio, otherwise, the moiré fringe spacings are too small (possible resolution problems) or too large, making the method useless for analyzing smaller deposit islands.

---

† Recently several accessories for performing microdiffraction in TEMs have become available commercially. They include scanning microscopy attachments for "scanning microdiffraction" (JELCO, SIEMENS; irradiated diffraction area claimed to be several hundred angstroms in diameter), field-emission electron sources with small area (about 1000-Å diam) and high-diffraction resolution capability (Tonomura, 1974), and electron detectors in the final image plane combined with rocking specimen illumination (PHILIPS, van Oostrum *et al.*, 1973).

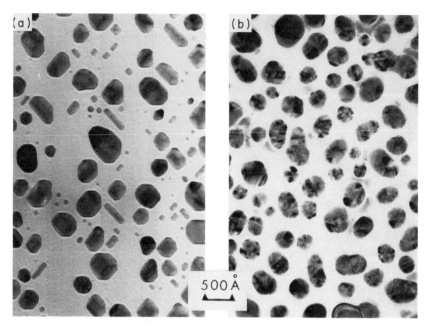

FIG. 17. Comparison of crystallographic habits of "epitaxial" silver overgrowth crystal-lites on single-crystal substrates. (a) Well-shaped particles of simple structure on MgO; (b) shapeless and heavily twinned deposit particles on mica.

(c)   The thin-film substrates used as reference crystals have to possess a high degree of crystallographic perfection.

(d)   When relatively large rotations of the deposit crystallites are involved, the interpretation of moiré patterns is often difficult, and the substrate lattice parameters have to be known with a high degree of accuracy.

(e)   Interfacial dislocations with a similar appearance as moiré fringes can sometimes interfere with the recognition of moiré fringes, and doubts have been registered (Matthews, 1972) that rotated moiré patterns in small islands are a unique indication of rotational misorientations of the overgrowth. Instead, they can sometimes also be interpreted as caused by anistropic stresses.

2. Lattice Imaging and Bragg Reflection Images

Of the more recently developed new methods for accurate orientation determinations of individual small deposit particles, we have discussed the most direct and also the most difficult approach in a previous section: the direct lattice-imaging method. The main difficulty connected with lattice imaging is, of course, the extremely high instrumental resolution required.

Another method that demands much lower resolution power and that is applicable to all epitaxial studies where the deposit crystallites have low-index lattice planes parallel to the imaging electron beam was recently introduced (Poppa *et al.*, 1971). The technique was named the "Bragg reflection images" (BRI) method and is based upon strong diffraction of the imaging electron beam in the entire deposit particle by lattice planes that are perpendicular to the substrate-specimen surface. Because of the spherical aberration of the objective lens, the dark-field images formed by these strongly diffracted (off-axis) electron beams are displaced in the image with respect to the location of the normal bright-field Gaussian image. Bright-shadow images that accompany the bright-field image of each particle result if one employs a large enough objective-lens aperture to permit the strongly diffracted beams to contribute to the image formation. These "shadow images" shift their position when changing the objective lens focus setting. As can be seen from Fig. 18, this specimen related shift δr is given by

$$\delta r = \lambda \, \delta z / d_{hkl} \tag{21}$$

as a function of the corresponding defocus δz. Therefore, d_{hkl} can be determined if the shift of a shadow image is measured. Figure 18 also shows another possibility of determining d_{hkl} of a deposit particle (suspended on a support film) that aids in finding the Gaussian image plane. (A contrast minimum in the phase structure background of an amorphous-carbon support film for an appropriate defocus setting can be readily recognized and this defines the Gaussian focus setting.) By registering the difference in focus setting δz_g between the Gaussian focus and the superposition of bright-field and shadow

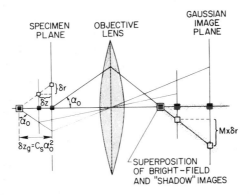

FIG. 18. Geometric relationships between the imaging parameters needed to deduce the equations used for determining the orientation of small deposit particles by the Bragg reflection image method (BRI method).

images of a particle, it is easily possible to determine d_{hkl} without making lengthy measurements in the EM image. The lattice spacing is now given by

$$d_{hkl} = \lambda(C_s/\delta z_g)^{1/2} \tag{22}$$

an quation that follows from the defocus relationship for spherical aberration $\Delta z = C_s \alpha_0^2$. The calibration of objective-lens defocus changes δz as a function of objective-lens current changes for a given spherical-aberration constant C_s that are necessary for applying Eqs. (1) and (2) is either provided by the instrument manufacturer or can be easily measured for each individual instrument. An in situ measurement procedure that uses evaporated MTP gold particles and Eqs. (1) and (2) was recently reported (Heinemann, 1971).

The shadow-image measurements necessary for applying the BRI method can be considerably facilitated by improving the often low image contrast of the shadows in the normal bright-field mode of microscope operation. This is accomplished by blocking the zero-order diffraction beam with a special objective-lens aperture. This aperture can either be a micromachined beam stop aperture (Heinemann et al., 1972) or a small-diameter conductive wire spanning a conventional circular objective-lens aperture.

Figure 19 shows as an example for the BRI method bright- and dark-field Bragg reflection images of evaporated gold particles most of which are multiply twinned and have {111} lattice planes parallel to the imaging electron beam. With gold particles grown on mica substrates, it was proven (Poppa et al., 1971) that the orientation of individual deposit particles as small as 20 Å in diameter can be determined. Furthermore, the extremely high azimuthal-measurement accuracies inherent in this technique make it an ideal tool for measuring the statistical distribution of rotational misalignments of over-growth crystallites. Only discrete misorientations, probably corresponding to different discrete minima of interfacial energy (van der Merwe, 1966) were found in the above-mentioned Au–mica study.

At present the full potential of this new analytical technique for studying the kinetics and microstructural mechanisms that control the introduction of epitaxial order during film formation has not yet been explored. The BRI technique should be particularly valuable in connection with systematic epitaxial studies of different stages of the nucleation and film-growth process.

3. Selected-Zone Dark-Field Microscopy (SZDM)

SZDM represents another most powerful new imaging approach for orien-tation studies of thin-film deposits. Its application is, however, not limited to deposits composed of separate crystallites as the BRI method but can also be very useful in measuring spatial orientation relationships in continuous films. The unconventional features of this new dark-field technique are: (a) suitability for mapping in microscopic detail the areal orientation distri-

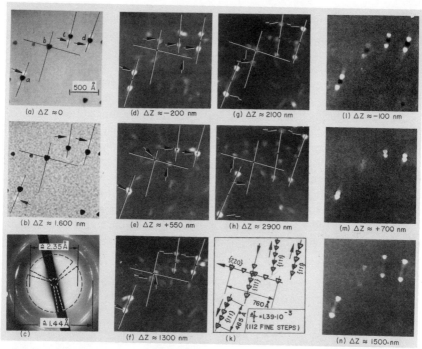

FIG. 19. Series of micrographs of epitaxial Au particles on a mica substrate demonstrating the essential features of the BRI method. (a), (b) Bright-field; (c) two possibilities of blocking the zero-order beam in the back focal plane for the central dark-field mode; (d)–(h) high-contrast central dark-field reflection images for various defocus settings; (k) schematic summation of results; (l)–(n) central dark-field BRI series without disturbing auxiliary lines for three different defocus values.

bution in thin deposits with great simplicity, speed, and efficiency, and without preference to azimuthal direction, and (b) implicitly high image resolution and image contrast. The principle ideas of SZDM were described in detail in one of the preceding sections on very high-resolution EM of specimen phase structures. Employing annular objective-lens apertures of different geometries it becomes possible to select only a well-defined number of Debye–Scherrer diffraction rings. Within one image, one can, therefore, map all specimen areas of random azimuthal orientation that diffract into the selected range of Bragg angles. Using fcc metal deposits as an example it is possible to select either the innermost (111) ring only, add the (200) ring, or choose a combination of (200) and (311) rings, etc.

A particularly suitable application of structure analytical power of this approach is the preferred nucleation of metal deposits at cleavage steps. Results of some investigators (Sato and Shinozaki, 1970) on alkali halide crystals have shown that simple well-oriented single-crystalline deposit particles nucleated and grew at the cleavage steps, while multiply twinned (MT) particles in less perfect registry with the single-crystal substrate were found in the substrate areas between the steps. Some far-reaching theoretical conclusions concerning generally applicable epitaxial mechanisms were later presented as the consequence of this evidence. It should be relatively easy to verify these experimental results by SZDM, because composite MT particles are usually characterized by crystallite portions with (111) lattice planes parallel to the electron beam. The conventional epitaxial particles are (100) oriented and the distinction between MT and epitaxial particles can easily be accomplished by combining SZDM with Bragg reflection imaging. Figure 20 shows Au particles grown on NaCl in bright field (a) and central dark field (b)–(d). The annular dark-field aperture was chosen so as to allow only {111} and {200} reflections. Part (b) was taken at 600 nm underfocus (superposition of {111} reflection images), and for (c) an underfocus of 1300 nm was

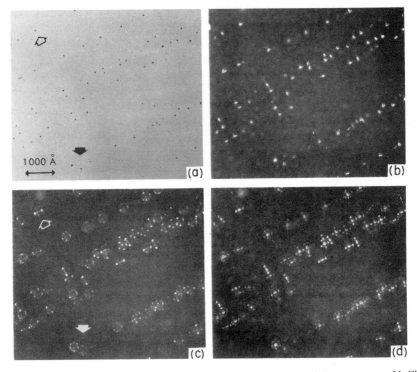

Fig. 20. Selected-zone dark-field imaging of epitaxial Au particles grown on NaCl. Bragg reflection images are used to distinguish between {100} oriented and multiply twinned particles. (a) Bright-field; (b) $\Delta z = 6000$ Å; (c) $\Delta z = 13000$ Å; (d) superposition of (b) and (c).

arbitrarily chosen. Part (d) is a superposition of micrographs taken with either defocus condition to demonstrate better the directions of shift of reflection images upon defocusing. The dark arrow in (a) indicates a group of multiply twinned particles; the light arrow in (a) points to a typical {100} epitaxial crystallite. The epitaxial particles are characterized by simple {200} Bragg images while the MT particles show a multitude of {111} reflection images making it easy to distinguish them from the epitaxial particles. The {111} oriented particles that are not multiply twinned cannot be identified in this way since those would feature {220} reflections, which are not allowed with the particular annular objective aperture used here. However, every single particle did show either {111} or {200} reflection images proving that no {111} oriented particles were lost during this analysis.

The results demonstrated in Fig. 20 leave no doubt that the proposed epitaxial processes of interfacial interactions of varying intensity (Sato, 1971) are not capable of satisfactorily explaining the experimental orientation observations on NaCl. Small differences in actual deposition conditions are always a possible source for the discrepancy of experimental results, but the validity of this hypothesis of epitaxial growth mechanisms must at least be appreciably more limited in scope than originally assumed.

III. In Situ Transmission Electron Microscopy

A. GENERAL FEATURES AND APPLICABILITY TO VAPOR-DEPOSITION STUDIES

"In situ" observation of a process generally means the direct observation of that process at the very time it is taking place. In the case of deposition onto substrates from the vapor phase, that means detecting and measuring the deposition products while they are being formed. When these products form as discrete small entities that are resolvable by TEM and when they possess sufficient image contrast to distinguish them from the substrate background, we can, in principle, count the number of nucleated deposit particles, possiby determine their individual orientation, and measure their growth and coalescence behavior. By doing so, one can hope to learn much about some of the important substrate-surface material parameters that very sensitively influence the heterogeneous nucleation process, one can explore in detail such basic surface physical mechanisms as the clustering of adatoms and the mobility of adatoms and clusters of different size on solid surfaces, and one can investigate and attempt to separate the various stages of film growth that influence the structure and physical properties of the end product: the continuous-deposit layer.

When compared with postdeposition type analytical TEM studies, it is obvious that particularly the kinetic film nucleation and growth processes, which are so difficult to study quantitatively, can be determined most directly, most accurately, and certainly most efficiently by in situ TEM. However, this favorable situation applies only if (a) the sophisticated experimental techniques necessary for realizing well-controlled deposition conditions, as used in conventional UHV equipment, can be successfully incorporated in a TEM, and (b) the specific constraints that are imposed upon the choice of suitable substrate and overgrowth materials by the presence of an image-producing electron beam can be fulfilled. It will be shown that the instrumental requirements have been met by combining modern UHV and vapor-deposition techniques with high-resolution TEM. However, only a few satisfactory substrate–overgrowth material combinations have been found to date, and investigators are still searching for other suitable "epitaxial" in situ systems.

It should also be pointed out that we are not justified in rating the usefulness of in situ TEM studies solely by their success in measuring genuine nucleation and growth phenomena. There are a fair number of applications of existing in situ techniques to other problem areas in the general field of thin-film deposition that should not be overlooked. Late growth and coalescence studies, for example, are much easier to perform than actual nucleation measurements, and a wealth of information pertaining to the eventual structure of continuous films can be obtained rather easily, particularly if measurements of deposit orientations are included in the in situ analysis. Another area of interest in this context is represented by studies conducted during postdeposition processing of discontinuous island films. Much light can be shed on various processes of transport of deposit atoms over substrate surfaces (e.g., Oswald ripening effects) and deposit selfdiffusion can be explored under varying environmental conditions. The possibility of observing structural and form changes in individual deposit islands and the interaction of islands that can be directly observed is an essential feature of the in situ technique. One also has to mention the obvious suitability of in situ TEM studies for directly investigating the influence of electron irradiation on nucleation and growth processes. This is most easily accomplished by comparing nucleation results inside and outside the electron-beam-illuminated specimen area. Since it is a well-known fact that electron-beam irradiation can produce quite dramatic and often desirable property changes in thin-film deposits, this is an area of significant practical value.

In the following chapters, a more detailed discussion will be presented of the present state of the art of in situ TEM investigations with particular emphasis on systematic well-controlled studies. The usually more qualitative and exploratory results of earlier in situ work have been thoroughly reviewed (Pashley, 1956, 1965a,b; Kenty, 1972; Stowell, 1972). For a comprehensive account of the historical development, the interested reader is also referred to the above reviews. Also, the role of misfit dislocations during the epitaxial layer growth of metals and chalcogenite compounds, which has been studied successfully by in situ TEM under less stringent vacuum conditions (Takayanagi *et al.*, 1974; Yagi *et al.*, 1971) will not be discussed.

B. INSTRUMENTATION AND PERFORMANCE

A number of basic instrumental requirements can be listed that have to be met by any in situ TEM system if the results obtained with it are not to be immediately subject to criticism. These are:

(a) background pressures at the site of the TEM substrate specimen in the low 10^{-9} Torr region with partial pressures of the active gases in the low 10^{-10} Torr range and better if possible,

(b) accurately controlled and measured flux of the depositing species and very carefully determined temperature of the thin-film substrates,

(c) means for reproducible in situ preparation of truly clean substrate surfaces,

(d) constant monitoring and assessing of the influence of the imaging electron beam and means for reducing this influence as much as possible while retaining reasonably high image resolution,

(e) the possibility of real time recording of rapid kinetic specimen changes.

In addition to these basic experimental prerequistites, it is in the interest of research expediency to provide somehow for a multiple-specimen capability. This is demanded by the time-consuming nature of modern UHV techniques even though the efficiency of information collection by in situ studies is much higher than that of the previously discussed step-kinetic studies on bulk substrates (Section II).

The goal of achieving an UHV environment at the site of the microscope specimen can be approached in two ways. The most obvious, but also the most difficult, approach is to provide an UHV environment in the entire electron microscope and add vapor-deposition facilities in the specimen-chamber region. This approach has been chosen by Braski (Braski *et al.*, 1968; Braski, 1970), who extensively modified the vacuum system of a commercial electron microscope and achieved pressures during deposition of the order of 10^{-8} Torr. Much lower pressures could not be obtained because of the impossibility of baking the entire microscope column for outgassing purposes.

The other approach to UHV in situ microscopy that has been pursued more often (Poppa, 1965; Valdre *et al.*, 1966, 1970; Barna *et al.*, 1967; Pashley *et al.*, 1968; Moorhead and Poppa, 1969) is based on the design philosophy of providing an UHV environment at the site of the TEM specimen only. Normal operating pressures of 10^{-5} Torr are maintained in the rest of the EM column (which remains essentially unchanged) and normal microscope operation conditions can be retained. The various designs reported differ by the type of oil-free pumping systems used, the extent of the low-pressure specimen and evaporation source surroundings, and by the facilities available for deposition control and specimen treatment. A potential disadvantage to this in situ approach is the existence of two molecular beams impinging upon the substrate because of the use of two conductance-limiting apertures through which the electron beam enters and leaves the UHV specimen chamber. This effect is negligible, however, for small acceptance angle apertures, and could in principle be entirely eliminated by covering the aperture holes with thin electron-transparent films that block the gas flow. Figure 21 shows as an example of the differential-pumping approach to in situ TEM the design of an UHV chamber as it is presently being used by the author (Moorhead and Poppa, 1969). The chamber is bakeable to about 100°C. A characteristic pump-down curve, which is the result of a carefully tested outgassing scheme, is

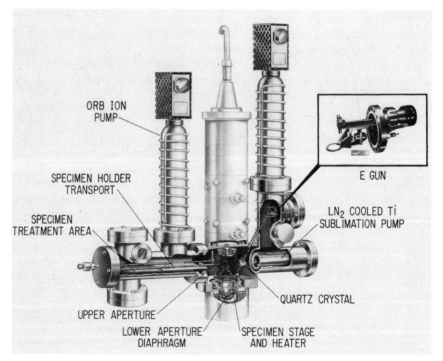

FIG. 21. UHV specimen chamber for in situ electron microscopy studies of vapor-deposition processes (with permission, North Holland Publ. Co.).

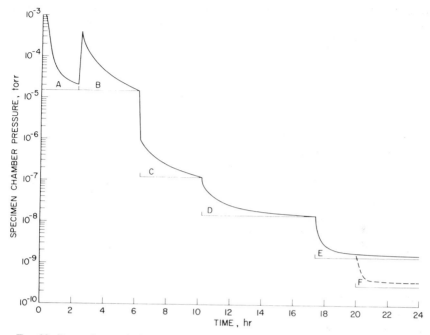

FIG. 22. Pump-down schedule of UHV-EM chamber (with permission of North Holland Publ. Co.). The points labeled are: A, evacuate to normal column pressure with EM vacuum system; B, self bake-out of "orb ion" pumps and initial bake of chamber; C, isolation of chamber from column vacuum, cooling of ion pumps, continuation of bake-out; D, cool down of chamber and final pump to base pressure for ion pumps; E, LN_2 cryopumping including LN_2-cooled Ti sublimation pump (normal configuration); F, LHe pumping, when utilized.

presented in Fig. 22. An ultimate base pressure of 5×10^{-10} Torr was achieved when an additional He cryopump was added to one of the spare ports on the left-hand side of the chamber. (These ports are provided for future specimen storage and treatment functions.) Usually, however, He pumping is not employed which results in a routine base pressure of $\sim 2 \times 10^{-9}$ Torr after about one day of pumping. During operation of the three-source e-gun evaporator the pressure in the specimen chamber as measured by an ionization gauge or by a quadrupole mass spectrometer rises to the middle of the 10^{-9}-Torr range. The evaporation rate is measured and feedback controlled by a stationary oscillating quartz crystal with a hole in the center. The specimen temperature can be raised to $\sim 1000°C$ in a rotation symmetric heater cone that is carefully calibrated. The upper beam-aperture diaphragm is LN_2 cooled and can be positioned from the outside when this is needed for better alignment of the EM. The lower aperture, however, is always stationary in the present design and accounts for the rather long working distance of the objective lens. Because of the long focal length ($f \approx 5$ mm), thermal drifts, and mechanical vibrations cause by LN_2 cooling of various chamber-pumping elements, the electron optical resolution of this in situ microscope is currently limited to approximately 10 Å.[†]

One of the most critical and application-limiting problems of the in situ TEM technique concerns the preparation, inside the chamber, of electron transparent substrates that are satisfactorily clean. If this problem can be successfully solved, the usually high substrate temperatures during film deposition and the relatively low pressures then combine to ensure sufficient time for valid nucleation and growth experimentation. Several experimental approaches are feasible for specimen-surface cleaning in an UHV chamber of the type described above.

Heating a previously prepared thin substrate specimen to high temperatures in an UHV environment for producing an initially clean surface is an obvious possibility but a wealth of LEED–Auger experience on bulk substrate surfaces of many materials proves that this is almost never a successful procedure unless very high temperatures are available and complex cleaning routines are pursued. It is for this reason that results of deposition studies on such well-known substrate materials as MoS_2 (Stowell et al., 1970; Stowell and Law, 1968) that were prepared by necessarily dirty cleavage procedures are suspect.

The chances of successfully cleaning substrate surfaces are somewhat improved when conducting the specimen heating in a controlled reactive gas environment that can be easily established in a UHV in situ chamber. One would be well advised, however, to simulate any doubtful cleaning procedure of a TEM specimen in a LEED–Auger apparatus before relying upon it for in situ studies. It would be even more advantageous if an Auger analyzer were directly attached to the in situ chamber for direct substrate-surface analysis after each cleaning step. This has been achieved recently (Heinemann and Poppa, 1974).

Such cleanliness control would also be desirable, of course, for any of the other cleaning methods feasible. Of these, repeated ion-bombardment cleaning and subsequent annealing is a technique well proven in LEED–Auger work for metal and semiconductor substrate materials, where it is possible to anneal out the accompanying heavy radiation damage (Venables, 1970). As reported recently (Heinemann and Poppa, 1973), a miniaturized ion gun can be incorporated in the in situ chamber of Fig. 21 with relative ease. Furthermore, such a gun can be used to remove previously deposited material following

---

[†]Recently a Siemens model 101 electron microscope with shorter focal length ($f = 2.1$ cm) has replaced the previously employed RCA as basic instrument. The resolution achieved now is approximately 5 Å.

completion of an in situ deposition test and thus provide one version of the often-mentioned multiple-specimen capability.

The usefulness of ion bombardment for substrate cleaning purposes is, however, restricted to metals and semiconductors. With the possible exception of a few selected metal oxides (Stroud, 1972; Kelly and Naguib, 1970) sputter cleaning cannot be applied to most of the important insulating epitaxial substrate materials because of the difficulty of restoring crystallographic order after bombardment. One, therefore, has to rely almost entirely on cleavage as a way of preparing clean surfaces, although in situ cleavage, in general, cannot be easily accomplished with thin substrate films and within the restricted space of an EM specimen chamber. Electron-beam cleavage of high-melting-point metal oxides such as MgO (Yagi and Honjo, 1964; Sato and Shinozaki, 1971) and PbO appears to constitute one quite satisfactory cleaning procedure. The resistance of MgO (Wertz et al., 1957; Palmberg et al., 1967) to low-intensity radiation damage by high-energy electrons (100 kV) seemed to make this material a very promising choice for in situ epitaxy investigations. Recently it was found, however, that MgO is also sufficiently damaged by the high-current-density pulses of the cleaving electron beam to render any reproducible quantitative epitaxial nucleation studies virtually impossible (Moorhead and Poppa, 1972).

It remains to discuss the suitability as a cleaning procedure of the deposition of fresh films of substrate materials by evaporation. For amorphous or very fine polycrystalline substrate surfaces, this is certainly the simplest way of reproducibly preparing in situ a clean substrate surface. A second material of usually higher electron-scattering power can be vapor-deposited from another vapor source in a subsequent low-rate deposition run. The nucleation and growth of this second material can then be properly studied under well-controlled conditions (Poppa, 1967, 1968; Hutchinson et al., 1972; Sherman, 1972). The extension of this technique to single-crystal in situ substrates presents significant problems, however, which can usually be ascribed to the difficulty of growing thin well-oriented epitaxial layers that are comparable in crystallographic and morphological perfection to clean bulk substrate surfaces. At the present time it must be concluded that a great deal of new and imaginative experimental work remains to be done in the area of preparing clean single-crystal substrate specimens for valid in situ epitaxy studies.

The only other problem area that is as critical as the state of cleanliness of the substrate surface is the susceptibility of the same surface to irradiation effects caused by the imaging electron beam. In principle, one has to consider at least three major sources of surface damage by medium-energy (100 kV) electrons. First, ionizing radiation damage can significantly change the chemical composition and/or the electronic nature of the substrate surface. This type of radiation damage has been reported for alkali halide (Rhodin et *al.*, 1968; Bauer *et al.*, 1966) and muscovite mica (Poppa and Elliot, 1971) substrate materials. The halide depletion of KCl surfaces and the surface compositional changes of mica surfaces were determined by AES. The actual atomistic nature of this damage is, however, still a matter of discussion (Lord and Prutton, 1974) as is the detailed mechanism by which the subsequent epitaxial processes are so significantly influenced (Green *et al.*, 1969, 1970). Next, the interaction of the imaging electron beam with gases or other foreign species adsorbed on the substrate surface has to be considered. These surface contaminants can either be preparation remnants or be due to adsorption from the gaseous environment at the substrate before and during deposition. The third type of beam effect can result either from the ionization of the depositing species (Green *et al.*, 1969) either in the gaseous phase in front of the substrate before condensation (Mihama and Tanaka, 1968; Postnikov *et al.*, 1971) or from charge effects after condensation on the substrate surface (Chopra, 1966, 1969). Electrical charging of deposit particles on insulating substrates by the electron beam also belongs in this category and has often been thought responsible for the unusually high mobility of deposit particles sometimes observed during in situ EM experiments. In spite of some theoretical estimates (Marcus and Joyce, 1971) that indicate the possibility of appreciable electrostatic forces between particles, more systematic and conclusive experimental work is definitely needed to assess the true significance of this effect (Murayama, 1972).

When attempting to select substrate materials that can be expected to be least susceptible to ionizing radiation damage, one faces a dilemma similar to that encountered during the search for suitable surface cleaning methods. The most desirable substrate materials of low free-surface energy, like, for instance, the alkali halides or other ionic crystals, are particularly radiation sensitive. The reason for this behavior is the high degree of ionic binding in these substances and the usually very low electronic conductances that afford no easy way of dissipating the energy of the impinging electron beam. Semiconductor materials with good electronic conductance and large percentages of covalent binding are a priori more suitable substrate materials in this respect. Because of their medium-high free-surface energies (~ 1200 ergs/cm^2 for Si, e.g.) semiconductors will, however, lead only to partly three-dimensional, partly two-dimensional nucleation and growth behavior according to the Stanski–Krastanov (Bauer, 1958) mechanism.

The exact degree of electron-beam influence is hard to predict in any substrate material. Even in mostly ionic compound crystals it ranges from total decomposition for the alkali and "earth alkali" halides to weak beam effects in metal oxides like MgO. In any in situ TEM, it is therefore very desirable to reduce a priori the intensity of the imaging electron beam as much as possible, and EM image intensifiers can fulfill a useful function in this connection. At present a large variety of intensifier devices is being offered commercially, ranging from relatively inexpensive channel-plate electron multiplier TV systems (Heinemann and Poppa, 1972) to expensive and sophisticated combinations of electric or magnetic intensifier tubes with very high-sensitivity closed-circuit TV systems (Hermann *et al.*, 1969). It is good to remember, however, that even the most efficient intensifier systems are limited to an approximate improvement of actual specimen beam exposure by a factor of 50 when compared with internal EM photography. The high image-intensification values often cited as necessary in this context are misleading (Heinemann and Poppa, 1972; Heinemann and Pound, 1972).

Another significant advantage of an image-intensifier system is the reduction of recording times for EM images. The average exposure time when using internal photography at higher electron optical magnifications is of the order of several seconds. However, rather fast EM specimen changes due to the kinetics of the deposition processes or due to specimen drift at elevated substrate temperatures and high magnifications cannot usually be avoided, and short recording times become a necessity. With an image-intensifier system one can easily realize photographic exposure times of 1/30 sec when recording off the TV screen. It becomes feasible, therefore, to combine with an intensifier system various facilities for efficiently recording kinetic specimen changes. One can use single-exposure automatic cameras, motion-picture facilities, or video-tape recorders, depending upon the specific type of in situ experiment conducted.

Finally, the great potential of an EM image scanning system for the real time reduction and evaluation of large volume nucleation and growth data has to be mentioned. The sequential scanning of image elements by the TV chain of an image intensifier can in principle be used for this purpose and can be quantitatively analyzed by a computer. It is, however, usually more efficient to analyze the optical output of the actual intensifier element with the help of one of the commercially available complete image analyzer systems (Section II, B, 3). These instruments use special-purpose computers to determine quantitatively and to record quickly such time-dependent image parameters as the number of deposit particles, their average size and their size distribution, the substrate coverage by the deposit, and many other deduced properties of island films that can be effectively used to measure details of the deposition process for comparison with theory.

C. Quantitative Kinetic Measurements

Quantitative and systematic in situ deposition studies have so far only been reported for very simple substrate–overgrowth combinations. An amorphous elemental substrate like evaporated carbon is an obvious choice because of its stability under the electron beam, and because of the ease of preparing clean substrate surfaces by vapor depositing a fresh film. The nucleation and growth of Ag and Bi on carbon (and SiO) substrates has been examined in detail (Poppa, 1967). Although the studies were conducted in what has to be considered a first-generation advanced in situ apparatus capable of only about 10^{-7} Torr background pressures, the results are assumed reliable. The exposure of the fresh substrate surface to the background gases before the start of nucleation was kept to a minimum and the results of more recent nucleation

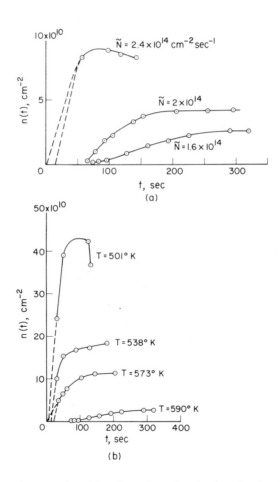

FIG. 23. Dependence on deposition time of number density of stable deposit particles of bismuth on amorphous-carbon substrates: (a) substrate temperature $T = $ const; (b) impinging flux $R = $ const (Poppa, 1967).

experiments of Ag on C at two orders of magnitude lower background pressures (Poppa *et al.*, unpublished results) generally confirm the salient features of the earlier measurements. They will, therefore, be discussed briefly.

By continually recording the number and size of the deposit particles in the same specimen area with increasing deposition time, the complete nucleation characteristics of the system and the growth kinetics of individual particles were determined as a function of the two major experimental parameters that determine the degree of supersaturation. These are the substrate temperature T and the impinging flux R. In this way $n(t)$ curves of the type shown in Fig. 23 were obtained.

The interpretation of this type of data in terms of classical or atomistic nucleation concepts (Pound and Hirth, 1963; Walton, 1962) was partly successful. The number of atoms in the critical nucleus ($n^* = 1$ or 2 or 3) was determined as was the energy of formation of the critical nucleus (ΔE_{n^*}) and approximate values for the free energy of desorption of adatoms from the substrate surface (ΔG_{des}). A refined interpretation of the same nucleation data, as carried out later by Lewis (Lewis, 1970a, b), further refined the probable "shape" of the critical nucleus in the classical model, and postulated the necessity for much greater than usual atom-cluster surface mobilities. These increased particle mobilities, required theoretically to explain the observed low values of maximum number densities of particles n_{max}, could later be confirmed by a careful reexamination of previously recorded micrographs. The exact comparison of different stages of deposition can best be accomplished by graphic superposition of in situ micrographs. In Fig. 24 this has been done for the growth of Bi on amorphous carbon ($T_s = 317°C$, $R = 1.67 \times 10^{14}$ cm^{-2} sec^{-1}). Several basic surface physical processes can be distinghished from Fig. 24 when considering that particles nucleate and grow during the deposition, that the particles at $t = t_3$ are shown shaded, and that the largest outline of all Bi particles must be assigned to the longest deposition time t_5. Figure 24 provides direct evidence for (a) high surface mobility of relatively large clusters of atoms (similar results have been obtained using different experimental methods by Metois *et al.*, 1972), (b) preferred directions of growth that are independent of the direction of the impinging flux for most crystallites, and (c) significant changes in position of very large crystallites during growth.

Direct measurement of the growth of individual particles as represented for one Ag crystallite in Fig. 25 can also be used to determine ΔG_{des} independent of nucleation kinetic measurements. For this purpose one has to measure the dependence of particle size D on deposition time t for various supersaturation conditions. Sigbee's (Sigsbee, 1971) detailed theoretical treatment of particle growth kinetics on substrate surfaces can then be used to determine ΔG_{des}. In situ work of this nature is presently in progress (Poppa and Moorhead, unpublished results; Elliot, 1974) and demonstrates as predicted the effect of adatom depletion by neighboring deposit particles upon the growth of individual deposit particles.

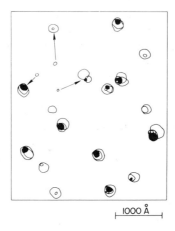

FIG. 24. Graphic superposition of five stages of an in situ deposition of Bi on amorphous carbon ($t_1 = 95$ sec, $t_2 = 127$ sec, $t_3 = 160$ sec, $t_4 = 192$ sec, $t_5 = 320$ sec; $T = 590°K$, $R = 1.67 \times 10^{14}$ cm^{-2} sec^{-1}). The particles at $t = t_3$ are shown shaded.

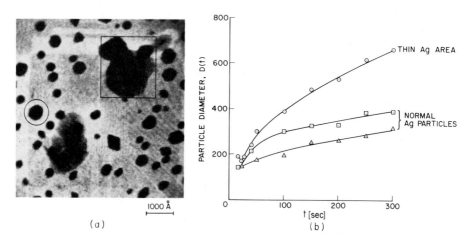

(a)

(b)

FIG. 25. Growth of individual silver particles on amorphous SiO at elevated temperature. Under the prevailing deposition conditions different growth behavior was found for "normal" Ag particles (O) and for large area thin Ag islands (□) (with permission, North Holland Publ. Co.).

D. AFTER-GROWTH EFFECTS

The detailed study by in situ methods of postdeposition microstructural changes in thin films is probably the area of in situ TEM techniques with the greatest potential for practical applications. On the other hand, studies of this kind can also shed appreciable light on some very basic surface physical processes (Robertson, 1973).

When the distribution of deposited matter as it is found on the substrate surface at the end of a deposition keeps on changing although the incoming flux has been stopped, it is obvious that various processes of material transport on surfaces must be operating. Surface self-diffusion of the deposit material and diffusion material across the substrate surface (Oswald ripening, see e.g. Geguzin *et al.*, 1969) have to be considered and interpreted in connection with their respective driving forces. These forces are mainly due to the reduction of specific substrate, deposit surface, and interfacial free energies. The dependence of these transport processes on such readily accomplished postdeposition treatments as prolonged annealing at elevated temperatures, cooling or warming to room temperature, or exposure to various gaseous environments can be studied by following changes in shape and size of individual deposit particles or clusters of particles. As an area of fundamental interest, one must also include in this context the very impressive in situ observations that can be made during the reevaporation of island films. These experiments permit the direct measurement of surface-curvature-dependent vapor pressures of small particles ranging in size from approximately 20-Å diameter to several hundred angstroms. All necessary data are collected within a few minutes while recording kinematographically the rate of shrinkage of deposit particles of different size. (For a silver deposit on carbon, reevaporation takes place with easily measureable rates at about 700°C.) A logical extension of these types of experiments would be to conduct the reevaporation measurements in different, well-defined, residual gas environments for determining the influence of adsorbed gas layers on the surface tension of various metals.

Postdeposition in situ TEM observations are, however, of particular importance to thin-film research from a practical and application-oriented point of view. All vapor-deposited or sputtered films are prepared in a certain vacuum environment under a specific and often very complex set of deposition parameters. In many cases, the as-deposited films will have different properties than will the same films after having been brought to room temperature (Bachmann and Hilbrand, 1966), after having been coated with another fixing film, and after removal from the vacuum system. Microstructural changes in well-defined areas of the deposit can be efficiently pursued and explored by in situ TEM during various postdeposition treatments, and these structural changes can be correlated with varying film properties. One problem area of particular relevance to the thin-film studies discussed in this chapter deserves mention in this context. It concerns the preparation of thin deposits prepared in a separate vacuum-deposition system for TEM or other analytical examinations. The reliability of the information obtained by these postdeposition studies (Section II,A,4) depends entirely upon the validity of the specimen preparation methods used. In an in situ TEM, one often has the unique opportunity of simulating each step of a specific specimen-preparation method if the substrate materials used can be obtained in electron transparent form (Poppa *et al.*, 1972). Specimen-preparation artifacts that would be extremely difficult to discover by any other experimental technique can be eliminated in this way.

E. STATUS AND FEASIBILITY OF VALID EPITAXIAL NUCLEATION AND GROWTH STUDIES

Introduction of an additional degree of order into a substrate–overgrowth system by substituting single-crystal substrate materials for amorphous substrates poses new problems experimentally but also opens up challenging opportunities for deeper understanding of vapor deposition and interface-formation processes. This applies in particular to the potential insights that could be gained concerning the operation of various epitaxial mechanisms by obtaining reliable epitaxial in situ data, and this potential has been the driving force behind the many attempts to solve the inherent severe experimental difficulties.

To the already sophisticated experimental requirements for reproducible in situ measurements, one now has to add a sensitive technique for the in situ determination of the average orientation of a deposit and, if possible, incorporate means for measuring the orientation of individual deposit particles. Analytical techniques with these capabilities would provide the investigator with a most powerful tool to examine systematically different stages of epitaxial film growth, and to determine whether the nucleation, the growth, or the coalescence phases of the deposition process exert the deciding influence on the generation of epitaxial order. The structure analytical methods for the determination of orientation in thin island films that were discussed in Section II,D can in principle be adapted for this purpose. The averaging diffraction methods will generally be found more useful for later growth stages featuring high particle densities, whereas the imaging methods are more suitable for the early stages of deposition. However, specific consideration must be given to the in situ requirements in that the orientation measurements must be made simultaneously with the other nucleation kinetic measurements. When simple SAD data are sufficient, this can sometimes be accomplished by automatically switching from the diffracting to the imaging mode of the TEM and vice versa

(Speidel, 1966). When high-diffraction sensitivities are required, one either has to conduct the deposition under selected-zone dark-field conditions (see Section II, D, 3) that are altered from run to run, or one can attempt to record the TEM images alternatively in bright- or selected-zone dark-field if the deposition process can be slowed down sufficiently by selecting conditions of very low supersaturation. Under favorable epitaxial circumstances, the orientation information can sometimes be found in the bright-field TEM image. This is the case when either moiré fringes or well-recognizable crystallographic habits of the deposition particles are present.

In spite of the appreciable fundamental and experimental difficulties encountered, a fair number of epitaxial in situ TEM studies are reported in the literature (see Section III, B), but not one study fulfills all of the very stringent requirements that have to be met for valid in situ work. However, some of the processes and mechanisms observed can still be considered representative and contribute in this sense to an increased understanding of the phenomenon of epitaxy. The best examples for this type of epitaxy study can be found in some of the pioneering work of Pashley and co-workers (Pashley, 1956, 1965; Stowell et al., 1970; Stowell and Law, 1968), who studied island coalescence processes and the introduction or elimination of lattice defects in later growth stages of Au on MoS_2. Other layer-structure substrate materials used for similar studies were graphite (Basset, 1960; Hutchinson et al., 1972; Sherman, 1972) and mica (Hutchinson et al., 1972; Sherman, 1972; Poppa et al., unpublished results). The only substrate material extensively used and not a layer material is MgO. This material was found suitable for in situ work because of the ability to cleave it through electron-beam-induced, highly localized thermal stresses (Section III, B). Unfortunately, however, it was found that MgO was not sufficiently resistant to electron bombardment damage (Moorhead and Poppa, 1972), and beam effects cannot be avoided entirely (see Fig. 26). On the other hand, it was surprising to find that the majority of the epitaxial results obtained for Au on bulk MgO (Palmberg et al., 1967) could be reproduced.

One other promising substrate material for well-controlled in situ deposition studies is PbSe. This, like most semiconductor materials, has predominantly covalent binding and exhibits a reasonable electronic conductivity. It therefore shows good resistance to radiation damage by the electron beam. It was also found that PbSe can be prepared in clean form by autoepitaxial vapor deposition of fresh films on epitaxial PbSe film substrates (grown previously outside the EM) if the original substrate films are subject to a predeposition heat treatment at about 300°C and if a substrate temperature of 150°C is maintained during the autoepitaxial deposition (Poppa and Moorhead, 1974) (see Fig. 27a). Care has to be exercised to avoid exposing the substrate films to electron-beam irradiation before deposition of the fresh

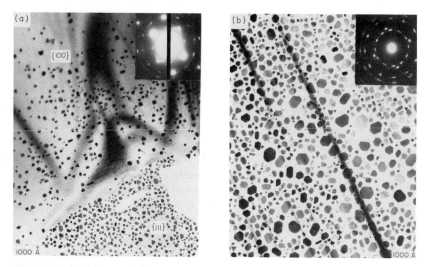

FIG. 26. Epitaxial growth of Au on MgO at $T = 400°C$. (a) Average orientation and habit of Au crystallites are different in {100} and {111} substrate areas. (b) A prevalence of multi-positioned Au crystallites in {111} orientation is found in some {100} substrate areas as a consequence of strong radiation damage by the cleaving electron beam (with permission, Claitor's Publ. Div.).

PbSe film because polycrystalline overgrowth results (Fig. 27b). As shown in Fig. 27c, the in situ deposition of Au on fresh epitaxial PbSe layers looks promising. However, the perfection of the autoepitaxial layer is not as good everywhere on the specimen as in the area of Fig. 22b and, consequently, good-quality epitaxial Au deposits are not reproducibly obtained at the present stage of investigation. Work on improving the perfection of the autoepitaxial PbSe layer and attempts using other chalcogonide substrate materials are in progress (Poppa and Moorhead, 1974).

F. CONCLUDING REMARKS

There is no doubt that in principle in situ TEM investigations offer very attractive research possibilities in many areas of surface and thin-film physics. The very nature of the technique is conducive to making microstructural changes directly visible and with high-image resolution while they are occurring. Since it is often said that "seeing is believing" and "a picture is worth a thousand words," a motion picture, which is the usual result of an in situ experiment, must be very worthwhile to study. For this very reason, a considerable amount of work has been done recently in this area with the goal of sufficiently improving the earlier in situ techniques so that they can be

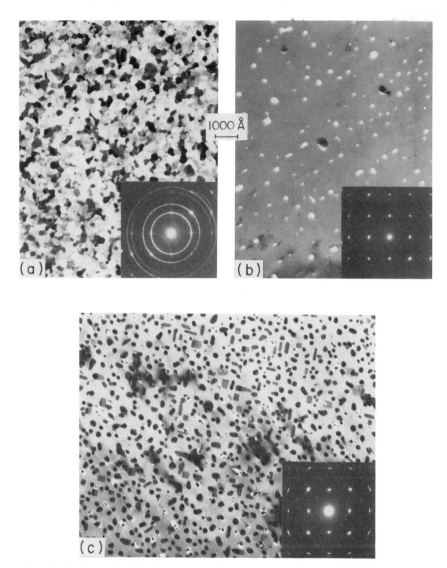

FIG. 27. (a) Growth of fresh polycrystalline lead selenide layer in preirradiated area of PbSe substrate film, (b) autoepitaxial growth of PbSe outside electron irradiated substrate area, and (c) epitaxial overgrowth of gold on fresh epitaxial PbSe film at $T = 250°C$ substrate temperature.

confidently applied. However, this effort has met with partial success only because of basic limitations due to the rarely negligible and complex effects of the ever-present imaging electron beam. At the present stage of development of the technique, it must be recognized that only very specific combinations of substrate and overgrowth materials seem to lend themselves to valid in situ TEM studies of epitaxial nucleation and growth of thin films. In spite of these limitations, in situ studies of thin-film growth have at times provided truly exciting results that could not have been obtained with any other analytical method. As an example, the last figure (Fig. 28) shows unequivocal evidence for the surprising ability of huge clusters of atoms to rotate spontaneously ($\Delta t < 1/30$ sec) on the substrate surface from one stable epitaxial position (indicated by triangular markers in Fig. 28) into a different equivalent position. For postdeposition investigations of microstructural changes in thin deposits, the limitations are less severe, and some of the best uses of in situ TEM tech-

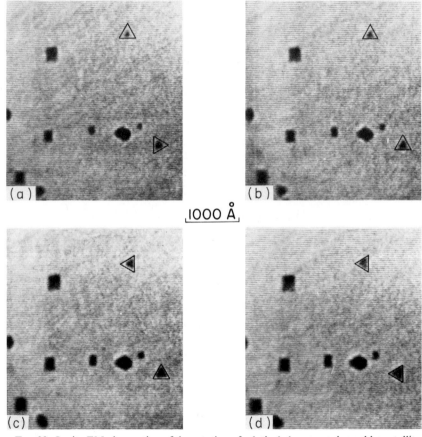

FIG. 28. In situ EM observation of the rotation of relatively large growing gold crystallites at high substrate temperature ($T = 500°C$). Different but equivalent epitaxial orientations on the {100}-oriented MgO substrate are assumed by the gold crystallites. (a) 650 sec; (b) 700 sec; (c) 750 sec; (d) 800 sec.

niques should be encountered while judiciously combining them with well-controlled, step-kinetic, vapor-deposition studies on conventional bulk substrates.

ACKNOWLEDGMENTS

Of my co-workers in the Materials Science Branch of NASA's Ames Research Center, the author would like to thank particularly Dr. Klaus Heinemann and Mr. R. Dale Moorhead. Mr. Moorhead was instrumental in designing the advanced in situ EM chamber. Dr. Heinemann provided the majority of the high-resolution electron micrographs used in this article, and contributed substantially to the sections dealing with phase contrast electron microscopy. Thanks are also due to Mr. Dell Williams who helped with the reading of the manuscript, and to Mrs. Olive Fordham for typing.

References

Abbink, H. C., Broudy, R. M., and McCarthy, G. P. (1968). *J. Appl. Phys.* **39**, 4673.
Abermann, R. (1972). Private communication.
Abermann, R., and Bachmann, L. (1969). *Naturwissenschaften* **56**, 324.
Allpress, J. G., and Sanders, J. V. (1967). *Surface Sci.* **7**, 1.
Anderson, J. C. (ed.) (1966). "The Use of Thin Films in Physical Investigations." Academic Press, New York.
Bachmann, L., and Hilbrand, H. (1966). *In* "Basic Problems in Thin Film Physics," p. 77. Vandenhoeck and Ruprecht, Goettingen.
Barna, A., Barna, P. B., and Pocza, J. F. (1967). *Vacuum* **17**, 219.
Bassett, G. A. (1958). *Phil. Mag.* **3**, 1042.
Bassett, G. A. (1960). *Proc. Eur. Conf. Electron. Microsc.* p. 270. Nederlandse Vereniging voor Electronenmicroscopie, Delft.
Bauer, E. (1958). *Z. Kristallogr.* **110**, 395.
Bauer, E. (1958). *Z. Kristallogr.* **110**, 372.
Bauer, E. (1969). *In* "Techniques of Metals Research," Part 2, p. 501. Wiley (Interscience), New York.
Bauer, E., Green, A. K., Kunz, K. M., and Poppa, H. (1966). *In* "Basic Problems in Thin Film Physics." Vandenhoeck and Ruprecht, Goettingen.
Beaman, D. R., and Isasi, J. A. (1972). Electron beam microanalysis. ASTM STP 506.
Bethge, H. (1962a). *Phys. Status Solidi* **2**, 3.
Bethge, H. (1962b). *Phys. Status Solidi* **2**, 775.
Bethge, H. (1969). *J. Vac. Sci. Technol.* **6**, 460.
Birjiga, M., Glodeanu, F., Popescu-Pogrion, N., Teodorescu, I., and Topa, V. (1972). *Thin Solid Films* **10**, 307.
Boersch, H. (1954). *Z. Phys.* **139**, 115.
Braski, D. N. (1970). *J. Vac. Sci. Technol.* **7**, 164.
Braski, D. N., Gibson, G. R., and Kobisk, E. H. (1968). *Rev. Sci. Instrum.* **39**, 1806.
Chakraverty, B. K., and Pound, G. M. (1963). *Proc. Dayton Int. Conf. Condensation Evaporation.* Gordon and Breach, New York.
Chopra, K. L. (1966). *J. Appl. Phys.* **37**, 2249.
Chopra, K. L. (1969). "Thin Film Phenomena." McGraw-Hill, New York.
Darby, T. B., and Wayman, C. M. (1970). *Phys. Status Solidi.* **1**, 729.

Dowell, W. C. T. (1963). *Optik* **20**, 535.

Downing, K. H. (1972). *Proc. Annu. Meeting EMSA, 30th* p. 562. Claitor's Publ., Baton Rouge, Louisiana.

Elliot, A. G. (1972). Ph.D. Thesis, Stanford Univ.

Elliot, A. G. (1974). *J. Vac. Sci. Technol.* **11**, 826.

Farnsworth, H. E. (1959). *In* "Surface Chemistry of Metals and Semiconductors" (H. C. Gatos, ed.). Wiley, New York.

Farnsworth, H. E., Schlier, R. E., George, T. H., and Burger, R. M. (1958). *J. Appl. Phys.* **29**, 1150.

Frankl, D. R., and Venables, J. A. (1970). *Advan. Phys.* **19**, 409.

Fuchs, E. (1966). *Rev. Sci. Instrum.* **37**, 623.

Geguzin, Y. E., Kaganovskii, Y. S., and Kalinin, V. V. (1969). *Sov. Phys.-Solid State* **11**, 203.

Green, A. K. (1973). NASA Contractor Rep. CR-2234.

Green, A. K., Bauer, E., and Dancy, J. (1969). *In* "Molecular Processes on Solid Surfaces," p. 479. McGraw-Hill, New York.

Green, A. K., Dancy, J., and Bauer, E. (1969). *J. Vac. Sci. Technol.* **7**, 159.

Green, A. K., Bauer, E., Peek, R. L., and Dancy, J. (1970). *Krist. Tech.* **5**, 345.

Green, A. K., Dancy, J., and Bauer, E. (1970). *J. Vac. Sci. Technol.* **7**, 159.

Haine, M. E., and Mulvey, T. J. (1954). *J. Sci. Instrum.* **31**, 326.

Hanszen, K. J., and Morgenstern, B. (1965). *Z. Angew. Phys.* **19**, 215.

Hanszen, K. J., and Morgenstern, B. (1967). *Naturwissenschaften* 126.

Hanszen, K. J., Morgenstern, B., and Rosenbruch, K. J. (1963). *Z. Angew. Phys.* **16**, 477.

Hass, G., and Thun, R. E. (eds.) (1964–1967). "Physics of Thin Films," Vols. 1–4. Academic Press, New York.

Hayeck, K., and Schwabe, U. (1971). *Z. Naturforsch.* **267**, 879.

Heinemann, K. (1971). *Optik* **34**, 113.

Heinemann, K. (1972). Private communication.

Heinemann, K., and Poppa, H. (1970). *Appl. Phys. Lett.* **16**, 515.

Heinemann, K., and Poppa, H. (1972a). *Appl. Phys. Lett.* **20**, 122.

Heinemann, K., and Poppa, H. (1972b). NASA Tech. Brief 72–10171 (ARC–10448).

Heinemann, K., and Poppa, H. (1972c). Unpublished results.

Heinemann, K., and Poppa, H. (1972c). *Proc. Annu. Meeting EMSA, 30th* pp. 610–612. Claitor's Publ., Baton Rouge, Louisiana.

Heinemann, K., and Poppa, H. (1973). *J. Vac. Sci. Technol.* **10**, 22.

Heinemann, K., and Poppa, H. (1974). To be published.

Heinemann, K., and Pound, G. M. (1972). Annu. Progr. Rep., NASA Grant NGR–05–020–569.

Hermann, K. H., Krahl, D., Kuebler, A., and Rindfleisch, V. (1969). *Siemens Rev.* **36**, 6.

Hirsch, P. B., Howie, A., Nicholson, R. B., Pashley, D. W., and Whelan, M. J. (1965). "Electron Microscopy of Thin Crystals." Butterworths, London and Washington, D.C.

Honma, T., and Wayman, C. M. (1965). *J. Appl. Phys.* **36**, 2791.

Hutchinson, T. E., and Sherman, D. (1970). *Proc. Annu. Meeting EMSA, 28th* p. 516. Claitor's Publ., Baton Rouge, Louisiana.

Hutchinson, T. E., Sherman, D. M., and Maa, Jer-shen (1972). Univ. of Minnesota Annu. Tech. Progr. Rep. IV–C00–1790.

Ino, S., Watanabe, D., and Ogawa, S. (1964). *J. Phys. Soc. Japan* **19**, 881.

Jacobs, M. H., Pashley, D. W., and Stowell, M. J. (1966). *Phil. Mag.* **13**, 121.

Jaeger, H., Mercer, P. D., and Sherwood, R. G. (1967). *Surface Sci.* **6**, 309.

Jaeger, H., Mercer, P. D., and Sherwood, R. G. (1969). *Surface Sci.* **13**, 349.

Jesser, W. A., and Kuhlmann-Wilsdorf, D. (1968). *Acta Met.* **16**, 325.

Kelly, R., and Naguib, H. M. (1970). *In* "Atomic Collision Phenomena in Solids," p. 172. North-Holland Publ., Amsterdam.

Kenty, J. L. (1972). Bibliography of EM in-situ Nucleation and Growth Studies, ARPA Contract DAAH01–70–C–13111, North American Rockwell, Anaheim, California.

Khan, I. (1970). *In* "Handbook of Thin Film Technology," p. 10–11. McGraw-Hill, New York.

Koch, F. A., and Vook, R. W. (1971). *J. Appl. Phys.* **42**, 4510.

Lee, E. (1974). Ph.D. Thesis, Stanford Univ.

Lenz, F. (1965). *Optik* **22**, 270.

Lenz, F. (1971). "Electron Microscopy in Materials Science" (U. Valdrè, ed.), p. 540. Academic Press, New York and London.

Lewis, B. (1970). *J. Appl. Phys.* **41**, 30.

Lewis, B. (1970a). *Surface Sci.* **21**, 273.

Lewis, B. (1970b). *Surface Sci.* **21**, 279.

Lord, D. G., and Prutton, M. (1974). *Thin Solid Films* **21**, 341.

Marcus, R. B. and Joyce, W. B. (1971). *Thin Solid Films* **7**, R3.

Markov, I., and Kashchiev, D. (1973). *Thin Solid Films* **15**, 181.

Matthews, J. W. (1960). *Proc. Eur. Reg. Conf. Electron. Microsc.* **1**, 276. Nederlandse Vereniging voor Electronenmicroscopie.

Matthews, J. W. (1972). *Surface Sci.* **31**, 241.

Metois, J. J., Gauch, M., Masson, A., and Kern, R. (1972). *Thin Solid Films* **11**, 205.

Mihama, K., and Tanaka, M. (1968). *J. Crystal Growth* **2**, 51.

Mihama, K., Shima, S., and Uyeda, R. (1974). *Proc. ICG4*, p. 100, Tokyo.

Moorhead, R. D., and Poppa, H. (1969). *Proc. Annu. Meeting EMSA, 27th* p. 116. Claitor's Publ., Baton Rouge, Louisiana.

Moorhead, R. D., and Poppa, H. (1972). *Proc. Annu. Meeting EMSA, 30th* p. 516. Claitor's Publ., Baton Rouge, Louisiana.

Murayama, Y. (1972). *Int. Conf. Thin Films, Conf. Abstr., Venice* p. 94.

Palmberg, P. W., Rhodin, T. N., and Todd, C. J. (1967). *Appl. Phys. Lett.* **10**, 122.

Palmberg, P. W., Rhodin, T. N., and Todd, C. J. (1967). *Appl. Phys. Lett.* **11**, 33.

Pashley, D. W. (1965a). *Advan. Phys.* **5**, 174.

Pashley, D. W. (1965b). *Advan. Phys.* **14**, 327.

Pashley, D. W. (1970). *Recent Progr. Surface Sci.* **3**, 23.

Pashley, D. W., Stowell, M. J., and Robinson, E. A. (1968). *Proc. Eur. Conf. Electron. Microsc.* p. 387. Tipografia Poliglotta Vaticana.

Palatnik, L. S., Kosevich, V. M., Zozulya, L. P., Zozulya, L. F., and Sorokin, V. K. (1970). *Sov. Phys.-Solid State* **11**, 2086.

Poppa, H. (1965). *J. Vac. Sci. Technol.* **2**, 42.

Poppa, H. (1967). *J. Appl. Phys.* **38**, 3883.

Poppa, H. (1968). NASA Tech. Note D–4506.

Poppa, H., and Elliot, A. G. (1971). *Surface Sci.* **24**, 149.

Poppa, H., and Moorhead, R. D. Unpublished results.

Poppa, H., and Moorhead, R. D. (1974). *J. Vac. Sci. Technol.* **11**, 132.

Poppa, H., Heinemann, K., and Elliot, A. G. (1971). *J. Vac. Sci. Technol.* **8**, 471.

Poppa, H., Heinemann, K., and Moorhead, D. R. (1972). *J. Vac. Sci. Technol.* **102**, 149.

Poppa, H., Moorhead, R. D., and Heinemann, K. (1972). *Nucl. Instr. Meth.* **102**, 521.

Postnikov, V. S., Margunov, V. N., Zolotukhin, I. V., and Tevleo, V. M. (1971). *Sov. Phys.-Solid State* **12**, 2399.

Pound, G. M., and Hirth, J. P. (1963). "Condensation and Evaporation," *Progr. Mater. Sci. II*. MacMillan, New York.

Prutton, M. (1971). *Met. Rev.* No. 152, 52.

Reimer, L. (1967). "Elektronenmiskroskopische Untersuchunges-und Praeparations-methoden." Springer-Verlag, Berlin and New York.

Rhodin, T. N., Palmberg, P. W., and Todd, C. J. (1968). *In* "Molecular Processes on Solid Surfaces." McGraw-Hill, New York.

Robertson, D. (1971). Ph.D. Thesis, Stanford Univ.

Robertson, D. (1973). *J. Appl. Phys.* **44**, 3924.

Robins, J. L., and Rhodin, T. N. (1964). *Surface Sci.* **2**, 346.

Robinson, V. N. E., and Robins, J. L. (1974). *Thin Solid Films* **20**, 155.

Routledge, K. J., and Stowell, M. J. (1970). *Thin Solid Films* **6**, 407.

Sato, H. (1971). Ford Sci. Res. Lab. Publ., Dearborn, Michigan.

Sato, H., and Shinozaki, S. (1970). *J. Appl. Phys.* **41**, 3165.

Sato, H. and Shinozaki, S. (1971). *J. Vac. Sci. Technol.* **8**, 159.

Scherzer, O. (1949). *J. Appl. Phys.* **20**, 20.

Schmeisser, H. (1974). *Thin Solid Films* **22**, 83.

Sherman, D. M. (1972). Ph.D. Thesis, Univ. of Minnesota.

Sickafus, E. N., and Bonzel, H. P. (1971). *In* "Progress in Surface and Membrane Science," p. 115. Academic Press, New York.

Sigsbee, R. A. (1971). *J. Appl. Phys.* **42**, 3904.

Speidel, R. (1966). *Optik* **24**, 298.

Stirland, D. J. (1966). *Appl. Phys. Lett.* **8**, 326.

Stowell, M. J. (1972). *Phil. Mag.* **26**, 349.

Stowell, M. J. (1972). *Thin Solid Films* **12**, 341.

Stowell, M. J., and Hutchinson, T. E. (1971a). *Thin Solid Films* **8**, 41.

Stowell, M. J., and Hutchinson, T. E. (1971b). *Thin Solid Films* **8**, 411.

Stowell, M. J., and Law, T. J. (1968). *Phys. Status Solidi* **25**, 139.

Stowell, M. J., Law, T. J., and Smart, J. (1970). *Proc. Roy. Soc. London* **A318**, 231.

Stroud, P. T. (1972). *Thin Solid Films* **11**, 1.

Takayanagi, K., Kobayashi, K., Yagi, K., and Honjo, G. (1974). *Thin Solid Films* **21**, 325.

Thomas, G. (1962). "Transmission Electron Microscopy of Metals." Wiley, New York.

Thon, F. (1965). *Z. Naturforsch.* **207**, 1.

Thon, F. (1971), "Electron Microscopy in Materials Science" (U. Valdrè, ed.), p. 570. Academic Press, New York and London.

Tonomura, A. (1974). *Optik* **39**, 386.

Toth, R., and Cicotte, L. (1968). *Thin Solid Films* **2**, 111.

Valdre, U., Pashley, D. W., Robinson, E. A., Stowell, M. J., Routledge, K. J., and Vincent, R. (1966). *Proc. Int. Conf. Electron. Microsc., Kyoto* **1**, 155. Maruzen, Tokyo.

Valdre, U., Robinson, E. A., Pashley, D. W., Stowell, M. J., and Law, T. J. (1970). *J. Phys.* **3E**, 501.

Van der Merwe, J. H. (1966). *In* "Basic Problems in Thin Film Physics," p. 122. Vandenhoeck and Ruprecht, Goettingen.

van Oostrum, K. J., Leenhouts, A., and Jore, A. (1973). *Appl. Phys. Lett.* **23**, 283.

Venables, J. A. (1970). *In* "Atomic Collision Phenomena in Solids," p. 132. North-Holland Publ., Amsterdam.

Venables, J. A. (1973). *Phil. Mag.* **27**, 697.

Vorobev, Yu. V., and Vyazigin, A. A. (1967). *Opt. Spectrosc.* **22**, 261.

Walton, D. (1962). *J. Chem. Phys.* **37**, 2182.

Wertz, J. E., Auzins, P., Weeks, R. A., and Silsbee, R. H. (1957). *Phys. Rev.* **107**, 1535.

White, J. R., Beer, M., and Wiggins, J. W. (1971). *Micron* **2**, 412.

Yagi, K., and Honjo, G. (1964). *J. Phys. Soc. Japan* **19**, 1892.

Yagi, K., Tokayanagi, K., Kobayashi, K., and Honjo, G. (1971). *J. Cryst. Growth* **9**, 84.
Zinsmeister, G. (1966). *Vacuum* **16**, 529.
Zinsmeister, G. (1968). *Thin Solid Films* **2**, 497.
Zinsmeister, G. (1969). *Thin Solid Films* **4**, 363.
Zinsmeister, G. (1971). *Thin Solid Films* **7**, 51.

3.2   X-Ray Topography

G. H. Schwuttke

IBM East Fishkill Laboratories
Hopewell Junction, New York

I. Introduction

X-ray diffraction topography techniques are most useful for the investigation of thick and nearly perfect crystals. The structural perfection of such crystals is conveniently displayed in transmission as well as reflection topographs. X-ray topography techniques are not directly applicable to the investigations of thin single-crystal foils. Very thin crystals appear featureless in transmission topographs. A minimal crystal thickness is necessary to see crystal defects in X-ray topographs. This minimum crystal thickness varies somewhat with the recording reflection and is for low-order reflections approximately 0.4ξ and for high-order reflections 0.15ξ (ξ is the extinction distance of recording reflection) (Tanner, 1972). In silicon, one obtains a minimum thickness of approximately 10 to 20 μm. Below 10-μm crystal thickness, silicon crystals appear featureless in X-ray topographs. Therefore, X-ray topography techniques are of limited use for the structural investigations of thin single-crystal films in the form of thin foils. However, the diffraction conditions are more favorable whenever thin films are deposited on perfect single-crystal substrates. Under such conditions, X-ray topography is very useful and can provide interesting information about thin-film properties as well as about the interaction between film and substrate

281

II. X-Ray Diffraction Topography

X-ray topography is in many ways similar to transmission electron micros-copy and can provide complementary and supplementary information about the defect state of single crystals. The diffraction of X rays like the diffraction of electrons is controlled by Bragg's equation, and local variations in diffraction conditions can cause variations of the X-ray intensity diffracted from a par-ticular crystal region. If these variations are recorded as a function of the coordinates of the diffracting region, an image can be obtained showing local variations in diffracting power due to dislocations, stacking faults, and other strain fields in the crystal. The contrast of such images recorded on photographic plates is called an *X-ray topograph* of the crystal and is analogous to the diffraction contrast observed in transmission electron micrographs. However, when comparing photomicrographs of X-ray topographs and transmission electron microscope pictures it is necessary to realize that it is customary to reproduce X-ray topographs as photographic negatives and electron micrographs as positive prints. This reverses the meaning of black and white contrast in terms of deficiency or excess intensity for the two techniques. In general, the contrast rules of X-ray topography are very similar to the rules of diffraction contrast in the electron microscope.

Since X rays cannot be manipulated by lenses or other magnification devices, X-ray topographic techniques have some severe limitations, specifically in the area of X-ray geometries and X-ray instrumentation. This has led to the development of many geometrically different methods. A number of useful techniques are listed and compared in Table I. The two more important tech-niques, the Lang (1958) and the scanning oscillator technique (SOT) (Schwuttke, 1965a; Van Mellaert and Schwuttke, 1972), are described in more detail in the following paragraphs.

If the crystal is not too thick and the defect density is not too large ($\sim 10^6$ dislocations/cm^2) the distribution of imperfections in the crystal is conveni-ently obtained by the Lang technique. In this method, a ribbon X-ray beam (coming from a point or a microfocus X-ray source) is sufficiently collimated so that only characteristic radiation is diffracted by the crystal. A stationary aperture allows the diffracted beam to impinge on a photo plate while the direct beam is stopped. A relatively large area/volume of crystal can be surveyed by scanning both crystal and photo plate together past the incident beam. This is shown in Fig .1a.

The Lang technique is excellent and has the following advantages: lower background than any other technique, less trouble with simultaneous reflec-tions because a strongly collimated beam is used, and a high resolution because the photographic plate can be placed close to the specimen. Disadvantages may arise, such as relatively long exposure times; therefore, the recording of

TABLE I

COMPARISON OF X-RAY DIFFRACTION MICROSCOPY TECHNIQUES

	Schulz (1954), Weissmann (1956), Guinier and Tennevin (1949)	Berg-Barrett (see Newkirk, 1959)	Double crystal (Bonse, 1958)	Wide-beam transmission (Schwuttke, 1965a)		Scanning transmission (Lang, 1958)	Scanning oscillating transmission (Schwuttke, 1965a; van Mellaert and Schwuttke, 1972)
				$\mu_0 t > 10$	$\mu_0 t < 1$		
Instrumentation	Simple	Simple	Complicated	Simple	Simple	Complicated	Complicated
available	No	Yes	No	Yes	Yes	Yes	Yes
exposure time	10–25 hr	Short (~1 hr)	Short (1 hr)	Long: 10–30 hr		0.1–1 hr/mm scan	0.1–0.5 hr/mm scan
Type of defect revealed	Grain misorientation	Subgrains Dislocations	Subgrains Dislocations Stacking faults	Dislocations	Subgrains Dislocations Stacking faults	Dislocations Stacking faults Segregation Precipitation	Dislocations Stacking faults Segregation Precipitation
Type of contrast	Dilation and tilt	Dilation and tilt Extinction	Dilation and tilt	Dynamical	Dilation and tilt Extinction	Extinction Dynamical	Extinction Dynamical
Sensitivity to: deformation	Low	Low	High	High	Low	Low	Low
sense of deformation	Tilts: yes Inhomogeneous deformation: no	Subgrains: yes Dislocations: no	Yes	Yes	No	Yes ($\mu t \sim 10$)	Yes ($\mu t \sim 10$)
Upper limit of dislocation density	—	$10^{16}/cm^2$	$10^5/cm^2$	$10^3/cm^2$	$10^6/cm^2$	$10^6/cm^2$	$10^6/cm^2$
Dislocation image width	—	1–5 μm	Up to 200 μm	~50 μm	5 μm	1–10 μm	1–10 μm

283

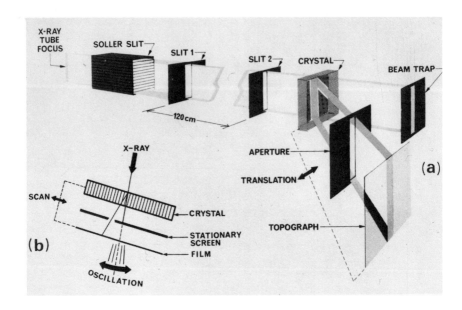

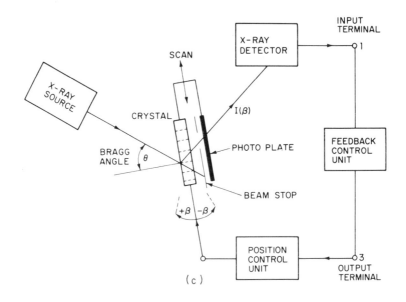

FIG. 1. Geometrical set-up for X-ray topography. (a) Lang technique; (b) scanning-oscillator technique; (c) automatic Bragg angle control (ABAC).

large-area topographs is not practical. Also, it is very difficult to record topographs in the presence of elastic strains. In many cases it is impossible. Standard Lang cameras provide a maximum size of the topograph of only approximately 25×25 mm.

The scanning oscillator technique (SOT) has all the advantages of the Lang technique, and in addition overcomes its shortcomings. It makes possible the recording of large-area topographs of entire crystals containing appreciable amounts of reversible elastic and/or frozen-in strains. Topographs of entire crystals 150 mm in diameter have been recorded. This technique is particularly suited for developing correlations between crystal imperfections and other properties of the crystal.

A possible geometry of the experimental set-up, based on the Lang technique, is seen in Fig. 1b. The line or point focus of a standard X-ray source is used to produce a ribbon-shaped beam. In addition, provisions have been made to oscillate crystal and film simultaneously while the crystal is being scanned (Fig. 1b). The angle of oscillation is chosen to cover the whole reflecting range of the crystal. Thus it becomes possible to record large-area topographs. A newer version of the SOT system is based on automatic Bragg angle control (van Mellaert and Schwuttke, 1972) and shown in Fig. 1c. Accordingly, the automatic Bragg angle control maintains the operation of the SOT system at all times at maximum X-ray intensity for exposing the photographic plate. The automatic Bragg angle control is accomplished by a continuous measurement of the angular derivative of the reflection curve. This measurement signal is applied to an automatic control system, which orients the crystal for operation at the peak of the reflection curve at all times while scanning. A significant reduction in exposure time and background scattering is thus obtained. Picture sharpness is also improved.

Topographs are recorded without magnification. Consequently, X-ray topographs are of very low resolution when compared to transmission electron micrographs. For evaluation, X-ray topographs are optically magnified. The useful magnification is limited in most cases to below 100 times. This limits X-ray topography to materials of low defect density and to single crystals of high perfection. On the other hand, this is one of the advantages of X-ray topography when compared to transmission electron microscopy, because it can supply a survey of large crystal areas.

III. Applications to Thin Films

In semiconductor technology, thin films are of great technological importance. Such films can vary in thickness from a few thousand angstroms to several microns and are used either for insulation or current conduction. The films are produced by different techniques, such as diffusion, epitaxy, vapor deposition, sputtering, and ion implantation. They are produced on semiconductor substrates, such as silicon, germanium, gallium arsenide, and others. The substrate crystals have a high degree of perfection, and this is the area where X-ray topography is most useful to the investigation of thin films. X-ray topography techniques provide information about defects in the film and in the substrate generated by the film deposition. In addition, X-ray topography is useful for the investigation of film adhesion to the substrate. It also provides information about local breakdown of film adhesion and generates information about tensile properties of films and how their properties change during semiconductor processing. Recently, X-ray topography has also been applied to the investigation of subsurface films as produced through ion bombardment and their annealing properties. Some specific examples demonstrating the application of X-ray topography to a variety of thin-film problems are discussed in the following sections.

A. SINGLE-CRYSTAL FILM–SUBSTRATE SYSTEMS

1. *Auto-Epitaxial Films*

Auto epitaxial film–substrate combinations are Si on Si, GaAs on GaAs, etc., and are primarily produced by epitaxy or diffusion. Dislocations and stacking faults are the most common types of defects produced in such films during processing. The defect generation in such films is primarily due to a mismatch in lattice parameters or to a mismatch in the coefficients of thermal expansion or both. Mismatch dislocations in epitaxial films on semiconductor substrates are readily investigated through X-ray topography and have been studied for a variety of systems (Schwuttke, 1965b; Meieran, 1967). Dislocation generation in epitaxial silicon deposited on silicon can occur for small differences in lattice constants, which in turn are the result of doping differences between epitaxial film and substrate. For systems like n-type film on n-type substrate or pp or even nn^+ or n^+n combinations, the lattice match is practically perfect. This can be different for np, pn, pp^+, or p^+ film–substrate combinations, such as phosphorus-doped substrates and boron-doped films or vice versa because the incorporation of phosphorus or boron into the silicon lattice leads to lattice contraction. Figure 2a is an example of the change of lattice constant of silicon with increasing boron doping. From these data a maximum lattice shrinkage of 0.2% can be estimated. If a low-resistivity boron-doped film is grown upon a high-resistivity substrate, dislocations are generated at the interface. The imperfections are straight lines oriented along [110] directions and vary in length from a few microns to several centimeters and are readily observed in X-ray topographs. A typical result is shown in Figs. 2b, c. Similar results have been reported for GaAs (Meieran, 1967).

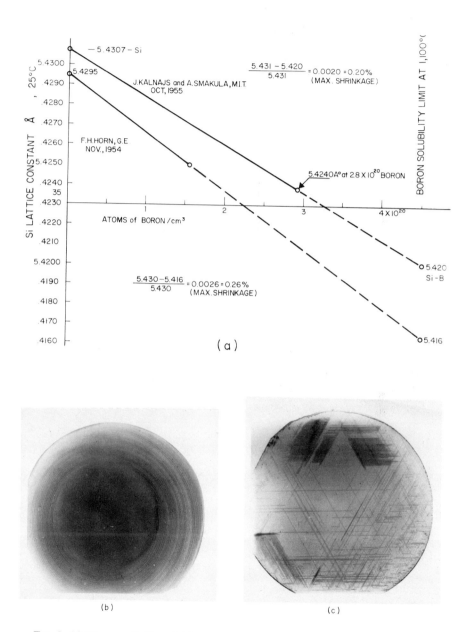

(a)

(b)

(c)

FIG. 2. (a) Change of silicon lattice constant with increasing boron doping. (b) X-ray topograph of 0.01 Ω-cm substrate, 220-type reflection. (c) X-ray topograph of 10 Ω-cm epitaxial film grown on substrate shown in Fig. 2b.

Stacking faults in epitaxial silicon films are usually too small for direct observation through X-ray topography and may escape detection in X-ray topographs whenever they are smaller than 5–10 μm. Nevertheless, stacking faults in epitaxial silicon were first discovered and analyzed through X-ray diffraction topography (Schwuttke, 1962; Schwuttke, and Sils, 1963) and subsequently confirmed through transmission electron microscopy (Queisser 1961). An example is shown in Figs. 3a, b.

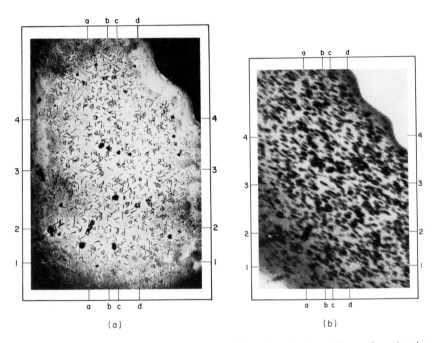

(a)   (b)

FIG. 3. (a) Optical photomicrograph of an (111) epitaxial silicon-film surface showing etch traces of stacking-fault structures. The crystal area shown is 3 × 4 mm. (b) X-ray topograph of film shown in Fig. 3a. Note stacking-fault contrast.

Dislocations in surface films produced by diffusion of boron or phosphorus into silicon have received a great deal of attention. Originally, such defects were theoretically predicted by Prussin (1961) and subsequently observed by X-ray topography (Schwuttke and Queisser, 1962). Numerous investigators have studied diffusion-induced dislocations by X-ray topography as well as other techniques including electron microscopy and optical microscopy (Julett and Lapierre, 1966; Ewing and Smith, 1968; Jaccodine, 1964; Joshimatsu and Kohra, 1968; Yoshida et al., 1968; Washburn et al., 1962; Lander et al., 1963; Blech et al., 1965). An example of diffusion-induced dislocations due to the diffusion of high concentration of boron into an n-type substrate is shown in Fig. 4. In this topograph the dislocations are

FIG. 4. Reflection X-ray topograph of boron-diffused silicon. Note diffusion-induced dislocation pattern.

quite well resolved. Unresolved dislocations are shown in the topograph of Fig. 5a recorded after a localized phosphorus diffusion into silicon. The excess diffraction contrast in the diffused area is due to a high-density dislocation network, which is shown resolved in the transmission electron micrograph of Fig. 5b.

2. *Heteroepitaxial Films*

Heteroepitaxial single-crystal film–substrate combinations are also of great interest in semiconductor technology and have become the focus of many new device concepts. Epitaxial deposition of pseudobinary alloys such as Ga(As, P) onto GaAs substrates (Howard and Cox, 1966; Howard and Dobrott, 1966) and also GaAs on Ge (Meieran 1967), GaAs–spinel and GaP–sapphire (McFairlane and Wang, 1972) are easily investigated through

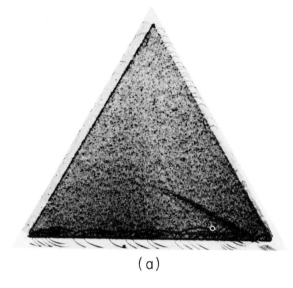

(a)

(b)

FIG. 5. (a) X-ray topograph of silicon wafer after localized phosphorus diffusion into triangular window. Note excess diffraction contrast inside window and emitter-edge dislocations around window periphery. (b) Transmission electron micrograph of phosphorus-diffused wafer similar to the one shown in Fig. 5a. Dislocation network in triangular diffused area (Fig. 5a) resolved.

X-ray topography. Interesting results have been obtained through the technique of compositional X-ray topography (Howard and Dobrott, 1966). This technique is an application of the scanning-reflection X-ray topographic method (Howard and Cox, 1966). The conceptual diagram of compositional X-ray topography for an idealized case of the GaAs–GaP heteroepitaxial system is shown in Fig. 6a. Accordingly, epitaxial layers A and B have been sequentially deposited on a GaAs substrate. Assuming that the A layer has the composition $GaAs_{1-x1}P_{x1}$ and the B layer has the composition $GaAs_{1-x2}P_{x2}$, then each layer and the substrate can be diffracted separately by utilizing the diffraction angle that corresponds to the lattice parameters (composition) of that layer. Thus separate topographs can be recorded for substrate and layers. This technique is useful for the investigation of compositional changes in sequentially deposited films and defects generated during the deposition in the films as well as for the investigation of strains in the films and the study of mismatch dislocations in the film interface. A sequence of compositional topographs for the system $GaAs$–$GaAs_{0.85}P_{0.15}$–$GaAs_{0.67}P_{0.33}$ is shown in Figs. 6b–d. The main defects in this system are hillocks that occurred during the film deposition. The topograph shown in Fig. 6d indicates that the hillocks have the same composition as the $GaAs_{0.85}P_{0.15}$ layer; consequently the growth defects were generated during the first phase of the film deposition and originate at the substrate–$GaAs_{0.85}P_{0.15}$ interface.

B. Nonsingle-Crystal–Substrate Systems

Other important film–substrate combinations in semiconductor technology are silicon oxide, silicon nitride, or metal film deposits primarily on silicon. The deposition of such films on single-crystal substrates leads to strain contrast visible in X-ray topographs of the substrates. Such artificially imposed elastic strains on crystals and their effects on X-ray topographs have been studied extensively (Schwuttke and Howard, 1968; Meieran and Blech, 1968). The results shown here are taken primarily from Schwuttke and Howard (1968).

Accordingly, the strain contrast is of particular interest along film edges or film discontinuities and is primarily activated by the mismatch in expansion coefficients. Since the thermal coefficient of expansion of silicon differs from its oxide, the substrate will contract more than the film upon cooling. Thus the slice shape is nearly parabolic and convex to the film surface. The compressive film stress generates a tensile stress located close to the substrate surface. The local stress is partially relieved along the sides of windows etched into the film; however, a strain gradient develops normal to the window edge or any other discontinuity in the film. The absence of diffraction contrast in topographs of the substrate recorded after removal of the film confirms the elastic nature of the strain. X-ray topography of strain fields in silicon due to the thin-film deposition has been extensively studied. Some examples are discussed in the following paragraphs.

1. Film–Substrate Adhesion

X-ray topography is an elegant technique for measuring the adhesion of thin films deposited on single-crystal substrates (Schwuttke and Howard,

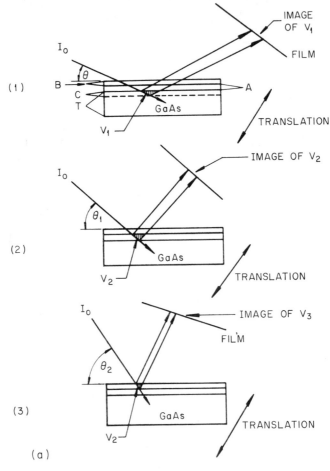

FIG. 6. (a) Conceptual diagram for compositional X-ray topography after Howard and Dobrott (1966). Each layer can be diffracted separately by utilizing the diffraction angle which corresponds to the lattice parameter (composition) of that layer. (b) The (440) topograph of GaAs substrate. Note hillocks visible as triangular regions of null contrast (courtesy of J. K. Howard). (c) The (440) topograph of the $GaAs_{0.67}P_{0.33}$ layer (courtesy of J. K. Howard). (d) the (440) topograph of the $GaAs_{0.85}P_{0.15}$ layer, only hillocks are in reflecting position (courtesy of J. K. Howard).

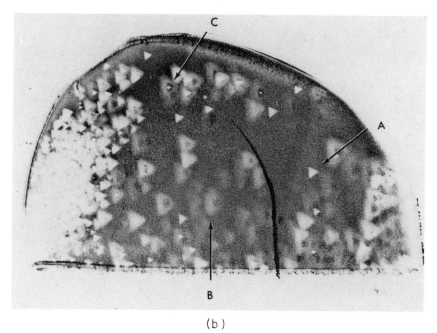

(b)

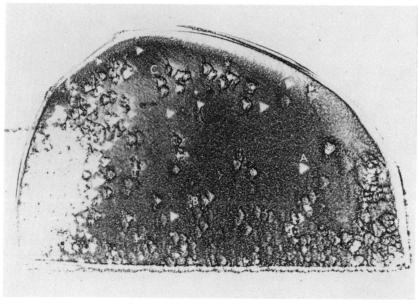

(c)

Fig. 6 (b,c).

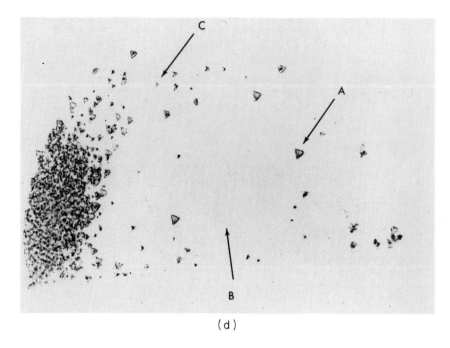

(d)

FIG. 6(d).

1968). Such measurements are possible and can be made nondestructively because any local loss of film adhesion results in a stress gradient across the good–no-film adhesion boundary line. Therefore, local failure of film adhesion is revealed in X-ray topographs through excess diffraction contrast. An example of such a measurement is seen in Figs. 7a, b. It represents the (220) topograph of a 1000-Å-thick silver-film deposit on a 0.25-mm-thick silicon substrate of 40-mm cross section. The optical appearance of the silicon film is mirrorlike and without any flaws. Its topograph was recorded with MoKα-radiation ($\mu_o t < 1$); consequently, the image contrast is strictly kinematical (Schwuttke and Howard, 1968). The circular good–no-film adhesion boundaries appear as two half-circles in the X-ray topograph (see Fig. 7b). This is in agreement with the contrast criterion (Schwuttke and Howard, 1968). The black spots in the topograph indicate areas of disturbed adhesion. A kinematical image is not influenced by the sign of the deformation; therefore, the image contrast in Figs. 7a, b could, for instance, indicate good adhesion inside the two half-circles and bad adhesion outside the half-circles, or vice versa. The adhesive-tape "strip test" was used on the silver film (Figs. 7) to identify the diffraction contrast as areas of no adhesion.

Film-adhesion measurements can be interpreted unambiguously when the topograph displays dynamical contrast (Schwuttke and Howard, 1968). In most cases, dynamical images can be obtained either by adjusting the substrate thickness or by choosing the radiation according to $\mu_0 t \geqslant 1$. For silicon, germanium, gallium arsenide, and other semiconductors, molybdenum, copper, or chromium radiation is a good choice, depending on the substrate thickness. An adhesion measurement that can be interpreted unambiguously is shown in Fig. 8. This topograph represents the $\bar{2}20$ recording of a molybdenum film on a 1-mm-thick silicon substrate of [111] orientation and of 40-mm diameter. The molybdenum film was prepared as being compressive. The optical appearance of the molybdenum film shown in Fig. 8 is mirrorlike. Interesting is its X-ray topographical appearance. The film contains strong buckling stresses over relatively large areas. These areas are marked with the letter b in Fig. 8. The compressive nature of the film is confirmed by its image contrast of the triangular window edges. An accumulation of black spots in the topograph (areas marked by the letter a) indicates adhesion loss. Note that the contrast morphology of these spots is identical to the contrast morphology of the window edges. The black spots are therefore "little windows," areas of no adhesion.

This technique was successfully applied to a variety of metal films on germanium and silicon substrate (Schwuttke and Howard, 1968). Adhesion measurements were made of silver, gold, rhodium, molybdenum, tungsten, nickel, and aluminum films. Silicon oxide films and silicon nitride films deposited under different conditions were also investigated through this method.

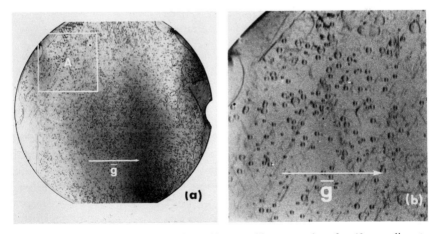

FIG. 7. X-ray topograph of a silver film on silicon-crystal wafer 40-mm diameter ($\mu_0 t \simeq 0.3$). (a) Low magnification showing entire wafer; (b) indicated part of (a) at higher magnification.

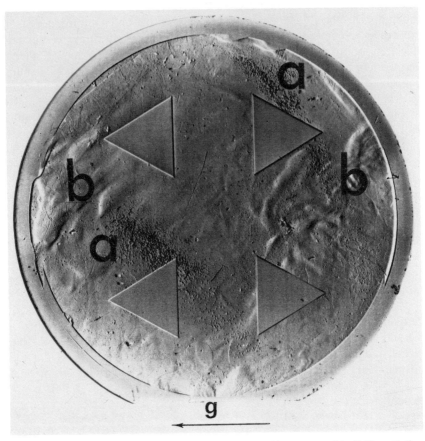

FIG. 8. X-ray topograph of molybdenum film on silicon, $\mu_0 t \simeq 14$, CuK$_\alpha$-radiation. (See text for discussion of labeling.)

The technique has also been applied to the study of electromigration and void formation in thin films on silicon (Howard and Ross, 1967).

It should be noted that adhesion measurements are best done through the scanning-oscillator technique (Schwuttke 1965a, van Mellaert and Schwuttke, 1972). Topographs of entire crystals under stain are easily obtained in that way. Topographs of 50-mm-diam crystal slices covered with films have been recorded in this way.

2. Stress Topography of Films

Surface stresses in substrates due to film deposition may have some other interesting consequences. For instance Si$_3$N$_4$ is frequently used in semiconductor technology as a masking material either alone or in conjunction with an intermittent layer of SiO_2. Interesting dislocation configurations are introduced into silicon wherever stress gradients are present, for instance, at diffusion-window edges. An example is shown in Fig. 9. This topograph shows dislocation generation at masking-window edges after 1200°C heat treatment of a Si wafer before diffusion (Westdorp and Schwuttke, 1969). The film composition on the wafer is as follows: Si_3N_4 film over 500 Å thermal SiO_2 on a 0.2-mm-thick [111] substrate. The wafer was annealed for 2 hr at 1200°C. For such a film combination the stress in the oxide film was reported to be compressive, while the stress in the silicon nitride film was reported to be tensile. The most probable source for the oxide stress is the difference in the thermal expansion coefficient of silicon and SiO_2. The tensile stresses in the silicon nitride film are intrinsic and are the result of the densification taking place at the deposition temperature of 1200°C. To account for the dislocation generation the following model was proposed: The maximum tensile stress σ_m generated in the silicon substrate due to Si_3N_4 film is $\sim 5 \times 10^7$ dyn cm^{-2}. This data is based on a 0.25-μm-thick nitride film deposited on a 900-μm-thick silicon wafer. The stress in the Si_3N_4 film is 1.2×10^{10} dyn cm^{-2}.

FIG. 9. X-ray topograph of silicon nitride film on silicon. Note dislocation generation around window edges.

Such a Si stress of 5×10^7 dyn cm^{-2} is unlikely to generate dislocations in the silicon because the empirical flow stress in silicon is 4×10^8 dyn cm^{-2} (Chaudhuri et al., 1962). This data is in agreement with the observation because perfect silicon nitride films, covering the silicon wafer completely, do not introduce dislocations into the substrate upon heat treatment. To form dislocations in the silicon, some stress concentration is necessary. Stress concentration occurs at Si_3N_4–Si interfaces formed when the nitride film is removed. Examples of these interfaces are cracks, diffusion windows or stress gradients at wafer edges. To estimate the stresses at crack fronts propagating into silicon, the Griffith criterion can be used (Westdorp and Schwuttke, 1969). Experimentally, it is found that after heat treatment, cracks penetrate into the silicon as deep as 20 μm. The stresses at these crack tips could be as high as the critical stress necessary to propagate the cracks further. For the case of plane stress, we estimated the critical stress as follows: $\sigma_c = (2ES/\pi C)^{1/2} \simeq 10^9$ dyn/cm^2, for $E_{111} = 1.9 \times 10^{12}$ dyn/cm^2 (Young's modulus in the $\langle 111 \rangle$ direction); $S_{111} = 1230$ erg/cm^2 (the surface tension of the crack); and $C = 20$ μm (the length of the crack), σ turns out to be about 10^9 dyn/cm^2. This is considerably higher than the flow stress of Si at the annealing temperature, which is about 10^8 dyn/cm^2. A large-scale relaxation of crack-tip stresses through dislocation nucleation thus appears reasonable.

The stress spikes at diffusion-window edges due to local relaxation of film–substrate constraints have not been calculated to date. It follows from our observations of dislocation-loop nucleation due to heat treatment at windows in Si_3N_4 and Si_3N_4–SiO_2 masks, that these stresses at the substrate level must be larger than 10^8 dyn/cm^2 at temperature. The discreteness of observed dislocation-loop families along diffusion-window edges is noteworthy (Fig. 9). It is reasonable to assume that the discrete-loop families generated along the window edges are caused by discrete microcracks introduced along the edges. Such microcracks may arise from local fracture of the thin Si_3N_4 membrane within the diffusion windows during the final stages of etching or from the causes enumerated for the continuous thick Si_3N_4 films. The microcracks are activated during annealing and relax their tip stresses through dislocation-loop nucleation. This model is also in agreement with the observation that such dislocation phenomena are caused through tensile Si_3N_4 films only and not through compressive Si_3N_4 films. Compressive films on substrates result in buckling stresses pointing away from the substrate; hence introduction of microcracks into the substrate is less likely.

3. Stress Jumping across Film–Substrate Interfaces

Stress measurements in films through X-ray topography have provided interesting information about "stress jumping." This effect can occur during

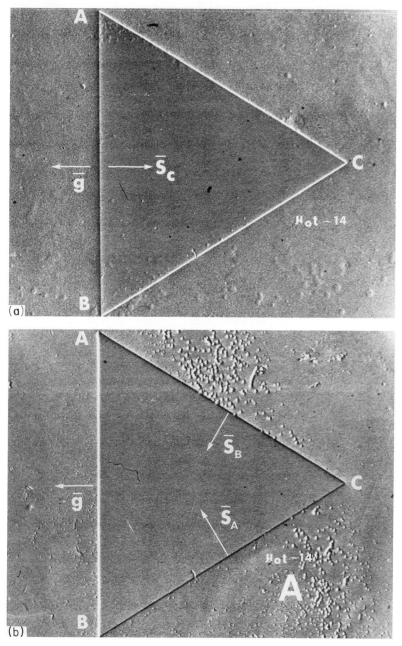

FIG. 10. X-ray topograph of molybdenum windows on silicon, $\mu_0 t = 14$. (a) Compressive film; (b) tensile film.

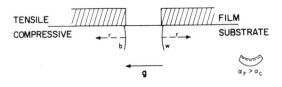

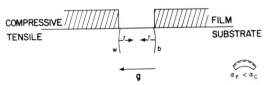

FIG. 11. Lattice curvature and stress relations due to film deposition on silicon.

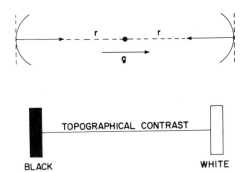

FIG. 12. Lattice curvature and topographical contrast due to film deposition in silicon.

semiconductor processing as a result of oxidation, diffusion, and metalization. These operations can change the magnitude of the film stress as well as the stress direction (sign of stress) in the film or in the diffused layer. The mechanism of stress jumping is activated whenever the stress across a film–substrate interface jumps from tensile to compressive or vice versa. This event was shown to have considerable influence on the defect state in the crystal as well as on the perfection of the film and was shown to be of great importance to the final semiconductor-device performance (Schwuttke, 1970). Through the technique of stress topography it is actually possible to obtain information about the relative stress magnitude and stress direction in planar semiconductor devices after each processing step. The technique exploits dynamical diffraction phenomena that occur along film–crystal interfaces (Schwuttke and Howard, 1968). Such diffraction phenomena are shown in the topographs of Figs. 10a,b for a set of molybdenum films deposited on silicon. One film (Fig. 10a) was prepared to exhibit tensile stress and the other film was prepared to exhibit compressive stress (Fig. 10b).

Note that the contrast at each interface edge is identically reversed for the windows cut into the tensile and compressive films. The topographical contrast displayed by the window edges is related to the sign of the film stress, which determines the curvature of the diffracting lattice planes. The lattice curvature–stress relations are shown schematically in Fig. 11 and the relationship between lattice curvature and topographical contrast is given in Fig. 12. Stress jumping in planar device structures can increase the stress in the substrate to well above the yield strength of the material. Consequently dislocations can form along the window edges. A most remarkable manifestation of stress jumping is the formation of emitter-edge dislocations in silicon. Stress topography has shown that the strain in a thermally grown silicon oxide on silicon is compressive (Fairfield and Schwuttke, 1968). Normally, phosphorus diffusion into silicon results in a tensile-strain effect around the window periphery. This is according to expectation, because a substitutional phosphorus atom in the silicon lattice has a smaller ionic radius than the silicon covalent radius, and consequently it must stretch the neighboring bonds. This strain acts in the same direction as the strain imposed on the silicon by the oxide window. However, under certain conditions a phosphorus-diffusion process may result in a compressive rather than a tensile-strain

Fig. 13. X-ray topograph of emitter-edge dislocations around edges of diffusion window.

effect around the window edge. This anomalous strain leads to stress jumping and is thus responsible for the generation of emitter-edge (EE) dislocations (Fig. 13). The influence of EE dislocations on device gain β has been shown to be considerable (Fairfield and Schwuttke, 1968).

C. Subsurface Films

Another interesting application of X-ray topography to the study of thin films is X-ray interference topography. This technique allows a study of subsurface amorphous films. Such films are produced through high-energy ion implantion into single crystals. If high-energy ions are implanted, for instance, into silicon a bicrystal is produced (Bonse *et al.*, 1969) as shown in Fig. 14. This bicrystal consists of a perfect bulk crystal and a perfect layer crystal. Both crystals are separated by a damage zone, which can be completely amorphous and lens-shaped. The crystals have the same crystallographic orientation and lattice constant but are separated by a small rigid-body displacement. Thus the "feat" of holding two crystals in "physical contact" with interferometric sensitivity is easily achieved. If X rays are diffracted from such a crystal it acts like a two-component interferometer. This is shown in Figs. 15a and 15b, which show, respectively, a wafer before ion implantation and a similar wafer 1-MeV C^+ implantation bombarded to a fluence of

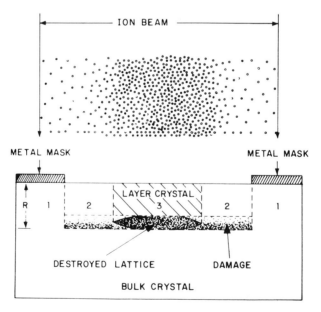

Fig. 14. Model of crystal after high-energy bombardment.

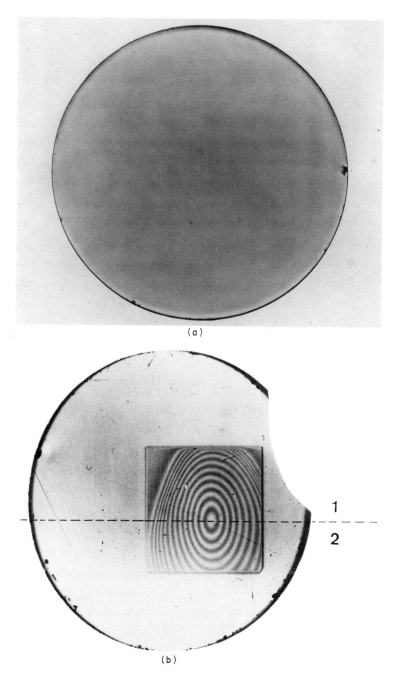

FIG. 15. (a) X-ray topograph of silicon wafer before high-energy ion implantation. X-ray topograph of silicon wafer after 1-MeV C^+ implantation. Note X-ray interference fringes.

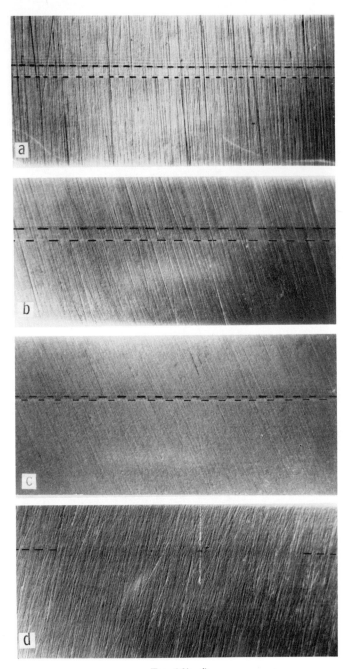

Fig. 16(a–d).

FIG. 16. Sequence of optical pictures of annealing cycles showing change of damage width in silicon after 1-MeV C$^+$ bombardment. (See text for discussion of figure parts.)

10^{16} C$^+$ ions/cm^2. X-ray interference fringes are clearly visible. Such fringes have some unique properties:

(1)   The number of fringes depends on the irradiation dose.

(2)   The number of fringes depends on the Bragg reflection.

(3)   The fringes are curves of constant $h \cdot c$ and thus contour lines of the displacement c over the beam cross section.

(4)   The variation δc per fringe obeys the equation $|\delta c| = (h \cdot z)^{-1}$.

(5)   The maximum displacement is equal to the total number of fringes times δc.

Such a monolithic two-crystal interferometer allows some interesting applications. Specifically, it can be used to study the annealing of the subsurface amorphous film (Brack *et al.*, 1973). An example is shown in Fig. 15b. This figure represents the interferometer topograph of a silicon wafer after 1-MeV C$^+$ implantation. One counts 13 interference fringes indicating a total displacement of ~ 70 Å due to a volume expansion of the damage zone. The interference fringes are very sensitive to dimensional changes of the subsurface film. Therefore, they can be utilized to study the annealing of the amorphous film.

The optical picture of the defect zone in the crystal shown in Fig. 15b after C$^+$ bombardment is shown in Fig. 16a. The subsurface damage layer is clearly visible on the bevel at a depth of 1.5 μm below the surface and identified by electron diffraction as amorphous silicon (Brack *et al.*, 1973). Changes in width due to annealing of the amorphous film are readily measured on such a bevel. Such annealing results are summarized in Figs. 16a–e. Accordingly, the damage width in the sample after bombardment is 3000 Å (Fig. 16a) and shows a small change after 500°C anneal (Fig. 16b). However, a considerable amount of shrinkage is apparent after the 600°C anneal

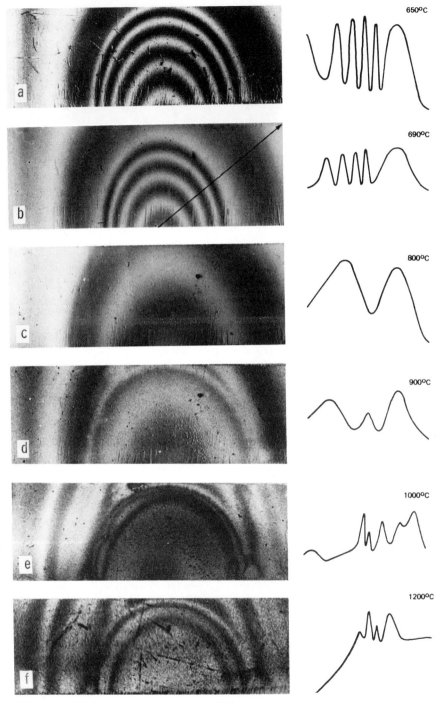

(Fig. 16c). The damage zone is not visible any more after 700°C anneal and appears again after 1200°C anneal. This is shown in Figs. 16d and 16e. All these changes are readily followed nondestructively with the two-crystal interferometer. A sequence of X-ray interference topographs recorded at different annealing cycles is shown in Fig. 17. The interference topographs were obtained from the crystal half labeled 1 in Fig. 15b. The other half, marked 2, was used for the bevel experiments summarized in Fig. 16. A quantitative evaluation of the X-ray interference patterns shown in Fig. 17 leads to the following conclusions: The amorphous-silicon film occupies a volume approximately 2% larger than single-crystal silicon. The volume change follows directly from the number of interference fringes counted after bombardment. As shown previously (Bonse et al., 1969), one fringe corresponds to 5.4 Å [111 reflection, (100) silicon, MoK α_1-radiation]. Therefore, 13 fringes indicate an expansion of ~ 70 Å. This is approximately 2.2% of the total damage zone measured to be approximately 3000 Å wide after implantation. During high-temperature annealing the width of the amorphous zone shrinks due to its crystallization to quasi-crystalline silicon, leading to a volume reduction; consequently, the number of interference fringes in the sample under discussion reduces from 13 to 2 after 800°C anneal. Higher annealing results again in a volume expansion, this time a direct result of silicon carbide formation, which is concluded after 1200°C annealing. These conclusions are supported by electron microscopy (Brack et al., 1973).

References

Blech, I. A., Meieran, E. S., and Sello, H. (1965). *Appl. Phys. Lett.* 7, 176.
Bonse, U. (1958). *Z. Phys.* 153, 278.
Bonse, U., Hart, M., and Schwuttke, G. H. (1969). *Phys. Status Solidi* 33, 361.
Brack, K., Gorey, E., and Schwuttke, G. H. (1973). *Crystal Lattice Defects* 4, 109.
Chaudhuri, A. R., Patel, J. R., and Rubin, L. G. (1962). *J. Appl Phys.* 33, 2736.
Ewing, R. R., and Smith, D. K. (1968). *J. Appl. Phys.* 39, 943.
Fairfield, J. M., and Schwuttke, G. H. (1968). *J. Electrochem. Soc.* 115, 415.
Guinier, A., and Tennevin, J. (1949). *Acta Cryst.* 2, 133.
Howard, J. K., and Cox, R. H. (1966). *Advan. X-ray Anal.* 9, 35.
Howard, J. K., and Dobrott, R. D. (1966). *J. Electrochem. Soc.* 113, 567.
Howard, J. K., and Ross, R. (1967). *Appl. Phys. Lett. Vol. II* 3, 85.
Jaccodine, R. J. (1964). *Appl. Phys. Lett.* 4, 114.
Juleff, E. M., and Lapierre, A. G. (1966). *Int. J. Electron.* 20, 273.

FIG. 17. Sequence of X-ray interferometer topographs showing annealing of damage due to 1-MeV C^+ bombardment.

Lander, J. J., Schreiver, H., Buck, T. M., Jr., and Matthews, J. M. (1963). *Appl. Phys. Lett.* **3**, 114.

Lang, A. R. (1958). *J. Appl. Phys.* **29**, 597.

McFairlane, S. H., III, and Wang, C. C. (1972). *J. Appl. Phys.* **43**, 1724.

Meieran, E. S. (1967). *J. Electrochem. Soc.* **114**, 292.

Meieran, E. S., and Blech, I. A. (1968). *Phys. Status Solidi* **29**, 653.

Newkirk, J. (1959). *Trans. A.I.M.E.* **215**, 483.

Queisser, H. J. (1961). *J. Appl. Phys.* **32**, 1776.

Prussin, S. (1961). *J. Appl. Phys.* **32**, 1876.

Schulz, L. G. (1954). *J. Metals* **200**, 1082.

Schwuttke, G. H. (1962). *J. Appl. Phys.* **33**, 1540.

Schwuttke, G. H. (1965a). *J. Appl. Phys.* **36**, 2712.

Schwuttke, G. H. (1965b). *Trans. Int. Vacuum Congr. 3rd* **2**, 301. Pergamon, Oxford.

Schwuttke, G. H. (1966). *J. Appl. Phys.* **37**, 2862.

Schwuttke, G. H. (1970). *Microelectron. Reliabil.* **9**, 397.

Schwuttke, G. H., and Howard, J. K. (1968). *J. Appl. Phys.* **39**, 1581.

Schwuttke, G. H., and Queisser, H. J. (1962). *J. Appl. Phys.* **33**, 1540.

Schwuttke, G. H., and Sils, V. (1963). *J. Appl. Phys.* **34**, 3127.

Tanner, B. K. (1972). *Phys. Status Solidi(a)* **10**, 381.

van Mellaert, L., and Schwuttke, G. H. (1972). *J. Appl. Phys.* **43**, 687.

Washburn, J., Thomas, G., and Queisser, H. J. (1962). *J. Appl. Phys.* **33**, 1540.

Weissmann, S. (1956). *J. Appl. Phys.* **27**, 389.

Westdorp, W. A., and Schwuttke (1969). "Thin Film Dielectrics" (F. Vratny, ed.), p. 546 Electron. Chem. Soc.

Yoshida, H., Arata, H., and Terunuma, Y. (1968). *J. Appl. Phys.* **7**, 209.

3.3 Low-Energy Electron Diffraction and Auger-Electron Spectroscopy

F. Jona and J. A. Strozier, Jr.

Department of Materials Science
State University of New York
Stony Brook, New York

I. The Problem

Perhaps the greatest obstacle on the way toward complete understanding of the epitaxial phenomenon is a widespread uncertainty about the details of the atomistic processes involved. While it is generally accepted that epitaxy occurs by way of nucleation first and postnucleation growth second, there is still no clear criterion that allows prediction of whether in a given system epitaxy will take place at all, in particular, whether the nuclei will be crystalline and with what orientation. In trying to establish such a criterion one would like to know precisely what happens when very few atoms of a given substance are deposited onto a well-defined and clean substrate surface at a given temperature. Assuming for a moment that the atoms remain on the surface or in its general vicinity (as contrasted to being evaporated away or diffusing deeply into the interior of the substrate), what one would like to know is whether they are distributed at random or with some kind of periodic order. If one has in mind the epitaxial growth of a film of the substance whose atoms are being deposited, the first question that one would like to ask, in fact, is: Must the first layer of atoms adsorbed on the substrate be ordered, or can epitaxy still take place even if the first layer is disordered? Intuition and, to a certain limited extent, the experiment seem to point toward an affirmative answer to the first part of the question, but there are no compelling reasons for denying the second part of the question a priori in all cases. Whatever the answer, the next question is likely to be: What controls the ordering of the first layer of adsorbed atoms? Immediately a number of other questions arise in the process of attempting an answer to this question. Namely: Is the dominant factor the mobility of the adsorbed atoms on the substrate surface? How is this mobility related to the usually unwanted diffusion of the adatoms into the substrate interior? Is it necessary for ordering that the substrate surface possess relatively strong adsorption sites with topographical distribution compatible with the interactions among adatoms? Is it necessary, and to what extent, that the "net" parameters of the substrate surface and of the would-be film be matched? And once a first layer of sorts has been established, what then controls the growth of a crystalline lattice *over* the first layer?

These questions, and others of similar nature, have motivated the work of many researchers in the field of epitaxy ever since the phenomenon was discovered. While it is true that, as we implied above, no complete and satisfactory answers have yet been given to most of these questions, it is fair to say that only recently has experimentation advanced to the point where substrate surfaces can be cleaned, kept clean, and characterized well enough for reproducible work. In addition, old techniques have been perfected and new techniques have been developed that allow meaningful insights into some of the questions listed above. Two such techniques are of concern to us here, namely, low-energy electron diffraction (LEED) and Auger-electron spectroscopy (AES). LEED gives information about the extent and nature of periodic order on surfaces and may eventually unveil the atomic and electronic structure of surfaces and overlayers. AES gives information about the types and, within certain limits, the concentrations of foreign atoms located at or in the immediate vicinity of a surface. In the following, we describe briefly the nature of the LEED and AES experiments and we discuss what in principle they can do for the scientist interested in epitaxy as well as what they have contributed in this field so far.

II. The LEED Experiment

The range of energies encompassed by the electrons used in LEED experiments extends from a few electron volts to, say, 500 eV, although most investigations make use of electrons with energies centered around 100 eV. Such electrons have very shallow penetration into solids, between two and five atomic layers, approximately, and are for this reason very suitable for the study of surfaces and surface phenomena. Being an experiment based on diffraction, LEED is sensitive to, and can give most information about, *order* (as contrasted to disorder) in the atomic arrangement on solid surfaces. Owing to the finite coherence width of the electron beams that are normally used in present-day LEED experiments, the lateral extent of ordered regions that can be examined with maximum resolution is approximately 100–200 Å. Of course, many such ordered regions (equivalent to, say, one-tenth of the area interrogated by the beam) are needed to produce an acceptable signal-to-noise ratio. Surface defects such as dislocations, steps, kink sites, etc., which do not in general occur in ordered arrays, cannot be resolved by LEED. The technique has therefore some rather disturbing limitations but seems nevertheless to be the only one capable of providing information about ordered atomic arrangements over and in the surface layer of crystalline solids. Unfortunately, it has not yet been developed to the point where it can be used as a routine tool for surface-structure analysis, in contrast, for example, to X-ray crystallography, but it can already provide very valuable information about such surface processes as, for example, epitaxy.

Let us briefly consider the main stages of the epitaxy process, as we understand it, and the extent of the contributions that LEED can make to their study. When we begin deposition of a given species of atoms on a given substrate surface we expect a certain number of such atoms to stick to (i.e., be adsorbed by) the surface and eventually to congregate into groups of atoms that will constitute the nuclei of the film to be grown. In this so-called "nucleation stage" the concentration of adsorbed atoms on the surface is often high enough to be detected by Auger spectroscopy but the nuclei are too small and too scarce to produce noticeable changes in the LEED pattern of the substrate surface. Later in the deposition process these nuclei may have grown to encompass a few hundred atoms, but if they are small in the directions parallel to the substrate surface, say, if they are roughly as high as they are wide, their effect on the LEED pattern will primarily be to cause an increase in background. If, on the other hand, the nuclei have grown predominantly in lateral dimensions, giving rise to what we may call "monolayer formation," then the changes in the LEED pattern may be significant. If the adsorbed layer is ordered, the LEED pattern may reveal a superstructure with periodicities determined by the combination of the nets of the adsorbed layer and the substrate surface (coincidence lattices or others, as discussed later in this chapter). If the adatoms are disordered, then LEED may only register a small and uniform increase in background. Figure 1 summarizes this situation in a schematic way.

In the intermediate stages of film growth, LEED may reveal the presence on the substrate surface of three-dimensional crystallites by displaying diffraction patterns of larger crystallite surfaces (parallel or inclined to the plane of the substrate surface) superimposed on the weakened pattern of the substrate. At this stage, information about the film–substrate interface is no longer available. If growth has occurred predominantly in the direction parallel to the first monolayer of adsorbed atoms, then some of the superstructure spots and some of the original substrate spots may still be visible, depending on the thickness of the grown film and the area of uncovered substrate surface, but the predominant features will be the LEED spots of the grown film surface. In the advanced stages of film growth, the latter spots will in fact be the only ones observable, and no information will be available about the substrate surface and the film–substrate interface. If the grown film is either amorphous or finely polycrystalline, then no LEED pattern can be observed.

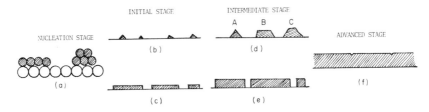

Fig. 1. (a) Clusters of 3 to 5 atoms randomly scattered on surface, generally not recognizable by LEED but noticeable by AES. (b) Clusters of 100 to 200 atoms, height approximately equal to lateral dimensions; LEED registers increase in background. (c) Monolayer formation, substrate surface visible between islands: (i) if adsorbed layer is disordered, LEED may register only increase in background; (ii) if adsorbed layer is ordered, LEED may reveal a superstructure (coincidence lattice or others). (d) Three-dimensional crystallites, substrate surface still visible; LEED reveals the presence of a deposit either through facets (A), diffraction spots from new grown surface (B), or high background (C). (e) Predominantly parallel growth, substrate surface hardly visible: (i) deposit amorphous, high LEED background, substrate spots almost obliterated; (ii) deposit crystalline, LEED may show weak superstructure spots and substrate spots, but the new surface spots are strong. (f) thick film, substrate surface no longer visible; if film is amorphous or thoroughly polycrystalline, LEED reveals only high, uniform background; if film is single crystalline, LEED reveals pattern of new surface alone.

It turns out, therefore, that LEED can be most profitably used in the initial and intermediate stages of film growth, provided, of course, that the film is single-crystalline. For this reason, LEED can be very useful in order to distinguish between monolayer formation and three-dimensional growth of nuclei and in order to determine the structure of the adsorbed layers relative to the substrate.

III. Auger-Electron Spectroscopy

Auger-electron spectroscopy (AES) is a nondestructive technique for chemical analysis with small depth resolution, which makes it particularly suited to the study of solid surfaces and thin films. The principle of the technique is as follows:

Suppose that a neutral atom is ionized in an inner shell S. (In most AES experiments the ionizing agent is a beam of electrons with energies of 2 to 3 keV that is made to impinge on the surface under study at near-grazing incidence.) The system consisting of both the ionized atom and the ejected electron (at rest at infinity) may be characterized by a positive energy E_S (measured relative to the energy of the neutral atom in its ground state). Now, a transition may occur in which the vacancy in shell S is filled by an electron from a higher level S', thereby liberating energy that may appear as radiation of frequency v, given by the equation $hv = E_S - E_{S'}$ (X rays, of course, are produced in this way). Alternatively, the transition may occur without emission of radiation: the liberated energy is communicated to another electron of the same atom and this electron is ejected from the atom. The effect is called the "Auger effect," and the ejected electron an "Auger electron" (from the work of the French physicist P. Auger). Calling E_i the ionization energy of the ejected Auger electron, we expect that its kinetic energy is given by $E_S - E_{S'} - E_i$. Thus, if we measure the energy of the Auger electron, we gain information about the electronic energy levels of the atom involved, and with this information we can identify the atom. AES consists therefore in first causing the emission of Auger electrons from a given surface and then carrying out an energy analysis of the emitted Auger electrons. The property that makes it particularly suitable for the study of surfaces is the very small escape depth of Auger electrons from any surface (only a few atomic layers).

It is clear, therefore, that AES gives only information about the chemical nature of the atoms on the surface investigated. It is a useful complement to LEED because it can tell us which atoms are in the surface layer, while LEED hopefully can tell us where they are located with respect to one anothr. The two techniques can make use of the same experimental arrangement in the vacuum chamber, and for this reason they are often used and mentioned together.[†] The quantitative aspects of the AES technique are less well known than its qualitative capabilities, but it is generally accepted that typical limits of detectability of impurities are of the order of 0.1% atomic concentration in the surface layers.

[†] The experimental arrangement for LEED is described in an article by Lander (1965) and that for AES in an article by Chang (1971a).

IV. Studies of Epitaxy by LEED

A. Scope of the Review

In view of the fact that studies of epitaxy by a variety of techniques are described in other chapters of this book, we are going to limit our discussion here to those cases and those systems in which epitaxy was established and examined by LEED, with or without the help of AES. By "epitaxy" we mean the *indefinite* growth of a macroscopic single-crystal film having a well-defined orientation with respect to the substrate. We exclude, then, discussion of the vast literature on chemisorption describing fractional and monolayer coverages of substrates, as observed by LEED, but reporting either unsuccessful attempts at epitaxy with prolonged deposition or no attempts at all. We will make some exceptions to this rule, however, because in some instances information as to why certain systems are *not* epitaxial can be just as useful as the information that they are. These restrictions greatly reduce the relevant literature, as we find that substantial use of the LEED technique for the study of epitaxy started only in the early 1960s.

We have cataloged the information gathered from the literature on various epitaxial systems in five groups, characterized by the type of overlayer material and the type of substrate used: (1) metals on metallic substrates; (2) metals on covalent substrates; (3) metals on ionic substrates; (4) covalent materials on metallic substrates; and (5) covalent materials on covalent substrates. This information is summarized in Table I and is discussed in some detail below.

B. Overlayer Structures

In a successful epitaxial experiment LEED will reveal at the beginning the pattern of the substrate surface and at the end the pattern of the grown film surface. In the intermediate stages LEED may show a variety of patterns that reflect the structural situation, i.e., the atomic arrangement, in the deposited overlayer or overlayers. These patterns may be quite complicated, not only and not always because the structures of the intermediate stages may in fact be complex but often also because there are strong interactions between the diffraction effects caused by the overlayer and those caused by the substrate structure. The interpretation of these patterns is often doubtful and will remain so until a complete solution of the problem of surface-structure analysis will be available. Nevertheless, it may be worthwhile, at this point in time, to classify the overlayer structures in two groups, which we may call "bulk" and "intermediate" structures.

In some cases, the first layer of adsorbed atoms already has the same structure as the corresponding plane of atoms in the macroscopic film to be grown. In these cases, we call the structure of the overlayer the "bulk" structure.

In other cases, the first few layers of deposited material have structures different from both the structure of the substrate surface and that of the macroscopic epitaxial film. These structures of the first few layers, which we call "intermediate structures," represent therefore a transition region, at the interface, between substrate surface and epitaxial film.

C. THE COINCIDENCE LATTICE

Coherent diffraction effects caused by the overlayer structure and the substrate structure may give rise to "coincidence lattices" defined as follows. Call a_{surf} the parameter of the substrate–surface net alone in a given direction parallel to the surface and a_{over} the parameter of the overlayer net alone in the same direction. If conditions are such that $na_{surf} = ma_{over}$, where n and m are integers not having any factors in common, then the translational symmetry of the overlayer–substrate combination has in fact the periodicity na_{surf}. This phenomenon causes the appearance of so-called $(1/n)$th order beams in the diffraction pattern of the substrate surface.

In actual cases that are believed to have been observed, n may vary approximately between 1 and 20, and thus the resulting LEED pattern is rather complicated, as it may show as many as $n-1$ spots between any two successive spots of the substrate pattern. It is worth noting, therefore, that the resulting LEED pattern is *not* the same as a simple superposition, i.e., the sum, of the two-dimensional reciprocal lattices of the substrate surface and the overlayer taken separately. The combination of overlayer and substrate nets is called a *coincidence lattice* because coincidences of the periodicities occur after an integral number of periods of the substrate net.

Despite the complexities of the associated LEED patterns, a coincidence lattice may represent a very simple physical situation, namely that in which, say, a monolayer of the adsorbate has the "bulk" structure and coincidences occur with the structure of the substrate surface. This is obviously the simplest of all epitaxial systems, namely, one in which the structure of the epitaxial film begins abruptly at the interface, at least in the first approximation, without transition regions. In a better approximation, a coincidence lattice may entail slight adjustments of the lattice parameters of the overlayer and/or substrate surface. It may also represent the result of mismatch between these parameters by amounts smaller than the experimental resolution of the LEED equipment (which is of the order of 0.2 Å).

D. DISCUSSION AND INTERPRETATION OF EXPERIMENTAL RESULTS

1. *Metals on Metallic Substrates*

Close-packed metals deposited on low-index planes of metallic substrates tend to form overlayers with close-packed structures and with periodicities practically identical to those of the bulk film material. In the initial stages of

TABLE I

PARTIAL LIST OF EPITAXIAL SYSTEMS STUDIED BY LEED[a]

Substrate	Overlayer	Structures observed	References
Al(100)	Si	$> \frac{1}{3}$L, disordered, RT	Jona (1971)
Al(110)	Si	$> \frac{1}{2}$L, 8×2, CL, RT	Jona (1971)
		$>$ 2–4L, disordered, RT	
Al(111)	Si	1L, 3×3, CL, RT	Jona (1971)
		$>$ 4–8L, disordered, RT	
$Al_2O_3(\bar{1}012)$	Si	Polycryst, $T <$ 600°C	Chang (1971b)
		Si(100) 2×1, Epi, $T >$ 650°C, film thickness $\simeq$ 1000 Å	
$Al_2O_3(11\bar{2}3)$	Si	Polycryst, $T <$ 600°C	
		Si(111)–7, Epi, $T >$ 700°C, film thickness $\simeq$ 1000 Å	Chang (1971b)
$Al_2O_3(0001)$	Si	Polycryst, $T <$ 600°C	
		Si(111)–7, Epi, $T >$ 725°C, film thickness $\simeq$ 1000 Å	Chang (1971b)
$Al_2O_3(11\bar{2}0)$	Si	Polycryst, $T <$ 600°C	
		Si(111)–7, Epi, $T >$ 750°C, film thickness $\simeq$ 1000 Å	Chang (1971b)
Be(0001)	Si	$> \frac{1}{3}$L, $\sqrt{3} \times \sqrt{3}$–30°, IS, RT	Jona (1973b)
		$>$ 1L, disordered, RT	
Cu(100)	Ag	1L, $c(10 \times 2)$, CL, $T = -195$–200°C	Jackson and
		$>$ 1L, Ag(111), Epi, $T = -195$–200°C	Hooker (1971); Palmberg et al. (1968); Bauer (1967)
	Au	1L, $c(14 \times 2)$, CL, $T < -50$°C	Jackson and
		$>$ 1L, Au(111), Epi, $T < -50$°C	Hooker (1971)
	Au	$c(2 \times 2)$, alloy, $>$ RT	Palmberg et al.
		Au(111), Epi	(1968a)
	Pb	$c(4 \times 4)$, "pseudo (100)", CL, $T \leqslant 400$°C	Henrion and
		$c(5 \times 1)$–45°, "pseudo (111)", CL, "dense monolayer"	Rhead (1972)
		Pb(100), Epi, Pb crystallites with "dense monolayer"	
	Te	$<$ 0.25L, $c(2 \times 2)$, RT	Andersson and
		0.3L, 6×6, CL, RT	Andersson (1968)
		$>$ 2L, Te crystallites $\approx$ 30° wrt substrate surface	

[a] The entries are listed in alphabetical order of the chemical symbol of the substrate surface [CL: coincidence lattice; Epi: epitaxy; IS: intermediate structure; L: monolayer equivalents (exposure in Langmuirs); PM: pseudomorphism; RT: room temperature; T: temperature]. The notation used is explained in the relevant text sections.

TABLE I (*continued*)

Substrate	Overlayer	Structures observed	References
Cu(110)	Pb	$c(2 \times 2)$, CL, $T \leqslant 400°C$ $p(5 \times 1)$, "pseudo (100)", CL, "dense monolayer"	Henrion and Rhead (1972)
Cu(111)	Ag	$\sim 1L$, CL $> 1L$, Ag(111), Epi	Bauer (1967); Bauer and Poppa (1972)
	Ni	1–5L, Ni(111), Epi, $T \simeq 200°C$	Hague and Farnsworth (1966)
	Pb	$p(4 \times 4)$, "pseudo (111)", CL, $T \simeq 400°C$ Pb(111), Epi, $T \simeq 400°C$	Henrion and Rhead (1972)
	Te	0.1L, $2\sqrt{3} \times 2\sqrt{3}$–30°, IS, RT 0.3L, $\sqrt{3} \times \sqrt{3}$–30°, IS (some disorder), RT–3000°C $> 1L$, Te(0001), Epi (surface reconstructed 2×1), RT	Andersson et al. (1968)
Ir(100)1×1	Xe	$\sim 1.5L$, Xe(111), Epi, $T \simeq 56°K$	Ignatiev et al. (1972)
Ir(100)1×5	Xe	$\sim 0.2L$, 3×5, CL, $T \simeq 56°K$ $\sim 1.5L$, Xe(111), Epi, $T \simeq 56°K$	
KCl(100)	Ag	Irradiate substrate with electrons (100 eV) $\sim 1L$, nuclei, $T < 80°C$ $> 1L$, Ag(100), Epi, $T \simeq 75°–350°C$	Palmberg et al. (1967a, 1968)
	Au	Au(100)–5, Epi, $T \geqslant 250°C$	
	Cu	Irradiate substrate with electrons (100 eV) $\sim 1L$, nuclei, $T < 80°C$ $> 1L$, Cu(100), Epi, $T \simeq 75–350°C$	Palmberg et al. (1968)
MgO(100)	Ag	Ag(100), Epi, $T \simeq $ RT–200°C	Palmberg et al. (1967b)
	Au	Au(100)–5, Epi, $T \simeq 100–200°C$	
Mo(110)	Al	$c(2 \times 2)$, $T \simeq 650°C$ after deposition Al(111) stressed, RT? Al(111), Epi, RT?	Jackson and Hooker (1971)
Nb(100)	Ar	$\sim 20L$, Ar(111), Epi, $T \simeq 20°K$	Dickey et al. (1970)
	Xe	$< 2L$, 7×7–45°, CL, $T = 7°K$ Anneal of the 7×7–45° structure yields: $> 2L$, Xe(111), Epi	Farrell et al. (1972)
Ni(100)	Na	$< 0.5L$, diffuse rings, Na uniformly spaced, RT 0.5–1L, $c(2 \times 2)$, IS, RT $> 1L$, disordered	Anderson and Pendry (1972); Gerlach and Rhodin (1969)
	Si	$\sim 1L$, $c(2 \times 2)$, $T \geqslant 300°C$ $> 1L$, disordered, $T \geqslant 300°C$	Jona (1973a)

TABLE I (*continued*)

Substrate	Overlayer	Structures observed	References
Ni(110)	Cs	Similar to Na/Ni(110)	Gerlach and
	K	Similar to Na/Ni(110)	Rhodin (1969)
	Na	$< \frac{1}{4}$L, complex coherent surface structures, $T = 200°C$	
		$> \frac{1}{4}$L, one-dimensional incoherent structures, $T = 200°C$	
		> 2L, disordered, $T = 200°C$	
Ni(110)	Si	~ 1L, $c(8 \times 2)$, CL, $T \gtrsim 300°C$	Jona (1973a)
		> 1L, disordered, $T \gtrsim 300°C$	
Ni(111)	Ag	> 0.1L, 6×1(?), CL, $T \simeq 150°C$	Feinstein *et al.*
		> 1L, Ag(111), Epi, $T \simeq 150°C$	(1970)
	Na	< 0.3L, diffuse rings Na uniformly spaced, RT	Gerlach and Rhodin (1969)
		0.3–0.5L, "incoherent close-packed hexagonal" structure, RT	
		> 2L, disordered, RT	
	NiO	NiO(111), Epi, $T = 200-300°C$, $O_2 \simeq 10^{-7}$ Torr	MacRae (1963)
	Si	$\gtrsim \frac{1}{3}$L, $\sqrt{3} \times \sqrt{3} - 30°$, IS, $T \gtrsim 300°C$	Jona (1973a)
		$\lesssim 2$L, 2×2, $T \gtrsim 300°C$	
		~ 2L, 2×12, CL, $T \gtrsim 300°C$	
		> 2L, disordered, RT	
PbS(100)	Ag	Ag(100), Epi, $T \simeq 200°C$	Green *et al.* (1971)
		Ag(111), Epi, RT	
	Au	Au(110), Epi, 3D crystallites, $T \simeq 200°C$	
Pd(100)	Ag	1L, 1×1, PM, $T = -50°C$	Palmberg and
		2L, 1×8, CL, $T = -70°C$	Rhodes (1968)
		4L, Ag(111), Epi ("fuzzy" disorder)	
	Au	1L, 1×1, PM, $T = -50°C$	
		~ 4L, 1×7, CL?, $T = -50°C$	
		15L, Au(100)–5, Epi, $T = -50°C$	
Si(001)	Si	Si(100), Epi, $T < 400°C$	Jona (1967)
Si(111)1×2	Cs	0.25L, complex CL, RT	Gobeli *et al.* (1966)
		0.3L, $\sqrt{3} \times \sqrt{3} - 30°$, IS, RT	
		1L, 1×1, RT	
		Cs (hcp), PM, RT	
Si(111)-7	Ag	0.3L, $\sqrt{3} \times \sqrt{3}-30°$, $T > 200°C$	Spiegel (1967)
		> 18L, Ag(111), Epi, $T \simeq 250°C$ ("delayed condensation")	
	Al	0.3L, α-Si(111)$\sqrt{3}$–Al, IS, $T > 500°C$	Lander and
		0.5L, β-Si(111)$\sqrt{3}$–Al, IS, RT	Morrison (1964)
		> 0.5L, Al(111), Epi, RT	

TABLE I (continued)

Substrate	Overlayer	Structures observed	References
Si(111)−7—cont.	Au	5×1, $T > 200°C$ $\sqrt{3} \times \sqrt{3}-30°$ 6×6 Au(111), Epi(?), ("delayed condensation")	Bauer and Poppa (1972)
	Pb	Pb(111), Epi, RT	Estrup and Morrison (1964)
	Si	Si(111)–7, Epi, $T \geqslant 400°C$	Jona (1967); Thomas and Francombe (1967, 1971)
	Sn	$> 1L$, disordered, RT	Estrup and Morrison (1964)
Ta(100)	Th	0.3L, $c(2 \times 2)$, islands, $T \simeq 950°C$ $\sim 1L$, 1×1, PM, $T \simeq 950°C$ 1–20L, Th(?), PM plus small islands, $T \simeq 950°C$	Pollard and Danforth (1968)
Ta(110)	Al	$L > 10$, Al(111), Epi, $T \simeq 600–670°C$	Jackson et al. (1967)
Ti(0001)	Cu	$> 1L$, Cu(111), Epi, $T < 100°C$ ($> 8L$ for complete coverage)	Schlier and Farnsworth (1958)
W(100)	Cs	1L, $c(2 \times 2)$, RT 1.5L, $p(2 \times 2)$, RT 2L, Cs close-packed hexagonal layer, RT	MacRae et al. (1969)
	Pd	Epi(?). orientation not specified	Gorodetskii and Yas'ko (1972)
	Th	(a) 0.5L, $c(2 \times 2)$, IS, $T \simeq 1400°C$ 1L, 1×1, PM, $T \simeq 1400°C$ 25L, Th(?), "duolayer" plus bulk islands, $T \simeq 1400°C$ (b) $> 1L$, Th(111), Epi, $\sim 700°C$	Estrup et al. (1966)
W(100)	Y	Epi, orientation not specified	Gorodetskii and Yas'ko (1970)
W(110)	Ag	1L, CL $> 1L$, Ag(111), Epi	Lo and Hudson (1972); Bauer and Poppa (1972)
	Cu	0.5L, 7×1, alloy, RT; 0.5L, CL, RT $> 1L$, Cu(111), Epi, RT	Taylor (1966); Moss and Blott (1969)
	Sc	Sc(0001), Epi	Gorodetskii and Yas'ko (1969)

deposition we usually have monolayer formation and we may observe complex LEED patterns corresponding to coincidence lattices of various multiplicities.

Taylor (1966) investigated the epitaxial growth of Cu on W(110). At room temperature, a coverage of about 0.5 monolayer (estimated from the temperature and pressure of a Knudsen-cell-type source) resulted in a 7 × 1 LEED pattern with the long periodicity along the W($\bar{1}$10) direction.[†] This was attributed to the formation of a Cu–W alloy which, upon continued deposition of Cu, converted to a Cu(111) epitaxial film with the [$\bar{1}\bar{1}2$]Cu direction parallel to the [$\bar{1}$10]W direction. At this stage, the interaction between Cu film and W substrate was such as to give rise to a coincidence-lattice pattern with large multiplicity, but this was found gradually to disappear upon growth of a thicker Cu film. Moderate heating of the substrate with a Cu film approximately ten layers thick caused diffusion of copper into the interior of the W substrate and consequent reappearance of the 7 × 1 LEED pattern. The same epitaxial system was reexamined more recently by Moss and Blott (1969), who made use of AES to monitor the chemical composition of the surface. The LEED observations made by Moss and Blott were identical to those made by Taylor but the interpretation was somewhat different. The 7 × 1 pattern was attributed not to a Cu–W monolayer alloy, but to a close-packed monolayer of Cu in a coincidence lattice in which the copper periodicity is identical to that of tungsten along the W[001] direction and smaller than that of tungsten along the W[$\bar{1}$10] direction. The authors speculate that this copper layer relaxes into the configuration of a normal Cu(111) plane as soon as it is covered by another copper layer. The coincidence between the normal Cu(111) configuration and the W(110) substrate surface then gives rise to the high-multiplicity coincidence lattice already reported by Taylor. According to Moss and Blott, however, heating of the substrate–film combination does not lead to diffusion of copper into the bulk tungsten but only to clustering of the copper film into islands.

Feinstein et al. (1970) examined the system of Ag on Ni(111) and observed epitaxial growth of Ag(111) at 150°C. There is evidence for the fact that the Ag film has bulklike periodicities already in the earliest possible detectable stages (at surface coverages equivalent to about 0.1 monolayer). Under these conditions, a coincidence-lattice pattern is expected and, in fact, one is observed

[†] Superstructures are usually referred to with a short-hand notation as follows. If the apparent periodicities of the overlayer (a_{over} and b_{over}) along two suitable directions in the surface plane are related to the periodicities of the substrate surface (a_{surf} and b_{surf}) along the same directions as $a_{over} = \alpha \times a_{surf}$ and $b_{over} = \beta \times b_{surf}$, with α and β rational numbers, then the overlayer structure is referred to, in short, as an $\alpha \times \beta$ structure. If the net is primitive, the notation may sometimes be preceded by a p, thus: $p(\alpha \times \beta)$; if the net is centered, the notation is preceded by a c, thus: $c(\alpha \times \beta)$. A detailed discussion of the vocabulary of surface crystallography was given by Wood (1964).

as soon as the area of the Ag film is sufficiently large for multiple scattering contributions to become visible (about 0.5 monolayer coverage). No evidence for alloying or pseudomorphism was found.

Henrion and Rhead (1972) studied the systems of Pb on Cu(111), Cu(100), and Cu(110). At approximately one monolayer coverage the former system produces a $p(4 \times 4)$ LEED pattern, which was interpreted as a coincidence lattice between the Cu(111) substrate and the Pb(111) overlayer. The mismatch of 2.5% between the two nets was not detectable in the LEED pattern. On Cu(100), the coincidence lattice produced a $c(4 \times 4)$ and a $c(5 \times 1)$–45° pattern, depending on coverage. On Cu(110), the corresponding patterns were $c(2 \times 2)$ and $p(5 \times 1)$, respectively. On all three surfaces it appears that Pb forms dense close-packed monolayers with nearest-neighbor atomic spacings close to those of bulk Pb, but only on Cu(111) was epitaxial growth of a Pb(111) macroscopic film observed. On Cu(100) and Cu(110), bulk Pb crystallites grow with random orientations on top of the Pb monolayer and no LEED patterns from the Pb crystals are observed.

Schlier and Farnsworth (1958) investigated the epitaxy of Cu(111) on Ti(0001), and Haque and Farnsworth (1966) that of Ni(111) on Cu(111). In both cases no coincidence lattices were reported, although the observations were made at low coverages (approximately one monolayer). In both cases the overlayers were observed to have the bulk lattice parameters, i.e., no pseudomorphism occurred. However, Gradmann (1969) has disputed Haque and Farnsworth's claim that the first layer of Ni on Cu(111) has the bulk Ni spacing. Gradmann argued on the basis of RHEED (reflection high-energy electron diffraction) data, and the fact that the misfit is only 2.5%, that pseudomorphism occurs in this system for the first and possibly more layers.

Jackson et al. (1967) reported epitaxy of Al(111) on Ta(110) at 600–670°C. The mismatch in this case is small, less than 0.5% for the two orientations of Al(111) rotated by 10° as revealed by the LEED observations. No extra beams were reported but the films were fairly thick (thicker than about ten layers), and therefore if the coverage was reasonably uniform, no coincidence lattice would be expected to be observable.

The system of Al on Mo(110) was studied by Jackson and Hooker (1971). In the early stages of evaporation, at room temperature, a streaked LEED pattern was observed that was attributed to a strained Al(111) layer. The mismatch in rotation is of 5° and in lattice spacing of 10%. Prolonged deposition of aluminum produced a film with an ideal Al(111) surface.

Palmberg et al. (1968b) investigated the epitaxial growth of Ag and Au on Cu(100) and Pd(100), while Bauer (1967) and Bauer and Poppa (1972) studied the growth of Ag and Au on Cu(100) and Cu(111). At approximately one monolayer coverage of the Cu(100) surface, Ag was observed to give rise to a $c(10 \times 2)$ LEED pattern and Au to a $c(14 \times 2)$ pattern. They were both attributed to coincidence lattices resulting from slight expansion of the overlayers in one direction and compression in another. For both Ag and Au, prolonged depositions led to epitaxial (111) films. On the Pd(100) substrate, on the other hand, both Ag and Au gave rise to 1 × 1 patterns at approximately one monolayer coverage, then to coincidence-lattice patterns at higher coverages (1 × 8 for Ag and 1 × 7 for Au) and finally to epitaxial films. The film surfaces are Ag(111) for silver and Au(100)–1 × 5 for gold. The latter is thought to be due to a hexagonal close-packed reconstruction of the top gold-surface layer.

A number of metal–metal systems have also been studied, of course, that do *not* appear to produce epitaxial films. In the following, we discuss briefly some of the literature reports on such systems. We consider first the cases of Th on Ta(100) and Th on W(100). Pollard and Danforth (1968) reported observing a $c(2 \times 2)$ LEED pattern for 0.3 monolayer coverage of Th on Ta(100) at 950°C, and then a pattern indicative of exact registry of the Th film with the Ta(100) substrate (i.e., exact pseudomorphism) after deposition of 1 to 20 monolayer equivalents of Th. However, electron-microscopic observations revealed the presence of rectangular crystallites (probably of Th) randomly oriented over the surface. The authors suggested that Th forms a pseudomorphic monolayer but any excess Th clusters into small islands of random orientation so that there is no epitaxy. Pollard (1970) also reported a study of the system Th on W(100) at 1400°C. The results were similar to those obtained in the system Th/Ta(100), namely, a fractional coverage of about 0.5 monolayer yielded a $c(2 \times 2)$ LEED pattern, increased coverage gave rise to a Th monolayer pseudomorphic with the W(100) substrate, but prolonged deposition (up to 25 monolayer equivalents) generated Th islands with random orientations. These results seem to be very similar to those obtained by Henrion and Rhead (1972) for the Pb/Cu(100) and the Pb/Cu(110) systems, as discussed above. The Th/W(100) system had been studied earlier by Estrup *et al.* (1966) with the conclusion that Th formed an epitaxial close-packed film at coverages greater than one monolayer, but Pollard (1970) attributed the observed LEED patterns to a CO-stabilized thorium layer grown over the initial pseudomorphic thorium monolayer.

The systems involving Na, K, and Cs on Ni(111), Ni(100), and Ni(110) were investigated by Gerlach and Rhodin (1969); the system Cs/W(100) was studied by Lander and Morrison (1969). In general, coverages approaching one monolayer showed a tendency toward formation of irregular close-packed structures, but higher coverages (greater than about two layers) caused obliteration of the corresponding LEED patterns and were thus identified as amorphous films. This behavior was attributed first to the fact that the ionicity of the adsorbate introduced large repulsive forces between adatoms (thus giving rise to irregular hexagonal layers) and then to the fact that epitaxial growth of non-close-packed materials such as the alkali metals is obviously difficult to initiate on irregularly close-packed monolayers.

2. Metals on Covalent Substrates

Up to coverages of the order of one monolayer, metals deposited on covalent substrates tend to form intermediate structures in the sense defined above in Section 4, B. Such structures may be needed in order to bridge the gap between the directional character of the covalent bonds in the substrate and the nondirectional character of the metallic bonds in the deposited film. The number of metal–covalent systems that were investigated with the LEED and AES techniques is considerably smaller than that of the metal–metal systems discussed above. One practical difficulty encountered in LEED studies of covalent surfaces is that these are, in general, more difficult to clean satisfactorily than some of the metal surfaces and, most important, that the structures of these clean surfaces are not simple but often reconstructed. Silicon surfaces, for example, exhibit superstructures with long periodicities, such as $Si(111) \, 7 \times 7$, which is often referred to more briefly as $Si(111)$–7.

The $Si(111)$–7 structure has been most often chosen as the substrate in a number of epitaxial experiments involving metal adsorbates such as Pb, Ag, Al, Sn, and In. The Pb–$Si(111)$ system was studied by Estrup and Morrison (1964). At low coverages, the occurrence of extra diffraction spots in the LEED pattern was interpreted as due to the formation of a $Pb(111)$ layer with bulk lattice spacing superimposed on the $Si(111)$–7 structure. At higher coverages, the patterns were found to consist only of features attributed to the Pb layer, and epitaxial growth was confirmed. This behavior is therefore very similar to that of close-packed metals on metallic substrates. Silver and aluminum, on the other hand, exhibit a different behavior in the sense that they give rise, at low coverages, to intermediate structures and produce, with increasing coverage, a number of complex LEED patterns that are not easily interpreted as coincidence lattices. For example, in the system $Ag/Si(111)$–7 at temperatures higher than 200°C Spiegel (1967) observed a so-called $\sqrt{3} \times \sqrt{3}$–30° pattern at about 0.3 monolayer coverage but confirmed epitaxy of $Ag(111)$, with the concomitant disappearance of all extra diffraction spots, at higher coverages. Lander and Morrison (1964) reported three stages in the successful epitaxial growth of $Al(111)$ on $Si(111)$–7. If 0.3 monolayer equivalents are deposited on the silicon substrate above 500°C, the LEED pattern exhibits a so-called α-phase [$Si(111) \, \sqrt{3} \times \sqrt{3}$–30°–Al]. Continued deposition up to one monolayer equivalent onto the α-phase at room temperature produces the β-phase [$Si(111) \, \sqrt{3} \times \sqrt{3}$–30°–Al], which yields a LEED pattern geometrically identical to that of the α-phase but with different intensities. Further deposition onto the β-phase at room temperature leads finally to epitaxial growth of $Al(111)$. Thus, it appears that for both the $Ag/Si(111)$–7 and the Al/Si(111)–7 systems the observed $\sqrt{3} \times \sqrt{3}$–30° structure plays the role of an intermediate structure that provides continuity between covalent and metallic bonding. The In/Si(111)–7 system exhibits at least eight (Lander and Morrison, 1964) and the Sn/Si(111)–7 at least five (Estrup and Morrison, 1964) different and complicated phases but neither produced epitaxy. Similarly, three different phases were observed in the system Ce/Si(111)–7, ending presumably with a close-packed hexagonal array at one monolayer coverage (Gobeli et al., 1966), but further deposition was not attempted and hence epitaxy cannot be claimed.

The growth of Au and Ag on cleaved PbS surfaces was investigated by Green et al. (1971). Despite the fact that Ag and Au have nearly identical lattice constants, they were found to assume very different epitaxial orientations. While Au was found to grow in the form of three-dimensional crystallites with two equivalent (110) orientations, Ag was observed to form (100) epitaxial films at temperatures higher than 200°C. The systems involving Al and P on diamond (111) surfaces were studied by Lander and Morrison (1966). Complex LEED patterns were observed in both cases but no macroscopic epitaxy was reported.

3. Metals on Ionic Substrates

The few examples in the literature in which LEED was applied to the study of metals on ionic substrates show that it was used primarily as a tool for the determination of the occurrence of epitaxy "in situ." Palmberg et al. (1967a, 1968a) employed LEED to discover that Ag, Au, and Cu are more readily grown epitaxially on electron-irradiated KCl surfaces than on surfaces freshly cleaved in situ. However, on MgO perfect epitaxy of Ag(100) and Au(100) was observed without prior electron damage (Palmberg et al., 1967b). In any case, it seems well established that metals deposited on alkali halide surfaces tend to develop small three-dimensional crystallites rather than mono- and multilayer structures, so that LEED would not seem to be very useful in these cases, except perhaps to reveal the occurrence or the failure of epitaxy in the later stages of film growth.

4. Covalent Materials on Metallic Substrates

If, as we mentioned above, intermediate structures are indeed needed in the epitaxial process that relates metallic to covalent crystals, then a tendency toward such structures ought to be encountered also in systems involving covalent materials deposited on metal substrates. Andersson et al. (1968) investigated the system Te/Cu(111) and found, in fact, a succession of structures very much like those observed in the system Al/Si(111)–7 discussed above. In particular, at room temperature, a $2\sqrt{3} \times 2\sqrt{3}$–30° LEED pattern was observed at about 0.1 monolayer coverage, a $\sqrt{3} \times \sqrt{3}$–30°

pattern at about 0.3 monolayer coverage, and finally a pattern corresponding to epitaxial Te(0001) at coverages in excess of one monolayer. Andersson and Andersson (1970) studied the system Te/Cu(100) but found that no epitaxy occurs. At very low coverages, a $c(2 \times 2)$ LEED pattern was observed, while at approximately 0.3 monolayer coverage a 6×6 pattern was observed that was attributed to a coincidence lattice. However, further deposition was found to lead to the formation of three-dimensional Te crystallites with no parallel growth. Therefore, it appears that in the Te/Cu system intermediate structures capable of bridging the gap between substrate and overlayer toward epitaxy can readily be formed on the (111) but not on the (100) surface.

Other cases in which epitaxy was attempted but not attained are found in the Si/Al, the Si/Ni, and the Si/Be systems (Jona, 1971, 1973a, b). Intermediate structures were found for various degrees of fractional coverage on Al(110), Al(111), Ni(100), Ni(110), Ni(111), and Be(0001) surfaces, but prolonged deposition at room temperature resulted in amorphous silicon in all cases. The explanation offered for the failure to attain epitaxy at room temperature was that since silicon does not grow epitaxially on itself at temperatures lower than about 400°C (see below) it can hardly be expected to do so at room temperature on intermediate structures formed over metallic substrates. At higher temperatures, on the other hand, the intermediate structures cannot be maintained on the surfaces because the silicon atoms diffuse rapidly into the interior of the substrates, thus preventing growth of any silicon film.

Epitaxial orientations were observed for some oxides grown onto clean metal surfaces at higher temperatures, for example, TaO(111) on the reconstructed Ta(110)–5 surface (Boggio and Farnsworth, 1965) and NiO(111) on Ni(111) (MacRae, 1963). The need for increased temperatures in order to attain epitaxy in these cases seems justified by the argument that the grown materials are anisotropic and hence not prone to assume close-packed structures. Higher mobility is therefore needed in order to facilitate proper atomic arrangements in intermediate structures and in the epitaxial film itself.

5. Covalent Materials on Covalent Substrates

The most thoroughly investigated system involving deposition of a covalent substance onto a covalent substrate is doubtlessly the system of silicon on silicon. The LEED studies of this system, however, are rather limited in number and did not contribute much to the understanding of the epitaxial process, primarily because of the ignorance of the complicated reconstructed structures of clean silicon surfaces. Jona (1967) used LEED to establish that silicon grows epitaxially on the Si(111)–7 structure at temperatures higher than about 400°C, but on the Si(100)–2 structure at somewhat lower temperatures. Thomas and Francombe (1967, 1971) reached essentially similar conclusions.

The epitaxy of silicon on several surfaces of sapphire [namely, ($\bar{1}$012), (11$\bar{2}$3), (0001), and (11$\bar{2}$0)] was examined with LEED and AES by Chang (1971b). The epitaxial relationships and the electrical properties were found to depend on deposit–substrate interactions, substrate temperature, and surface structure. A range of epitaxial temperatures was identified for each of the substrate orientations studied within which the proper surface structure could be maintained during film growth. However, detailed understanding of the atomistic processes involved in this system does not seem possible at the present stage, primarily because the structures of the substrate surfaces are very complicated and not well known.

V. Conclusions

Despite the respectable age of the LEED technique, its use in the study of epitaxial processes has just begun, and therefore the number of epitaxial systems that have been scrutinized with LEED is rather small. For this reason, it is premature to draw general conclusions from the available results. It is possible, however, to recognize some trends and directions, and to make a few preliminary statements.

It seems to be well established by the literature reports that LEED is indeed a powerful technique for the study of monolayer formation in the early stages of epitaxy. Many examples can be found in Table I in which coincidence lattices or intermediate structures have been observed at very low surface coverages. The mere observation of these ordered structures indicates that monolayer formation does indeed occur. The other initial mode of growth (namely, isolated three-dimensional nuclei) is, in fact, not likely to be accessible to observation by LEED. For example, neither coincidence lattices nor intermediate structures have been reported in the systems of Au and Ag on KCl and MgO, where three-dimensional nuclei are known to constitute the initial growth phase. The reason is simple: If the three-dimensional nuclei are sufficiently large in the plane parallel to the surface to be "seen" (i.e., to scatter low-energy electrons coherently), they are also large enough in the direction perpendicular to the surface effectively to hide the interfacial region. Of course if the nuclei are very small, each consisting of a few atoms only, but very numerous, then indeed the interfacial region could be accessible to examination by LEED, and superstructures would be observable, but in this case the distinction between monolayer formation and three-dimensional nucleation is likely to be irrelevant.

The first question that we asked at the beginning of this chapter concerned the role played by the first layer of deposited atoms: Must this monolayer be ordered for epitaxy to occur? The present review of the literature does not allow the conclusion that an ordered first monolayer is *necessary* for further epitaxial growth but does seem to suggest that in the majority of epitaxial systems investigated the first monolayer *is* in fact ordered. Some authors claim that in certain systems epitaxy is possible even on disordered overlayers (Spiegel, 1967; Distler *et al.*, 1966, 1968a,b), although in most cases these claims have been disputed (Chopra 1970), and only one claim seems to be based upon LEED observations (Spiegel, 1967). We may therefore conclude, tentatively, that in general an ordered first monolayer of adatoms is likely to be necessary, although certainly not sufficient, for the subsequent epitaxial growth of the deposited material.

Some answers may be also suggested for the questions that were asked subsequently, primarily for the one about the factors that control the ordering of the first monolayer deposited. It appears certain that the temperature of the substrate must be high enough to allow sufficient surface mobility of the adatoms, but not so high as to cause rapid diffusion of such atoms into the interior of the substrate. In addition, on metallic substrates a tendency of the deposited atoms toward close-packing is likely to favor ordering of the first monolayer and subsequent epitaxial growth. In nearly all cases that have been reported in the literature, close-packed metals were found to grow epitaxially on almost any metal surface with their close-packed planes parallel to the plane of the substrate surface. This observation suggests that the function of the substrate, in these cases, is primarily that of providing a flat table on which the deposited atoms are free to arrange themselves in two-dimensional close-packed arrays. The orientation of the array may be achieved by a slight bias generated by the surface atoms in the vicinities of the nucleation sites. The metals that do not grow epitaxially are either not close-packed and/or are ionized to a certain extent by the substrate. This seems to be the case, for example, for the alkali metals and thorium on tungsten and tantalum. Obviously, ionization causes mutual repulsion and thus prevents long-range close-packed order. Thus, coverages approaching unity result in irregular close-packed structures that are not conducive to further ordered growth.

If the interactions between adsorbate and substrate are stronger than the adsorbate–adsorbate interactions (as in the systems involving metals on covalent substrates and metals on ionic substrates), then the first monolayer is also likely to be ordered, but the resulting order tends to reflect some of the characteristics of the substrate. Whether or not such an ordered monolayer can sustain further ordered growth cannot be said at the outset. Steric and chemical considerations are probably of paramount importance in this process. Metal–covalent systems, for example, often seem to form intermediate structures at the monolayer level, as if to bridge the bonding and structural gap between substrate and adsorbate, but only in a very few cases has extended epitaxial growth been found to occur in subsequent stages.

In any case, mismatch of lattice parameters between substrate and overlayer does not seem to play an important role in the process of monolayer formation and, eventually, epitaxial growth. This statement, of course, has been made before (Pashley, 1970). Indeed, when the mismatch is small (smaller than about 5%), pseudomorphism may occur, but the instrumental resolution of modern LEED equipment is not good enough to provide a definite answer. When the mismatch is large, coincidence lattices are formed by slight adjustments of the lattice parameters of the initial monolayer.

The question about the mechanism of growth of the second layer of deposited atoms over the first, and of subsequent layers after that, is the most obscure one at the moment. In a zeroth approximation, one might be tempted to suggest that the factors governing this stage of growth are closely related to those that control homoepitaxy and crystal growth in general. Thus, successful completion of this stage of epitaxial growth could only be achieved if at least those conditions that allow homoepitaxy to occur were fulfilled. There is, however, no direct experimental evidence for this supposition, and no reliable statement can be made on the subject at the present time.

Note added in proof: Owing to the rapid expansion of the field covered in this chapter, it may be useful to point out that the text and the table were prepared in the summer of 1972.

ACKNOWLEDGMENT

One of the authors (F.J.) would like to express his deep appreciation to the Air Force Office of Scientific Research, Air Force Systems Command, for the continued support given to his studies of epitaxy by LEED and AES under Grant AFOSR-72-2151.

References

Anderson, S., and Pendry, J. B. (1972). *J. Phys.* C **5**, L41.
Andersson, D. E., and Andersson, S. (1970). *Surface Sci.* **23**, 311.
Andersson, S., Marklund, I., and Martinson, J. (1968). *Surface Sci.* **12**, 269.
Bauer, E. (1967). *Surface Sci.* **7**, 351.
Bauer, E., and Poppa, H. (1972). *Thin Solid Films* **12**, 167.
Boggio, J. E., and Farnsworth, H. E. (1965). *Surface Sci.* **3**, 62.
Chang, C. C. (1971a). *Surface Sci.* **25**, 53.
Chang, C. C. (1971b). *J. Vac. Sci. Technol.* **8**, 500.
Chopra, K. L. (1970). *Surface Sci.* **20**, 201.
Dickey, J. M., Farrell, H. H., and Strongin, M. (1970). *Surface Sci.* **23**, 448.
Distler, G. I., and coworkers (1966). *Nature* (*London*) **212**, 807.
Distler, G. I., and coworkers (1968a). *J. Crystal Growth* **2**, 45.
Distler, G. I., and coworkers (1968b). *Naturwissenschaften* **3**, 132.
Estrup, P. J., and Morrison, J. (1964). *Surface Sci.* **2**, 465.
Estrup, P. J., Anderson, J., and Danforth, W. E. (1966). *Surface Sci.* **4**, 286.
Farrell, H. H., and Strongin, M. (1972). *Phys. Rev.* **B6**, 4711.
Farrell, H. H., Strongin, M., and Dickey, J. M. (1972). *Phys. Rev.* **B6**, 4703.
Feinstein, L. G., Blanc, E., and Dufayard, D. (1970). *Surface Sci.* **19**, 269.
Gerlach, R. L., and Rhodin, T. N. (1969). *Surface Sci.* **17**, 32.

Gobeli, G. W., Lander, J. J., and Morrison, J. (1966). *J. Appl. Phys.* **37**, 203.

Gorodetskii, D. A., and Yas'ko, A. A. (1969). *Sov. Phys.-Solid State* **10**, 1812.

Gorodetskii, D. A., and Yas'ko, A. A. (1970). *Sov. Phys.-Solid State* **11**, 2028.

Gorodetskii, D. A., and Yas'ko, A. A. (1972). *Sov. Phys.-Solid State* **14**, 636.

Gradmann, U. (1969). *Surface Sci.* **13**, 498.

Green, A. K., Dancy, J., and Bauer, E. (1971). *J. Vac. Sci. Technol.* **8**, 165.

Haque, C. A., and Farnsworth, H. E. (1966). *Surface Sci.* **4**, 195.

Henrion, J., and Rhead, G. E. (1972). *Surface Sci.* **29**, 20.

Ignatiev, A., Jones, A. V., and Rhodin, T. N. (1972). *Surface Sci.* **30**, 573.

Jackson, A. G., and Hooker, M. P. (1971). *Surface Sci.* **28**, 373.

Jackson, A. G., Hooker, M. P., and Haas, T. W. (1967). *J. Appl. Phys.* **38**, 4998.e

Jona, F. (1967). *In* "Surfaces and Interfaces" (J. J. Burke, N. L. Reed, V. Wiss, eds.), Vol. I, Chemical and Physical Characteristics, p. 399. Syracuse Univ. Press. Syracuse, New York.

Jona, F. (1971). *J. Appl. Phys.* **42**, 2557.

Jona, F. (1973a). *J. Appl. Phys.* **44**, 351.

Jona, F. (1973b). *J. Appl. Phys.* **44**, 4240.

Lander, J. J. (1965). *Solid State Chem.* **2**, 26.

Lander, J. J., and Morrison, J. (1964). *Surface Sci.* **2**, 553.

Lander, J. J., and Morrison, J. (1966). *Surface Sci.* **4**, 241.

Lander, J. J., and Morrison, J. (1969). *Surface Sci.* **14**, 465.

Lo, C. M., and Hudson, J. B. (1972). *Thin Solid Films* **12**, 261.

MacRae, A. U. (1963). *Appl. Phys. Lett.* **2**, 88.

MacRae, A. U., Muller, J., and Lander, J. J. (1969). *Phys. Rev. Lett.* **22**, 1048.

Moss, A. R. L., and Blott, B. H. (1969). *Surface Sci.* **17**, 240.

Palmberg, P. W., and Rhodin, T. N. (1968). *J. Chem. Phys.* **49**, 134.

Palmberg, P. W., Rhodin, T. N., and Todd, C. J. (1967a). *Appl. Phys. Lett.* **10**, 122.

Palmberg, P. W., Rhodin, T. N., and Todd, C. J. (1967b). *Appl. Phys. Lett.* **11**, 33.

Palmberg, P. W., Todd, C. J., and Rhodin, T. N. (1968). *J. Appl. Phys.* **39**, 4650.

Pashley, D. W. (1970). *Recent Progr. Surface Sci.* **3**, 23–69.

Pollard, J. H. (1970). *Surface Sci.* **20**, 269.

Pollard, J. H., and Danforth, W. E. (1968). *J. Appl. Phys.* **39**, 4019.

Schlier, R. E., and Farnsworth, H. E. (1958). *J. Phys. Chem. Solids* **6**, 271.

Spiegel, K. (1967). *Surface Sci.* **7**, 125.

Taylor, N. J. (1966). *Surface Sci.* **4**, 161.

Thomas, R. N., and Francombe, M. H. (1967). *Appl. Phys. Lett.* **11**, 108.

Thomas, R. N., and Francombe, M. H. (1971). *Surface Sci.* **25**, 357.

Wood, E. A. (1964). *J. Appl. Phys.* **35**, 1306.

3.4　High-Energy Electron Diffraction

D. B. Dove

Department of Materials Science and Engineering
University of Florida
Gainesville, Florida

I. Introduction

The classic experiments of Davisson and Germer (1927) on the diffraction of low-energy electrons and the almost simultaneous demonstration by Thomson and Reid (1927) of the diffraction of high-energy electrons transmitted through thin foils laid the foundation for the extremely powerful electron-diffraction techniques employed in the examination of surfaces and thin films. This early work verified the essential correctness of the de Broglie hypothesis that particles may exhibit wavelike properties. An interesting account of early electron-diffraction studies on metal foils and deposited films may be found in the book by Thomson and Cochrane (1939). For electrons the effective wavelength, given by Planck's constant/momentum, is typically a fraction of an angstrom for electron energies sufficient to allow penetration through a thin foil. Figure 1 shows a plot of wavelength λ in angstrom units (Å) for electrons accelerated through a potential V:

$$\lambda = 12.25/(V + 10^{-6}V^2)^{1/2}$$

With energies in the range 20–100 keV commonly employed in HEED investigations, wavelengths are smaller than prominent crystalline interplanar spacings, resulting in Bragg angles of a fraction of a degree to several degrees.

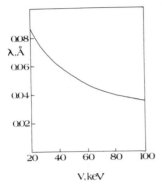

FIG. 1. Plot of electron wavelength λ versus accelerating voltage V.

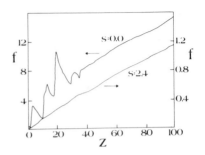

FIG. 2. Atomic elastic scattering cross sections for electrons versus atomic number for zero angle of deflection ($s=0$) and for scattering at $s = 2\sin\theta/\lambda = 2.4$.

Compared with X rays and neutrons as a radiation for the examination of materials, electrons possess the very important property that they may be deflected and focused by electrical or magnetic fields. Electron-diffraction experiments are therefore readily carried out with well-collimated beams that may be brought to focus on an observation screen or photographic plate. A second consequence of the existence of the electronic charge is that scattering cross sections for electrons are many orders of magnitude larger than those of X rays or neutrons, permitting diffraction patterns to be obtained from layers as thin as a few tens of angstroms or less. Examination of tabulated atomic scattering amplitudes shows (see "International Tables for X-Ray Crystallography." Kynoch Press, Birmingham, England) that the scattering probability at large angles increases with the Z number of the scattering atom, while the amplitudes at small angles do not behave so regularly. High-energy electrons are deflected by the atomic coulombic field. Thus the strong fields near the nucleus produce a large deflection, whereas scattering at small angles depends on the field resulting from both nuclear charge and atomic electron screening effects. Figure 2 shows a plot of scattering amplitudes at small and large angles as a function of atomic number. The electron intensity scattered by an atom into a constant solid angle is proportional to the square of the scattering amplitude. In the HEED case, intensities decrease rapidly with angle of deflection 2θ, or more precisely with scattering parameter $2\sin\theta/\lambda$.

The theory of electron diffraction has received considerable attention in recent years due to the desire to understand diffraction contrast effects observed in the electron microscopy of defects in crystalline foils, imaging and diffraction effects being, of course, intimately related.

The electron microscope provides an extremely versatile facility for HEED observations and has, to a large extent, superceded the purely diffraction instrument. Modern diffraction instruments are available with various additional capabilities such as ultrahigh vacuum or electronic intensity read-out.

II. Diffraction Instruments

A basic arrangement for transmission HEED is shown in Fig. 3a. Electrons emitted from a hot filament held at a high negative potential, within a vacuum of the order of 10^{-4} Torr or better, are accelerated by an anode at ground potential. One or more magnetic or electrostatic lenses and apertures form a well-collimated beam that is brought to a focus on a fluorescent screen or photographic plate. A thin foil specimen is placed in the beam at some distance L in front of the screen. Diffraction of the beam may occur through a deflection 2θ, giving rise to a beam striking the screen at a distance R from the location of the spot due to the unscattered beam. Evidently $R/L = \tan 2\theta$, and hence the angle of deflection is readily found. For the short wavelengths employed in HEED, θ is usually small so that the approximations $\theta = \sin\theta = \tan\theta$ are valid, and diffracting planes are nearly parallel to the incident beam. In an alternative configuration shown in Fig. 3b, reflection HEED, the beam is allowed to strike the specimen at grazing incidence, and the beams diffracted out of the surface form a diffraction pattern on the screen. In most cases the pattern formed in reflection HEED actually arises from the penetration of the beam through aspherities on the surface and is very similar to a transmission pattern. A very different pattern is obtained upon diffraction from an atomically flat surface as illustrated in Figs. 4a, b. The reader is referred

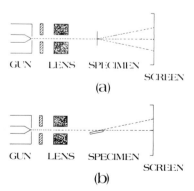

FIG. 3. Basic configuration for Heed investigations: (a) in transmission; (b) in reflection.

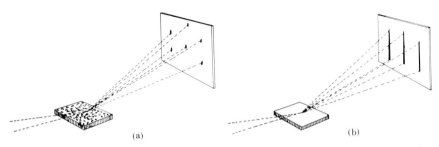

FIG. 4. Diffraction at grazing incidence may give rise to a transmission type of pattern due to surface irregularities as in (a) or to a streak pattern that is extremely sensitive to surface condition as in (b).

to a review by Bauer (1969) for an extensive discussion of the theory and practice of HEED in reflection.

The excellent electron-optic properties of the electron microscope make it an exceedingly versatile diffraction instrument for both transmission and reflection. Description of techniques for obtaining patterns from very small areas of specimens and for obtaining highly magnified diffraction patterns may be found in the work of Beeston *et al.* (1973). To overcome the very slight radial distortion of the pattern due to the image-forming lenses, some microscopes have a special high-precision diffraction position in which the specimen is located beneath the projector lens. The electron microscope however is generally badly limited with regard to quality of vacuum and may be hampered by lack of space surrounding the specimen.

An instrument designed for high-quality electron-diffraction observation is typified by the unit of Balzer A. G., Lichtenstein. This system has a well-developed electron-optic column with 120-keV capability, double-condenser lens system that can reduce beam size at the specimen to 1 μm, and large specimen space with provision for specimen heating, cooling, or other treatments. The vacuum system includes trapped diffusion pumps, and pressure at the specimen may be brought to below 10^{-7} Torr with differential pumping, while 10^{-6} Torr is maintained in the rest of the system.

A unit by Varian Associates, Palo Alto, California, has a more basic electron-optic column but is designed for complete ultrahigh vacuum operation. The system is ion-pumped, bakeable, with large specimen space, and with 10^{-10}-Torr capability. The pattern is formed on a fluorescent screen and is recorded by an external camera.

Ultrahigh vacuum is especially important for reflection work since the pattern is seriously degraded by contamination buildup under ordinary vacuum conditions. A system by Vacuum Generators (distributed by Veeco, Inc., Plainview, New York) combines a basic electron-optic column and UHV operation with a more elaborate system for the recording of electron intensities. In this instrument a pair of coils between the specimen and a fluorescent screen deflects the pattern in any desired direction across a small aperture in the screen. Electrons passing through the aperture enter an electrostatic filter that rejects electrons that have lost more than a few electron volts in energy during interaction with the specimen. The elastically scattered electrons are transmitted by the filter and strike a small second fluorescent screen. The brightness of this screen is measured by a photomultiplier tube, providing a very sensitive method of measuring electron intensity. Detector output is recorded automatically versus scan coil current, providing a series of intensity profiles across the diffraction pattern. Figure 5 shows schematically a scanning electron-diffraction arrangement, using magnetic deflection of the pattern. The design of such systems has been described by Grigson (1968) and Grigson and Tillet (1968). A method for direct electronic measurement of intensities using a translatable Faraday cage has been described by Burggraf and Goldsztaub (1962).

The electron-diffraction system employing magnetic deflection and retarding field analyzer possesses two useful features. The calibration between current in scan coil and deflection to a given value of $2 \sin \theta / \lambda$ is insensitive to electron energy. In addition, the retarding field for the filter may be derived from the electron-gun voltage source, and hence a power supply of somewhat less than ultrahigh stability may be employed if necessary. A scanning electron-diffraction system with energy filter is available as an accessory for the AEI electron microscope (A.E.I. Scientific Apparatus Ltd., Harlow, Essex, England). This attachment is located beneath the microscope column and has an auxiliary oil pump to maintain vacuum in the filter during specimen change.

A system combining a high-quality electron-optic column with ultrahigh vacuum and/or electronic intensity read-out does not yet appear to be commercially available.

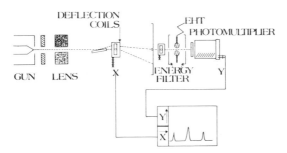

FIG. 5. Arrangement for electron intensity measurement with energy filtering of the scattered beams (scanning electron diffraction).

III. Analysis of Diffraction Patterns

The theory of the diffraction of electrons has been extensively developed in recent years particularly with regard to the treatment of diffraction effects in electron microscopy. Reviews have been given, for example, by Cowley (1968), Gevers (1970), and Howie (1970). As in the case of X rays, a kinematic, i.e., geometric wave interference, theory often suffices.

In general, a diffraction pattern contains information on (1) the symmetry and orientation of crystalline layers, (2) the position of atoms within the crystalline unit cell, and (3) the physical size of the volume of scattering material. The analysis of the geometric array of diffracted beams has provided a valuable means of establishing epitaxial relationships. An extensive discussion of the interpretation of spot patterns with regard to crystallite identification, and orientation determination, has been given in the book by Andrews *et al.* (1967). This work contains many patterns together with their analysis. Kikuchi patterns frequently provide a means for precise orientation of thicker foils. Thomas (1970) has reviewed the application of Kikuchi patterns and has presented extensive Kikuchi maps for various crystal structures.

Determination of crystal structures using electron diffraction has only been carried out in rather special cases and requires careful measurement of relative spot intensities for a number of crystal orientations. A treatment of structure analysis using electrons has been given by Vainstein (1964) with particular emphasis on Fourier transform methods.

Information on the size of scattering regions may be obtained from a measurement of the intensity profile of diffracted beams. Typically crystallites on the order of only a very few unit cells in size lead to gross spot broadening, while spot width for large scattering volumes is determined by the quality of the electron-optic column rather than by scattering volume. The interpretation of diffuse patterns, particularly as arising from diffraction by glassy or amorphous films, has been reviewed by Dove (1973).

While the theory of the diffraction of electrons in transmission has been extensively studied, comparatively little work has been carried out for the grazing-incidence reflection case. A two-beam dynamical calculation has been given by Harding (1937), and calculations and measurements have been reported by Colella (1972), Colella and Menadue (1972), and Moon (1972). Reflection diffraction patterns from highly perfect surfaces in ultrahigh vacuum frequently consist of long streaks normal to the plane of the specimen. In this condition, high-order reflections from ordered surface structures normally only observed by low-energy electron diffraction may be seen (Sewell and Cohen 1965). Recent work by Dove *et al.* (1973) suggests that such patterns may be interpreted in terms of surface-layer, i.e., two-dimensional, diffraction and that much useful information may be obtained from a simple kinematic interpretation in the absence of an easy to apply dynamical theory. With further experimentation under UHV conditions and with development of dynamical theory, reflection HEED may provide a powerful technique for detailed atomic structural analysis of surfaces and of very thin surface layers.

References

Andrews, D. W., Dyson, D. J., and Keown, S. R. (1967). "Interpretation of Electron Diffraction Patterns." Plenum Press, New York.

Bauer, E. (1969). *In* "Techniques of Metals Research" (R. F. Bunshah, ed.), Vol. 2. Wiley (Interscience), New York.

Beeston, E. E. P., Horne, T. W., and Markham, R. (1973). "Electron Diffraction and Optical Diffraction Techniques." American Elsevier, New York.

Burggraf, C., and Goldsztaub, S. (1962). *J. Microsc.* **1**, 441.

Colella, R. (1972). *Acta Cryst.* **A28**, 11.

Colella, R., and Menadue, J. F. (1972). *Acta Cryst.* **A28**, 16.

Cowley, J. M. (1968). "Crystal Structure Determination by Electron Diffraction." Pergamon, New York.

Davisson, D. J., and Germer, L. H. (1927). *Nature (London)* **119**, 558.

Dove, D. B. (1973). *In* "Physics of Thin Films," Vol. 7 (M. Francombe and R. W. Hoffman, eds.), p. 1. Academic Press, New York and London.

Dove, D. B., Ludeke, R., and Chang, L. L. (1973). *J. Appl. Phys.* **44**, 1897.

Gevers, R. (1970). *In* "Modern Diffraction and Imaging Techniques in Material Science" (S. Amelinckx, R. Gevers, G. Remaut, and J. Van Landuyt, eds.), p. 1. North-Holland, Amsterdam.

Grigson, C. W. B. (1968). *Advan. Electron. Electron Phys. Suppl.* **4**, 187.

Grigson, C. W. B., and Tillett, P. I. (1968). *Int. J. Electron.* **24**, 101.

Harding, J. W. (1937). *Phil. Mag.* **23**, 271.

Howie, A. (1970). *In* "Modern Diffraction and Imaging Techniques in Material Science" (S. Amelinckx, R. Gevers, G. Remaut, and J. Van Landuyt, eds.), p. 285. North-Holland, Amsterdam.

Moon, A. R. (1972). *Z. Naturforsch.* **27A**, 390.

Sewell, P. B., and Cohen, M. (1965). *Appl. Phys. Lett.* **7**, 32.

Thomas, G. (1970). *In* "Modern Diffraction and Imaging Techniques Material Science" (S. Amelinckx, R. Gevers, G. Remaut, and J. Van Landuyt, eds.). North-Holland, Amsterdam.

Thomson, G. P., and Reid, A. (1927). *Nature (London)* **119**, 890.

Thomson, G. P., and Cochrane, W. (1939). "Theory and Practice of Electron Diffraction." MacMillan, New York.

Vainshtein, B. K. (1964). "Structure Analysis by Electron Diffraction" (translation by E. Feigl and J. A. Spink). Pergamon, Oxford, and MacMillan, New York.

3.5   X-Ray Diffraction

Richard W. Vook

Department of Chemical Engineering and Materials Science
Syracuse University
Syracuse, New York

I. Introduction

X-ray diffraction techniques are useful tools for the microstructural investigation of thin films (Chopra, 1969; Mader, 1970). In contrast to high-energy electron diffraction, the diffracted X-ray beam tends to originate in a relatively large volume of the film. Consequently, average microstructural information is generally obtained. In this chapter a review of the main experimental methods is given along with some of the principal applications. The techniques (Bertin, 1969) generally apply to both monocrystalline (Borie and Sparks, 1966) and polycrystalline (Vook, 1968) films.

Structural and microstructural information on thin films can be obtained from three kinds of X-ray diffraction measurements: the position of the diffraction line's peak, its profile as given by a rocking curve, and the total energy diffracted as the crystalline material is rotated through the Bragg reflecting position. Each of these methods as they apply to thin films will be discussed in detail in the following sections. Not all X-ray methods will be covered, but the emphasis will be on those primarily concerned with diffraction phenomena.

II. Peak Position Measurements

A. Crystal Structure

The assumption is made in this chapter that the crystal structure of the material being investigated is known or that it can be determined for simple structures. In the latter case, the structure can be identified by indexing the observed Bragg reflections and comparing them with the allowed ones as calculated from the structure factor or by using standard powder-diffraction data tables (Powder Data File, 1967).

B. Lattice Parameter—Uniform Strain

The interplanar spacings of any real crystal are not uniform throughout the crystal. Even in the most perfect crystal, variations in the interplanar spacings will occur as a result of thermal atomic vibrations and surface relaxations. One can, however, speak of a "time-averaged interplanar spacing" or "lattice parameter." If such an average value deviates from the normal value obtained from a thick, unstrained, defect-free sample, then that deviation is a measure of the uniform strain in the crystal specimen. Thick specimens are defined as those in which the coherently diffracting domains exceed about 1500 Å in size. Such a uniform strain gives rise to a diffraction-line peak shift but produces no line broadening.

When the lattice parameters of the unit cells making up the real crystal are not all the same, then a nonuniform strain is present. To a first approximation, one can consider the corresponding broadened diffraction line as being formed by the superposition of unbroadened diffraction lines from a distribution of small perfect crystal regions in the sample. Each such region has undergone a uniform strain that may vary from region to region. The collection of such uniformly strained regions comprises the nonuniform strain or microstrain distribution in the material (Cullity, 1956).

Thus uniform strain gives rise to a peak shift and no diffraction line broadening, while microstrains inevitably produce such broadening. Further-more, in the latter case, the occurrence of peak shifts depends upon the actual distribution of lattice constants. If that distribution is symmetric with respect to the average crystal-lattice parameter, then no peak shift occurs. If it is asymmetric, an additional uniform strain and corresponding displacement of the diffraction line is introduced.

Measurements of lattice constants from the diffraction line peak positions involve distinguishing between random and systematic errors. The basic techniques are given in the book by Cullity (1956). Briefly, these errors are minimized by using a least-squares analysis fit to a linear extrapolation-function plot. In such a graph for cubic crystals, a lattice parameter a_{hkl} is calculated from the interplanar spacing d_{hkl} (obtained from the Bragg law, $\lambda = 2d \sin \theta$) for each hkl reflection. The different a_{hkl}s are then plotted versus an extrapolation function of the Bragg angle θ. A straight line is then fit to these points using the least-squares method. The applicable extra-polation function is determined by using tungsten powder (tungsten is elastically isotropic). A good fit to a straight-line plot should be obtained. Typical extrapolation functions are the Nelson–Riley function (NRF) $(\cos^2 \theta/\sin \theta + \cos^2 \theta/\theta)/2$ (Nelson and Riley, 1945) and $\cos^2 \theta/\sin \theta$, where θ is the observed Bragg angle.

The average lattice parameter is the value a_0 obtained by extrapolating the linear fit to the lattice parameters a_{hkl} to the point where the extrapolation function is zero. It is illustrated for tungsten powder in Fig. 1 (Selleck, 1972), where the extrapolated $a_0 = 3.1642$ Å. This value is the average lattice para-meter of the tungsten powder at room temperature.

The above discussion holds for cubic crystals in which certain other effects, such as stacking faults and thermally induced strains, are not present. These imperfections produce *systematic* peak shifts that in general are different for each reflection. As a result, a least-squares analysis does not apply. When systematic peak shifts are absent, and for any crystal structure (either poly-crystalline or monocrystalline), one can determine the interplanar spacing d_{hkl} for the (hkl) reflecting planes from a minimum of two orders of a reflection (e.g., 111 and 222), and the more orders present, the better. A similar extra-polation-function plot is made. In general the determination of lattice constants for noncubic crystals is more difficult than for cubic crystals (Cullity, 1956). This chapter will be concerned primarily with analyses applicable to cubic crystals.

The uniform strain ε_u^{hkl} in the $\{hkl\}$ reflecting planes is defined as $\varepsilon_u^{hkl} = (d^{hkl} - d_0^{hkl})/d_0^{hkl}$, where d_0^{hkl} is the $\{hkl\}$ interplanar spacing of the unstrained crystal and d^{hkl} refers to the measured crystal. For cubic crystals $\Delta d/d = \Delta a/a$, since $d^2 = a^2/(h^2 + k^2 + l^2)$; and the uniform strain in the lattice para-meter $\varepsilon_u = (a - a_0)/a_0$. Obviously the measured value of ε_u for a given thin

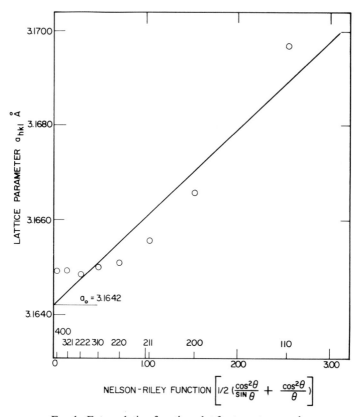

FIG. 1. Extrapolation function plot for tungsten powder.

film depends upon many factors, prime among which are the deposition conditions and annealing treatments. Typical results for polycrystalline films have been given by Suhrmann *et al.* (1960), Vook and Witt (1965a, 1966), Vook *et al.* (1967, 1972), Walker and Goldsmith (1970), Wedler and Wissmann (1968b), and Witt *et al.* (1968). Uniform strains in epitaxial monocrystalline Cu_2O (Borie *et al.*, 1962) and NiO (Cathcart *et al.*, 1969) films have also been reported. Values ranging from tenths of one percent to a few percent have been observed.

C. STACKING FAULTS IN fcc METALS

Hcp, bcc, and fcc metals may be thought of as being composed of layer structures in which deviations from the normal stacking sequence may occur. Such deviations are called "faults" of one kind or another. They give rise to diffraction effects because additional phase changes are introduced in waves that traverse them. The observed diffraction effect depends not only on the phase change but on how the faults are arranged in the material being examined. This section will be concerned primarily with peak shifts due to stacking faults in polycrystalline fcc metals. However, a short introduction to the diffraction effects from the various faults in hcp and bcc metals is also given. The normal faults in these latter two structures do not give rise to peak shifts in powder diffraction patterns.

Hexagonal close-packed metals may contain stacking (deformation) faults (Christian, 1954) and growth faults (Wilson, 1942) on {002} planes. If the atomic stacking sequence on {002} layers in hcp crystals is normally ABAB, then a stacking fault is illustrated by the sequence ABABCACA, with the fault lying between the adjacent B and C layers. A growth fault gives rise to the sequence ABABCBCB (Warren, 1969). These faults in hcp metals give rise to no peak shifts and to no diffraction line asymmetries. However, stacking faults produce changes in peak intensities for the various diffraction lines, while the line broadening is the same for all lines. For growth faults the line broadening depends on which planes are reflecting. These facts have been used to calculate fault densities in hcp metal filings (Stratton and Kitchingman, 1966).

Body-centered cubic metals may contain stacking (deformation) faults and twins on {211} planes (Hirsch and Otte, 1957; Guentert and Warren, 1958). Neither fault introduces peak shifts in powder patterns. Small peak shifts, however, are possible due to small changes in lattice spacing at the fault (Wagner et $al.$, 1962). Twin faults introduce small diffraction line asymmetries that are too small to give a significant measure of the twin fault density. Some information on the composite density of deformation faults and twins can be obtained, but it is difficult to separate the two (Warren, 1969).

Intrinsic and extrinsic stacking faults in fcc metals (ISFs and ESFs) produce systematic peak shifts that depend on which planes are reflecting (Paterson, 1952; Johnson, 1963; Warren, 1963; Holloway and Klamkin, 1969). The magnitudes of the shifts are equal in the two cases but their directions ($\pm \theta$) are opposite. The net density of stacking faults in the sample can be determined from the magnitude of the peak shifts. In an extrapolation-function plot of the lattice parameter obtained from each Bragg reflection, the vertical deviation of a point from the straight line for a faulted crystal is given by

$$(\Delta a)_{hkl} = (G_{hkl} a_0) \alpha_{hkl} \qquad (1)$$

where a_0 is the average lattice parameter and α_{hkl} is the stacking fault density in the {hkl} oriented grains (Wagner, 1957; Smallman and Westmacott, 1957). The calculated values of G_{hkl} are as follows: -0.0345, $+0.0689$, -0.0345, $+0.0125$, $+0.0172$, -0.0345, -0.0073, $+0.0069$, 0, $+0.0029$, for the 111, 200, 220, 311, 222, 400, 331, 420, 422, and $333 + 511$ reflections, respectively

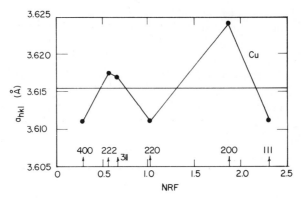

FIG. 2. Extrapolation function plot of a copper film having an ISF density α equal to 3.5×10^{-2} with $a_0 = 3.6153$ Å (courtesy Western Periodicals, North Hollywood, Cal.).

(Wagner *et al.*, 1962). Figure 2 illustrates the method and the systematic peak shifts obtained in one particular case (Vook, 1968). It is a theoretical plot based upon Eq. (1). In a polycrystalline sample, and especially a thin film, it is possible for the stacking-fault density to vary from crystallite to crystallite.

Experimental plots usually differ somewhat from the ideal case (Vook and Witt, 1965a, 1966; Light and Wagner, 1966; Marchetto *et al.*, 1968; Witt *et al.*, 1968). In the usual diffractometer experiment (Bragg–Brentano focusing) a flat polycrystalline thin film is used as the specimen (see Fig. 3). Thus the *hkl* diffraction line arises almost entirely from {*hkl*} oriented grains, that is, crystallites that have their {*hkl*} planes parallel to the surface of the film. For example, the 111 diffraction line comes almost entirely from {111} oriented grains. Consequently, a variation in stacking-fault density from one set of {*hkl*} oriented grains to another would cause a nonideal set of systematic peak

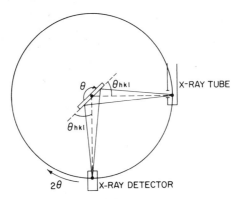

FIG. 3. Bragg–Brentano focusing on an X-ray diffractometer.

shifts for the various diffraction lines. Indeed, it seems likely that during film growth, such a nonuniform fault density arises in the individual grains comprising the film.

Stacking faults also cause diffraction line broadening because the size of the coherently diffracting domains is reduced. ISFs produce a symmetric line broadening (Paterson, 1952), while ESFs produce complex asymmetric line broadening more like that obtained from twins (Johnson, 1963; Warren, 1963).

In Cu, Au, and Ag films deposited on substrates near 80°K in high vacuum, ISF densities α ranged from about 0.01 to 0.06 (Vook and Witt, 1965a, 1966; Witt et al., 1968) with α expressed in units of number of faults per $\{111\}$ plane. Thus a value of 0.01 means that on the average there is one fault every 100 $\{111\}$ planes in the sample. Stacking-fault densities have been determined for films made under other experimental conditions as well (Hofer and Hintermann, 1965; Light and Wagner, 1966; Wedler and Wissmann, 1968b).

D. THERMALLY INDUCED STRAINS

1. Thermal Expansion Strains

Cubic materials have isotropic thermal expansion coefficients. Thus when their temperature is changed, the corresponding changes in interplanar spacings are independent of crystallographic direction. Such is not the case for crystals whose thermal expansion properties are anisotropic.

Consider the case of a "rigid" thin film (or a nonadhering thin film lying on a rigid substrate) having anisotropic thermal expansion coefficients. Let $\varepsilon_3' = (\Delta d/d)_{hkl}$ be the thermal expansion strain in the $\{hkl\}$ reflecting planes as measured in the normal Bragg–Brentano X-ray diffractometer focusing experiment (Fig. 3). Thus $(\Delta d/d)_{hkl}$ is the fractional change in d-spacing due to thermal expansion of the (hkl) reflecting planes. In crystalline materials with anisotropic expansion coefficients, ε_3' depends upon hkl. [See Fig. 4, where this result has been illustrated for nonadhering indium films cooled from 293 to 80°K (Witt and Vook, 1969).] The abscissa is the usual Nelson–Riley extrapolation function. In cubic crystals the points corresponding to the allowed reflections would lie on a horizontal straight line. The systematic individual deviations can, of course, be used to measure the thermal expansion coefficients.

Thus a nonadhering thin film lying on a rigid substrate may experience thermal expansion strains when its temperature is changed. These strains in anisotropic crystals are dependent on which $\{hkl\}$ planes are giving rise to the observed X-ray reflection. They thus cause systematic diffraction line peak shifts.

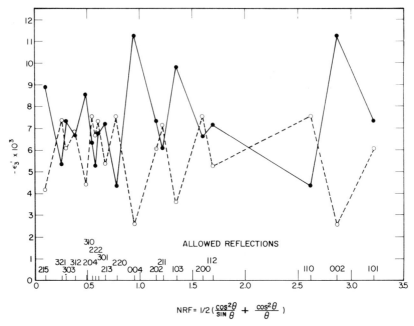

FIG. 4. Normal and substrate-induced thermal expansion strains illustrated for non-adhering (○) and adhering (●) indium films on glass (293–80°K).

2. Substrate-Induced Thermal Strains

Most crystalline materials are elastically anisotropic. When they are stressed, the resulting strains depend strongly on crystallographic direction. Thin films are frequently attached to substrates having different thermal expansion coefficients and therefore can experience substrate-induced strains when the temperature is changed. For polycrystalline films of elastically anisotropic materials, the thermally induced strain in a given crystallite depends strongly on the orientation of that crystallite on the substrate (Vook and Witt, 1965b; Witt and Vook, 1968, 1969).

Using again the normal X-ray diffractometer Bragg–Brentano focusing method, one will observe systematic diffraction line peak shifts from such a film–substrate combination when the temperature is changed. These peak shifts can be converted to the corresponding substrate-induced strains by again using the formula $\varepsilon_3' = (\Delta d/d)_{hkl}$. Typical results for adhering indium films on glass substrates are shown in Fig. 4, which is based on calculations by Witt and Vook (1969) for films formed on glass at 293°K and cooled to 80°K. It is to be noted that substrate-induced thermal strains are quite different from the thermal expansion strains that would be observed for nonadhering indium films. Thus a basis for distinguishing between adhering and nonadhering thin films has been developed.

Not only have substrate-induced strains been calculated as a function of crystallographic orientation, but also the corresponding strain energies per unit volume U_{hkl} (Vook and Witt, 1965b; Witt and Vook, 1968, 1969). These energies appear to be useful in interpreting relative grain growth in thick polycrystalline films deposited at low temperatures and annealed (Vook and Witt, 1965a, 1966; Witt et al., 1968).

3. Film–Substrate Adhesion

Figure 4 shows that the thermally induced strains for adhering and non-adhering indium films on glass are quite different. A similar plot for tin films on glass was given earlier (Vook, 1968). In both cases, the two patterns of relative peak shifts tend to be rough opposites of each other. Thus by measuring the peak shifts and calculating the corresponding thermally induced strains ε_3' for a particular film–substrate combination and temperature cycle, one can readily distinguish adhering from nonadhering films.

Moreover, one can in principle detect adherence in the individual grains of a polycrystalline film. For example, if the {103} oriented grains in indium, i.e., grains with their {103} planes parallel to the surface of the film, suddenly lost total or partial adherence during thermal cycling, their corresponding peak shift would tend to deviate from the value calculated for adherence. It would tend to move toward the value calculated for nonadherence. Lateral strains induced by neighboring grains would probably prevent a complete reversal of strain. Thus, in principle, the adherence of individual sets of {khl} oriented grains during thermal cycling could be investigated.

E. SURFACE RELAXATION PHENOMENA

1. Experiments

Recent X-ray diffractometer measurements of the Bragg angles from thin, highly oriented polycrystalline, uniformly thick Sn, In, and Au films have shown that this angle is thickness dependent (Vook and Otooni, 1968a,b; Vook et al., 1967, 1972). Precise X-ray measurements on these films in the 100–1000-Å-thick range are made possible by their strong fiber textures.

The fact that the films are "uniformly thick" is deduced from the (a) photographic or (b) diffractometer detection of the sharp fringes in the X-ray interference function corresponding to the hkl Bragg reflection (Croce et al., 1961, 1962, 1966; Vook and Witt, 1965a; Vook and Otooni, 1968a; Vook et al., 1967, 1972). See Figs. 5 (Selleck, 1972) and 6 (Otooni and Vook, 1968b) for typical examples. The formulas for determining film thickness in these two

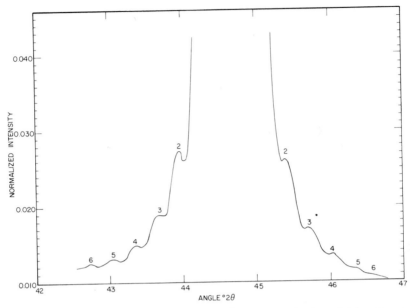

FIG. 5. Diffracted intensity obtained for a 370-Å-thick gold film by point counting over the 111 X-ray reflection. The precision is given approximately by the width of the trace. Subsidiary fringes indicated by number.

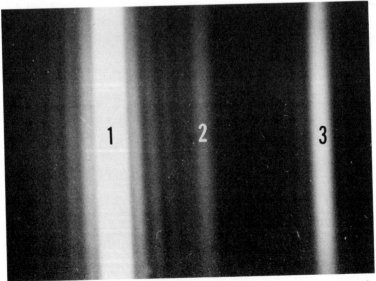

FIG. 6. Photographic recording of the X-ray interference function from a 325-Å-thick gold film. The three patterns are due to CuKα (1), WLα (2), and CuKβ (3).

cases have been derived earlier (Croce *et al.*, 1962; Vook *et al.*, 1967):
(a) diffractometer trace

$$t = \lambda/2 \, \Delta\theta \cos\theta = d \tan\theta/\Delta\theta \qquad (2)$$

where λ is the X-ray wavelength, θ the Bragg angle, and $2\Delta\theta$ the angular spacing in radians between successive minima; and (b) photographic film

$$t = (\lambda/\cos\theta)(L/L_1) \qquad (3)$$

where the photographic film is placed perpendicular to the diffracted beam, L is the distance from the axis of the diffractometer to the photographic film, and L_1 the distance on the film between the fringe minima on one side of the pattern. An appreciable variation in film thickness in these films would tend to wash out the fringes. This method for film-thickness determination is precise to within a few percent in the range 100–1000 Å.

The film-thickness dependence of the observed 100, 101, and 111 reflections from Sn, In, and Au films, respectively, is illustrated in Fig. 7. These reflections correspond to the observed fiber textures. In the figure, $\Delta d/d$ is the fractional change in interplanar spacing as calculated from the measured Bragg angles. The value of d in the denominator is the extrapolated value for thick films. A $t^{-1/2}$ least-squares straight-line fit to the Sn data was used (Vook *et al.*, 1967; Vook, quoted in Chopra, 1969), where t is the film thickness. Both In (Vook and Otooni, 1968b) and Au (Vook *et al.*, 1972) were fit with $1/t$ least-squares straight lines. The quantity $\Delta d = d_{hkl} - d$, where d_{hkl} is the measured interplanar spacing of the $\{hkl\}$ reflecting planes.

While Sn has the advantage of not adhering strongly to the glass substrate (Bourgault and Vook, 1967), it has the distinct disadvantage of forming an oxide on its surface in air. Thus the results may be influenced by such oxide-induced strains.

The In films were formed on glass near liquid-nitrogen temperature and annealed to 15°C. Heating indium above 15°C results in a transformation in which the previously uniformly thick film tends to agglomerate into small particles (Ehrhart and Marrud, 1964). When this transformation occurs, the X-ray interference fringes disappear. Thus the films had to be cooled during the X-ray investigations. Again the results for In may possibly have been influenced by substrate and/or surface oxide-induced strains. Moreover, the proximity to the agglomeration transformation temperature during X-ray measurement may have been significant. It has been shown for small particles that, as a result of surface tension phenomena, the particles tend to be compressed. Consequently they exhibit smaller than bulk interplanar spacings and the effect is larger the smaller the particle (Komnik, 1964; DePlanta *et al.*, 1964; Mays *et al.*, 1968; Vermaak and Kuhlmann-Wilsdorf, 1968).

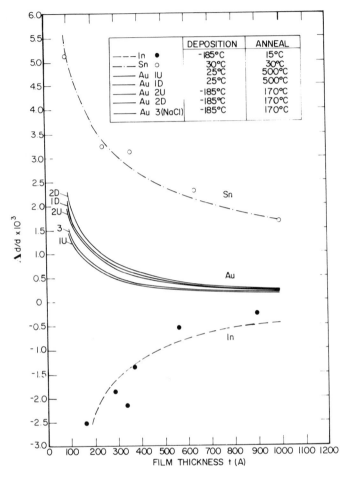

FIG. 7. Variation of the fractional change in interplanar spacing as a function of film thickness.

2. Theory

Two analyses of the Au lattice parameter–thickness data have been carried out (Vook *et al.*, 1972; Shiah and Vook, 1973). In the first it was assumed that, to a first approximation, the interplanar spacing $\bar{d}\ (=d_{hkl})$ corresponding to the measured Bragg angle is the average of all the interplanar spacings d_i in the reflecting planes. The measured change in interplanar spacing is then attributed to an outward relaxation of the first few atom layers near the surface. Each d_i is assumed to be dependent upon bulk (B) and surface (S) strains such that

$d_i = d_0(1 + \alpha_i^B + \alpha_i^S)$, where α_i^B and α_i^S are the fractional changes in d_0, and d_0 is the unstrained bulk interplanar spacing. Thus the relative interplanar spacing $\bar{\delta}_p = \bar{d}/d_0$ for a film of p layers and constant α_i^B ($=\alpha^B$), is given by

$$\delta_p = (1 + \alpha^B) + \frac{q}{p}\sum_{i=1}^{N}\alpha_i^S = 1 + \alpha^B + \frac{q}{p}(\alpha^S) \qquad (4)$$

where N is a small integer equal to the number of layers effectively contributing to the surface relaxation. When only one surface is free and therefore relaxed, $q = 1$. For two free surfaces, i.e., for a film that is detached from its substrate, $q = 2$. Since the film thickness $t \simeq pd_0$, the measured interplanar spacing $\bar{d}$ varies linearly with $1/t$.

Vook et al. (1972) showed that by comparing the measured data on Au with Eq. (4), it was possible to evaluate α^B and α^S. Moreover, the individual α_i^S values could be estimated by making reasonable assumptions for the ratios $\alpha_{i+1}^S/\alpha_i^S$. Values of α^S in the range of 3 to 5% were obtained for the (111) Au surface.

The above analysis rests on the assumption that one can average the d-spacings in a set of planes and that the average d-spacing is the one obtained from the Bragg law. As a result one finds that the interplanar spacing depends linearly on the reciprocal of the film thickness. A subsequent analysis (Shiah and Vook, 1973) was based on the kinematic theory of X-ray diffraction (James, 1965). In this method, a distribution of d-spacings in the reflecting planes was assumed (exponential decrease from the surface) and the resulting exact diffraction line peak shift calculated.

A good approximation to the exact peak shift was also derived and the equivalent expression is given in Eq. (5):

$$\bar{d} = d_0 + C(1/t)^2 \qquad (5)$$

where C is a constant that depends upon α^S and the assumed constant ratio between the magnitude of successive interplanar spacings near a free surface.

Thus the exact analysis shows that when the surface layers are relaxed outward, the shift in the experimental diffraction line peak position indicates an increased interplanar spacing and the magnitude of this shift varies linearly with $(1/t)^2$. The precision of the experimental X-ray data for Au is such that it does not allow one to distinguish between a $1/t$ or $(1/t)^2$ dependence.

Applying this kinematic theory analysis to the Au data gives relatively large values of α^S when the surface strains are assumed to decrease rapidly as one goes away from the surface. Reasonable values are obtained only when a slowly decreasing strain distribution is assumed. Thus any residual nonuniform strain distribution in the films would clearly affect the results obtained from this analysis. Indeed, those Au films for which α^B was smallest gave the lowest and most reasonable α^S values. Furthermore, a $(1/t)^2$ relation will be more sensitive to surface phenomena in thin films than a $1/t$ relation. Thus strains due to surface oxide or adsorption layers would have a bigger effect on a $(1/t)^2$ analysis. Clearly more work is needed in this area. Nevertheless, it has been shown that X-ray diffraction techniques applied to thin films of varying thickness can be useful in studying surface phenomena.

3. Film–Substrate Adhesion

Another way of using X-ray diffraction methods to measure film–substrate adhesion (see Section II,D,3) was given recently (Vook *et al.*, 1972). The method is based on Eq. (4). For a film that is detached from its substrate, $q = 2$. For a film still on its substrate and adhering "perfectly" to it, $q = 1$, assuming that there are no substrate-induced interfacial strains in the reflecting planes. Thus by comparing the lattice parameter versus thickness data for detached and undetached films, it was possible to determine what fraction of the undetached film adhered to the substrate. It was shown that Au films deposited on glass and annealed in vacuum at 500°C had much greater adhesion to the substrate than films deposited at −185°C and annealed at 170°C.

F. X-Ray Reflection

This subject has been thoroughly discussed recently by Bertin (1969), and only a brief summary will be given here. Using a highly collimated, monochromatic X-ray beam that strikes the surface of a smooth film at near the angle of total reflection ($\lesssim 1°$), one can obtain two sets of fringes, one on the high-angle and one on the low-angle side of the specularly reflected beam (Kiessig, 1931; Parratt, 1954; Sauro *et al.*, 1966). One set is due to reflected X rays (Wainfan and Parratt, 1960) and the other to scattered X rays (Sauro *et al.*, 1963). In each case the fringes arise from the interference between waves originating at the film–air interface and waves coming from the film-substrate interface. Recently Umrath (1967) has shown that a polychromatic X-ray beam can be used instead of monochromatic radiation.

Studies of these fringe patterns may be used in at least four ways for thin-film investigations. By comparing the experimental curves with theory, one can obtain information on the electron density at the reflecting interface. Interface roughness will also affect the shape of the experimental curve by tending to smear out and obliterate the pattern. Third, film thickness can be calculated from the angular positions of the fringes. Sauro *et al.* (1963) and Umrath (1967) report accuracies of 5 and 1.5%, respectively. And finally the average film density can be calculated from the experimentally measured critical angle for total reflection. Films in the range of ten to several thousand angstroms thick may be investigated with these methods.

III. Diffraction Line Profile Analyses

A. PARTICLE SIZE AND NONUNIFORM STRAIN

Initial qualitative information on thin films frequently can be obtained using elementary X-ray diffraction techniques: pinhole, Laue, and Debye–Scherrer (Barrett and Massalski, 1966; Cullity, 1956). A low-angle ($\sim$5–10°) variant of the Laue technique has been used by Heinzel and Walker (1966) and Walker (1970) for the quick evaluation of a film–substrate sample. Isherwood and Quinn (1967) and Quinn (1970) have reported on an edge-irradiated glancing angle X-ray diffraction film technique that is applicable to thin films and that is a variant of the cylindrical camera method. In this section, however, we will be concerned primarily with a quantitative analysis of diffraction line broadening from polycrystalline films. However, as mentioned earlier, the same methods apply to single-crystal films (Borie, 1960; Borie and Sparks, 1966).

In a coarse-grained film giving rise to spotty pinhole diffraction patterns, it is possible to determine the average crystallite size by counting spots in the Debye rings (Warren, 1960; Taylor, 1961; Wagendristel et al., 1969). However, when the thickness of a coherently diffracting domain measured in a direction perpendicular to the reflecting planes decreases to around 1500 Å or less, detectable diffraction line broadening occurs. This result follows directly from the kinematical expression for the diffracted intensity $J(\varepsilon)$ from a crystal having p reflecting planes (James, 1965):

$$J(\varepsilon) = |q|^2 \sin^2(pB\varepsilon)/\sin^2(B\varepsilon) \qquad (6)$$

where $B = 2\pi d(\cos \theta_0)/\lambda$, q is the reflection coefficient for each (hkl) reflecting plane, and d is the interplanar spacing. The angle ε is given by $\theta = \theta_0 + \varepsilon$, where θ_0 is the Bragg angle. Thus $J(\varepsilon)$ is a symmetric function of ε centered at the Bragg angle θ_0, i.e., where ε is zero. Figures 5 and 6 are the experimental intensity distributions corresponding to Eq. (6). For various reasons, the experimentally observed lines from typical samples do not reveal this fringed function but rather give the usual bell-shaped curve. By use of an Ewald-type construction shown in Fig. 8, it can be seen that the line broadening $\Delta\varepsilon$ corresponding to the half-width of the principal maximum of Eq. (6) is related to the thickness t ($=pd$) of the coherently reflecting domains approximately as shown in Eq. (7). In diffractometer work, one measures $\Delta 2\varepsilon$, so that we have

$$\Delta(2\varepsilon) = 2\pi/pB = \lambda/t \cos \theta_0 \qquad (7)$$

Thus the line broadening is inversely proportional to t. In a similar manner, the mosaic spread in monocrystalline films has been defined as equal to the angular width at half the maximum intensity of the observed rocking curves (Cathcart et al., 1967, 1969).

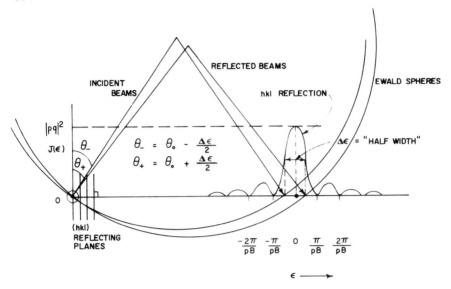

FIG. 8. Ewald construction corresponding to the intensity $J(\varepsilon)$ in reciprocal space.

One can also show that particle-size line broadening is independent of the reflection, that is, it is the same for all hkl reflections. To prove this statement requires a conversion of Eq. (7) to reciprocal space coordinates where it can be more conveniently handled. A consideration of the Ewald construction (James, 1965) shows that it is the intensity distribution in reciprocal space that corresponds directly to the measured intensity distribution. Actually we will be concerned only with one set of (hkl) reflecting planes and the various orders of a reflection that the structure factor allows. Thus if $\mathbf{g}_{hkl}$ is the vector from the origin of reciprocal space to the hkl reciprocal lattice point (relp), then $|\mathbf{g}_{hkl}| = g_{hkl} = 1/d_{hkl} = 2(\sin\theta)/\lambda$. Then the line broadening in reciprocal space is given by

$$\Delta g_{hkl} = \cos\theta\,(\Delta 2\varepsilon)/\lambda = 1/t \tag{8}$$

Equation (8) means that the half-width of the intensity distribution in reciprocal space equals the reciprocal of the thickness of the coherently reflecting planes. Thus in fcc crystals, for example, the particle-size line broadening in a 111 reflection is the same as in the 222, 333, etc., reflections.

On the other hand, the line broadening due to nonuniform strain depends strongly on the reflection and is zero at the origin of reciprocal space. If $\Delta d/d = \varepsilon$ is a measure of the fractional spread in interplanar spacings d for a given set of reflecting planes in the crystal, then $\Delta g/g = -\Delta d/d = -\varepsilon$. Thus the line broadening Δg in reciprocal space is proportional to the magnitude of the reciprocal lattice vector to the relp responsible for the reflection.

Consequently, comparing the line broadening in at least two orders of a reflection makes possible the separation and evaluation of the two broadening effects. Included in particle-size line broadening is, for example, the symmetric broadening arising from intrinsic stacking faults and the asymmetric line broadening produced by extrinsic stacking faults and twin faults, all in fcc materials. All such faults tend to reduce the size of the coherently diffracting domains. In the next two sections on Fourier analysis and integral breadth, these two most common methods for carrying out this separation are discussed.

B. FOURIER ANALYSIS

This method for separating particle size from nonuniform strain line broadening is generally called the "Warren–Averbach method" (Warren and Averbach, 1950, 1952). It has been discussed thoroughly by Warren (1969) and Wagner (1966b) for polycrystalline materials and by Borie (1960) and Borie and Sparks (1966) for monocrystals.

Basically the method consists of Fourier-analyzing at least two orders of a diffraction line, e.g., 200 and 400 in fcc crystals. After suitable corrections are made, such as for example for $K\alpha_1 - K\alpha_2$ overlap (Rachinger, 1948), instrumental effects (Stokes, 1948), and polarization, the resulting Fourier coefficients are used to obtain $D_e(F)$, the effective thickness of the coherent domains perpendicular to the reflecting planes, and $\varepsilon(F)$, the rms nonuniform strain averaged over the dimensions of the effective particle size. The procedures are fairly involved so that computer programs have been found to be very useful, if not essential (DeAngelis, 1966; Hoff and Kitchingman, 1966; Wagner, 1966a, 1969; Rashid and Allstetter, 1970).

In thin films having a large amount of diffraction line broadening, the higher-order lines tend to be very weak. Moreover in fcc polycrystals, the 222 line tends to overlap the 311 line. A similar situation may exist between 111 and 200. Thus before being able to accurately Fourier analyze a given line, one must separate its intensity from that of any neighboring line. No very good method exists for performing this operation. Consequently integral breadth methods that are in principle less accurate but simpler to use are often employed when the line broadening is severe. Thus when the line broadening is not large, as for the case of films deposited or annealed at relatively high temperatures (Segmüller, 1963; Light and Wagner, 1966), or in some electro-deposited films (Hofer and Javet, 1962; Eichkorn and Fischer, 1964; Hofer and Hintermann, 1965), the Fourier analysis method is preferred. However, recently the accuracy of the integral breadth particle size analysis method was investigated using annealed gold films (Atasagun and Vook, 1970). It was shown that the integral breadth analysis was accurate to better than 10% for particle sizes varying from 100 to 700 Å.

C. INTEGRAL BREADTH

Use of an integral breadth method to analyze line broadening requires the employment of theoretical expressions for the various components of the observed diffraction line. The integral breadth β is defined as

$$\beta = \int J(2\theta) \, d(2\theta)/J_{max} \qquad (9)$$

where $J(2\theta)$ is the diffracted intensity. Four commonly used expressions are Cauchy, Cauchy–Gaussian mean, Gaussian, and monocrystal (Schoening *et al.*, 1952; Klug and Alexander, 1954; Taylor, 1961; James, 1965). The latter distribution is given by Eq. (6). Generally particle-size line broadening tends to follow a Cauchy distribution, while nonuniform strain may be approximated by the Cauchy–Gaussian mean or the Gaussian distribution.

The initial step in the integral breadth analysis is to determine the true diffraction line integral breadth β, i.e., instrumental effects must be removed from the broadened line. The method generally used is to compare the observed broadened line with one having only instrumental broadening. This latter line will be called the "standard line." It may arise from an annealed specimen consisting of the same material as that being investigated or some other annealed material having lines near those of the sample under investigation. To make this comparison, one first separates the $K\alpha$ diffraction line into its $K\alpha_1$ and $K\alpha_2$ components, usually by using the Rachinger (1948) method.

One then obtains the integral breadths b of lines from the standard sample and B from the sample being analyzed. The best theoretical fits for the lines of the standard sample are then made. That is, it is determined whether the standard lines are best approximated by the Cauchy, Cauchy–Gaussian mean, or Gaussian distributions. Then the true diffraction line integral breadths β are calculated using graphical methods that relate β/B to b/B (Klug and Alexander, 1954; Taylor, 1961; Ruland, 1968). Schoening (1967) reported such graphs for monocrystalline films or their equivalent (fiber texture perpendicular to the film).

Once the diffraction breadth β has been obtained, it is then necessary to separate the particle size from the nonuniform strain breadths. Schoening (1965) developed a graphical method for the 111–222 and 200–400 fcc reflections. He considers two cases; in both he assumes particle-size broadening to be Cauchy. Strain broadening, however, was assumed to be either Cauchy–Gaussian mean or Gaussian. Use of his graphs gives D_e, the effective particle size, and $\bar{\varepsilon}$, the average nonuniform strain in the reflecting planes. An approximation to Schoening's (1965) method has been given by Halder and Wagner (1966).

When Schoening's (1965) method is not applicable, then other graphical methods may be used to determine the particle-size integral breadth β_p and the nonuniform-strain integral breadth β_s (Klug and Alexander, 1954; Taylor, 1961). Finally D_e and $\bar{\varepsilon}$ are calculated from the following formulas (Cullity, 1956):

$$D_e = K\lambda/(\beta_p \cos\theta) \tag{10}$$

$$\bar{\varepsilon} = \beta_s/(2 \tan\theta) \tag{11}$$

The constant K is a "shape" factor of order 1 and has been calculated for a number of cases (Klug and Alexander, 1954; Taylor, 1961). It depends on the indices of the reflection, the crystal structure, and the "shape" of the particle.

A reasonable approximation to use in this situation is to solve the problem for two extreme cases. One assumes that particle size and strain broadening have either both Cauchy or Gaussian line shapes. Then D_e and $\bar{\varepsilon}$ are calculated for each case and the respective values averaged, since the true value most likely lies somewhere in between.

When the particle-size and microstrain line broadening are assumed to be Cauchy, one may add the integral breadths:

$$\beta = \beta_p + \beta_s \tag{12}$$

$$\beta = \frac{\lambda}{D_e \cos\theta} + 2\bar{\varepsilon} \tan\theta \qquad \text{(Cauchy)} \tag{13}$$

where the shape factor K in Eq. (10) was taken as unity. A plot of ($\beta \cos\theta$) vs $\sin\theta$ for two orders of a reflection gives a straight line with slope $2\bar{\varepsilon}$ and whose intercept at $\sin\theta = 0$ is λ/D_e.

When both lines are taken as Gaussian, one can add the squares of the breadths:

$$\beta^2 = \beta_p^2 + \beta_s^2 \tag{14}$$

$$\beta^2 = \left(\frac{\lambda}{D_e \cos\theta}\right)^2 + (2\bar{\varepsilon} \tan\theta)^2 \qquad \text{(Gaussian)} \tag{15}$$

A plot of ($\beta \cos\theta$)^2 vs $\sin^2\theta$ is made. Again the slope $(2\bar{\varepsilon})^2$ and the $\sin^2\theta = 0$ intercept $(\lambda/D_e)^2$ give the $\bar{\varepsilon}$ and D_e.

Schoening's (1965) integral breadth method has been compared to the Warren–Averbach Fourier analysis method for a powder by Halder and Wagner (1966). The Fourier analysis effective particle size $D_e(F)$ is the thickness of the coherent domain in a direction perpendicular to the reflecting planes. The integral breadth particle size $D_e(\beta)$ for a powder is proportional to the cube root of the volume of the particles, the proportionality factor giving the shape of the particle. In general $D_e(\beta) \geqslant D_e(F)$.

The Fourier analysis strain $\bar{\varepsilon}(F)$ is the rms strain averaged over the dimension of the effective particle size. The integral breadth strain is $\bar{\varepsilon}(\beta)$. Experimentally Wagner and Aqua (1964) have shown for filings that $\bar{\varepsilon}(\beta) \simeq 1.25\bar{\varepsilon}(F)$.

Aqua (1966) compared measurements of particle size and strain by three techniques, namely Fourier analysis, integral breadth, and variance. He showed that for filings, the three methods yielded results that were of the same order of magnitude. Similar results were obtained by Langford (1968).

Only one check of the accuracy of an integral breadth method for measuring particle size in thin films has been reported (Atasagun and Vook, 1970). This work was based on the theoretical developments of Schoening (1965, 1967). In the experimental work the crystallites comprising the Au films extended all the way through the film. Thus it was known that the particle size equaled the film thickness. Carrying through the integral breadth analysis, it was found that $D_e(\beta)$ equaled the film thickness to better than 10% in the range 100–700 Å. Thus this particular integral breadth analysis was verified for the measurement of particle size.

Once the effective crystallite size D_e has been obtained, the various contributions to D_e must be sorted out. The line broadening due to intrinsic stacking faults and twin faults in fcc crystals results in particle sizes D_α and D_β, respectively (Wagner et al., 1962), which are determined from the respective fault densities α and β. By considering only the main contributions to D_e, one has

$$\left(\frac{1}{D_e}\right)_{hkl} = \frac{1}{D_{hkl}} + \frac{1}{D_\alpha} + \frac{1}{D_\beta} \tag{16}$$

where D_{hkl} is the true crystallite size in a direction perpendicular to the (hkl) reflecting planes (Warren, 1969).

A number of experimental investigations of the line broadening from thin films using integral breadth methods have been reported (Vook and Witt, 1965a, 1966; Witt et al., 1968; Wedler and Wissmann, 1968b; Walker, 1970). Experimental values of $\bar\varepsilon(\beta)$ range from a few tenths of a percent to a few percent depending upon the nature of the film, i.e., its deposition conditions and annealing treatment. Copper, silver, and gold films deposited on substrates near 80°K in high vacuum (Vook and Schoening, 1963) have D_{hkl} values equal to hundreds of angstroms (Vook and Witt, 1965a, 1966; Witt et al., 1968). It was found that a large proportion of the line broadening was due to faults and twins formed during deposition. Thus these analyses emphasized the importance of sorting out and evaluating the various contributions to diffraction line broadening.

D. Radial Distribution Function Method

When thin films are formed by liquid (Duwez et al., 1960; Duwez and Willens, 1962) or vapor quenching (Buckel and Hilsch, 1952; Buckel, 1954; Rühl, 1954; Mader, 1965) onto cold substrates, broad and diffuse rings may be observed in their electron and X-ray diffraction patterns. In such films the particle sizes are less than about 20 to 30 Å. Frequently, their structure has been intepreted qualitatively as being amorphous. However, one usually defines an amorphous structure as one with no crystallike ordering beyond nearest neighbor atoms; and even the nearest neighbor bonds are expected to be distorted. Consequently a more quantitative analysis of a diffraction pattern containing broad rings is required to establish whether such a film is truly amorphous or rather micropolycrystalline. Such a method is the radial distribution function (RDF) method.

The RDF is defined as $4\pi r^2 \rho(r)$, where $\rho(r)$ is the atomic distribution function, i.e., the weighted, local atomic density at a distance r from the center of an average atom or from an arbitrary origin (Guinier, 1963; Warren, 1969). The quantity $4\pi r^2 \rho(r)\, dr$ is then the average number of atoms between distances r and dr from the origin. The RDF is the Fourier transform of the interference function $I(K)$, which is given by the scattered X-ray or electron intensities.

It has been shown (Hosemann and Bagchi, 1962; Guinier, 1963; Wagner, 1968) that for an assembly of N atoms, the coherent, elastic X-ray scattering per atom $I_a^{coh}(k)$ in electron units (eu) is given by

$$I_a^{coh}(k) = \langle f \rangle^2 I_{sa}(k) + \langle f^2 \rangle - \langle f \rangle^2 + \langle f \rangle^2 I(k) \qquad (17)$$

where $k = 4\pi \sin\theta/\lambda$, $I_{sa}^{coh}(k)$ is the small angle scattering, $I(k)$ the interference function, $\langle f \rangle$ the average atomic scattering factor, and $\langle f^2 \rangle$ the mean-square scattering factor of the material.

The evaluation of particle sizes by the RDF method for X-ray diffraction using Eq. (17) has been carried through for vapor-quenched Ag–Cu (Wagner et al., 1968) and Ag–Ge (Light and Wagner, 1968) alloy films. The broad diffuse rings that were observed in the diffraction patterns were interpreted as due to microcrystals in the range of 12 to 45 Å in size. Thus these films cannot be characterized as amorphous but rather as micropolycrystalline.

The RDF method is thus a powerful tool for the quantitative analysis of amorphous and micropolycrystalline materials.

E. Twin Fault Densities in fcc Metals

Twin faults in fcc metals occur on {111} planes and produce asymmetric diffraction line profiles. The 111 line has excess intensity on its high-angle side, while the 200 line has excess intensity on its low-angle side. Measurements of this excess intensity in powder patterns are used to determine the twin fault density β, which is defined as the probability that a twin fault exists between any two {111} layers. Thus $1/\beta$ is the average twin plane spacing (Warren, 1969).

Several measurements of β in thin films have been reported. Appreciable densities of twins were measured in pure metals deposited at liquid nitrogen temperature in high vacuum (Vook and Witt, 1965a, 1966; Witt et al., 1968). Low densities were detected in films deposited at 130°C in poor vacuum (Light and Wagner, 1966). High densities were also reported in αCu–Ge alloy films by Marchetto et al. (1968). It is interesting to note that the twin and stacking fault densities reported for thin films formed in high vacuum are of the same order of magnitude as those reported on filings made at the same temperature.

IV. Integrated Intensity Measurements

A. Preferred Orientation

The X-ray integrated intensity is the total energy reflected as a crystal is rotated through the reflecting region (James, 1965). Essentially it is the area lying under a diffractometer trace of the line. Its magnitude is proportional to the volume of material giving rise to the reflection. Thus in a thin polycrystalline film, the integrated intensity for the hkl reflection is proportional to the volume of all crystallites that have their {hkl} planes lying parallel to the surface of the film.

A randomly oriented film has known relations between the relative integrated intensities of the various hkl reflections. These values have been tabulated in many cases (Powder Data File, 1967). For a given specimen, one compares the values obtained with the tabulated values for an unoriented, i.e., randomly oriented, powder. Differences are usually ascribed to preferred orientation.

A quantitative analysis of the preferred orientation in a particular specimen is best described by means of a pole figure. An {hkl} pole figure is a stereographic projection showing the density of {hkl} poles in a given direction in the sample. This pole density is directly related to the corresponding hkl integrated intensity (Cullity, 1956).

The preferred orientation of thin films has been measured quantitatively and recorded on a pole figure in a few cases (Witt et al., 1965; Schwartz and Witt, 1966; Wedlar and Wissmann, 1968a). A qualitative estimation of preferred orientation is much more frequently made (Wilkinson and Birks, 1949; Croce et al., 1961; Gandais, 1961; Segmuller, 1963; Vook et al., 1964; Vook and Witt, 1965a, 1966; Witt et al., 1968).

B. Grain Growth

When thin polycrystalline films are annealed, grain growth may occur.

Frequently such growth is accompanied by changes in texture. Measurements of the integrated intensities of the various *hkl* reflections allow one to measure changes in texture. Complete texture analyses require a determination of the pole figure at each stage of the anneal. None has been reported so far, although such a study would be very informative. Rather one usually looks only at the reflections coming from reflecting planes that are parallel to the surface of the film. Thus in this method, determining the integrated intensity from a 111 reflection, for example, gives a number that is proportional to the volume of crystallites having {111} planes parallel to the surface of the film. Changes in the 111 integrated reflection during annealing are thus proportional to changes in the volume of material in the {111} orientation.

Vook and Witt (1965a, 1966) and Witt *et al.* (1968) have made such measurements on thick films deposited on glass in high vacuum at 80°K and annealed to room temperature and higher. They showed that a correlation existed between the percent increase in volume of {*hkl*} oriented material and the corresponding substrate induced thermal expansion strain energies per unit volume (Vook and Witt, 1965b). However, for thin films, such a correlation did not exist, presumably because surface energies may be more important there than the volume strain energies.

C. FILM THICKNESS

Since the integrated intensity is a measure of the amount of material being sampled by the incident X-ray beam, it may be used to measure film thickness. Borie and Sparks (1961) developed such a method and applied it to the determination of the thickness of thin epitaxial Cu_2O films grown on a copper single crystal. Thicknesses of films considerably less than 100 Å (Cathcart *et al.*, 1969) have been measured with this technique.

Film thicknesses may also be determined by X-ray fluorescence and diffraction methods (Cullity, 1956). In the former, calibration of the system resulting in a plot of intensity versus thickness is required. In the case of diffraction methods one can examine the integrated intensity from the film and compare it with that obtained from an effectively infinitely thick sample. Another method is to use the peak intensity of a diffraction line from a crystalline substrate B supporting a film A. This intensity depends upon the absorption coefficient and thickness of the film A. Thus by comparing the peak intensities of a diffraction line from the substrate B in the case where the film A lies on B with the case of the substrate alone, one can calculate the thickness of the film. Methods involving use of the absorption of X rays to measure thickness usually require film thicknesses of several thousand angstroms for good precision.

ACKNOWLEDGMENTS

The author is grateful for the support of the National Science Foundation and expresses his appreciation to Western Periodicals, North Hollywood, California, for their permission to use Fig. 2.

References

Aqua, E. N. (1966). *Acta Cryst.* **20**, 560.

Atasagun, M., and Vook, R. W. (1970). *J. Vac. Sci. Technol.* **7**, 362.

Barrett, C. S., and Massalski, T. B. (1966). "Structure of Metals." McGraw-Hill, New York.

Bertin, E. P. (1969). *In* "Physical Measurement and Analysis of Thin Films" (E. M. Murt and W. G. Guldner, eds.), pp. 35–82. Plenum Press, New York.

Borie, B. S. (1960). *Acta Cryst.* **13**, 542.

Borie, B. S., and Sparks, C. J., Jr. (1961). *Acta Cryst.* **14**, 569.

Borie, B. S., and Sparks, C. J., Jr. (1966). *In* "Local Atomic Arrangements Studied by X-Ray Diffraction" (J. B. Cohen and J. E. Hilliard, eds.), pp. 341–349. Gordon and Breach, New York.

Borie, B. S., Sparks, C. J. Jr., and Cathcart, J. V. (1962). *Acta Met.* **10**, 691.

Bourgault, R., and Vook, R. (1967). Unpublished research.

Buckel, W. (1954). *Z. Phys.* **138**, 136.

Buckel, W., and Hilsch, R. (1952). *Z. Phys.* **132**, 420.

Cathcart, J. V., Peterson, G. F., and Sparks, C. J., Jr. (1967). *In* "Surfaces and Interfaces I: Chemical and Physical Characteristics" (J. Burke, N. Reed, and V. Weiss, eds.), pp. 333–346. Syracuse Univ. Press, Syracuse, New York.

Cathcart, J. V., Peterson, G. F., and Sparks, C. J., Jr. (1969). *J. Electrochem. Soc.* **116**, 664.

Chopra, K. L. (1969). "Thin Film Phenomena." McGraw-Hill, New York.

Christian, J. W. (1954). *Acta Cryst.* **7**, 415.

Croce, P., Gandais, M., and Marrud, A. (1961). *Rev. Opt.* **40**, 555.

Croce, P., Devant, G., Gandais, M., and Marrud, A. (1962). *Acta Cryst.* **15**, 424.

Croce, P., Devant, G., and Verhaeghe, M. F. (1966). *In* "Basic Problems of Thin Film Physics" (R. Niedermayer and H. Mayer, eds.), pp. 194–197. Vandenhoeck and Rupprecht, Göttingen.

Cullity, B. D. (1956). "Elements of X-Ray Diffraction." Addison-Wesley, Reading, Massachusetts.

DeAngelis, R. J. (1966). *In* "Local Atomic Arrangements Studied by X-Ray Diffraction" (J. B. Cohen and J. E. Hilliard, eds.), pp. 271–288. Gordon and Breach, New York.

DePlanta, T., Ghez, R., and Piuz, F. (1964). *Helv. Phys. Acta* **37**, 74.

Duwez, P., and Willens, R. H. (1962). *Trans. Met. Soc. AIME* **222**, 362.

Duwez, P., Willens, R. H., and Klement, W. (1960). *J. Appl. Phys.* **31**, 1136.

Ehrhart, J., and Marrud, A. (1964). *Rev. Opt.* **43**, 33.

Eichkorn, G., and Fischer, H. (1964). *Z. Metallk.* **55**, 582.

Gandais, M. (1961). *Rev. Opt.* **40**, 101, 306, 464.

Guentert, O. J., and Warren, B. E. (1958). *J. Appl. Phys.* **29**, 40.

Guinier, A. (1963). "X-Ray Diffraction." Freeman, San Francisco, California.

Halder, N. C., and Wagner, C. N. J. (1966). *Acta Cryst.* **20**, 312.

Heinzel, C. O., and Walker, G. A. (1966). *J. Appl. Phys.* **37**, 3809.

Hirsch, P. B., and Otte, H. M. (1957). *Acta Cryst.* **10**, 447.

Hofer, E. M., and Javet, P. (1962). *Helv. Phys. Acta* **35**, 369.

Hofer, E. M., and Hintermann, H. E. (1965). *J. Electrochem. Soc.* **112**, 167.

Hoff, W. D., and Kitchingman, W. J. (1966). *J. Sci. Instrum.* **43**, 654.

Holloway, H., and Klamkin, M. S. (1969). *J. Appl. Phys.* **40**, 1681.

Hosemann, R., and Bagchi, S. N. (1962). "Direct Analysis of Diffraction by Matter." North-Holland, Amsterdam.

Isherwood, J., and Quinn, T. F. J. (1967). *Brit. J. Appl. Phys.* **18**, 717.

James, R. W. (1965). "The Optical Principles of the Diffraction of X-Rays." Cornell Univ. Press, Ithaca, New York.

Johnson, C. A. (1963). *Acta Cryst.* **16**, 490.

Kiessig, H. (1931). *Ann. Phys.* **10**, 715, 769.

Klug, H., and Alexander, L. (1954). "X-Ray Diffraction Procedures." Wiley, New York.

Komnik, F. (1964). *Sov. Phys.-Solid State* **6**, 479.

Langford, J. I. (1968). *J. Appl. Cryst.* **1**, 48, 131.

Light, T. B., and Wagner, C. N. J. (1966). *J. Vac. Sci. Technol.* **3**, 1.

Light, T. B., and Wagner, C. N. J. (1968). *J. Appl. Cryst.* **1**, 199.

Mader, S. (1965). *J. Vac. Sci. Technol.* **2**, 35.

Mader, S. (1970). *In* "Handbook of Thin Film Technology" (L. I. Maissel and R. Glang, eds.), Chapter 9. McGraw-Hill, New York.

Marchetto, G., Vook, R. W., and Bourgault, R. (1968). *J. Vac. Sci. Technol.* **5**, 123.

Mays, C. W., Vermaak, J. S., and Kuhlmann-Wilsdorf, D. (1968). *Surface Sci.* **12**, 134.

Nelson, J. B., and Riley, D. P. (1945). *Proc. Phys. Soc. (London)* **57**, 160.

Otooni, M. A., and Vook, R. W. (1968). Unpublished research.

Parratt, L. G. (1954). *Phys. Rev.* **95**, 359.

Paterson, M. S. (1952). *J. Appl. Phys.* **23**, 805.

Powder Data File (1967). Amer. Soc. for Testing Materials, Philadelphia, Pennsylvania.

Quinn, T. F. J. (1970). *J. Phys. D. Appl. Phys.* **3**, 210.

Rachinger, W. A. (1948). *J. Sci. Instrum.* **25**, 254.

Rashid, M., and Allstetter, C. (1970). *J. Appl. Cryst.* **3**, 120.

Rühl, W. (1954). *Z. Phys.* **138**, 121.

Ruland, W. (1968). *J. Appl. Cryst.* **1**, 90.

Sauro, J., Fankuchen, I., and Wainfan, N. (1963). *Phys. Rev.* **132**, 1544.

Sauro, J., Bindell, J., and Wainfan, N. (1966). *Phys. Rev.* **143**, 439.

Schoening, F. R. L. (1965). *Acta Cryst.* **18**, 975.

Schoening, F. R. L. (1967). *Acta Cryst.* **23**, 535.

Schoening, F. R. L., van Niekerk, J. N., and Haul, R. A. (1952). *Proc. Phys. Soc. (London)* **65B**, 528.

Schwartz, M. and Witt, F. (1966). *Trans. Met. Soc. AIME* **236**, 226.

Segmüller, A. (1963). *Z. Metallk.* **54**, 247.

Selleck, F. (1972). Unpublished research.

Shiah, R. T. S., and Vook, R. W. (1973). *Surface Sci.* **38**, 357.

Smallman, R. E., and Westmacott, K. H. (1957). *Phil. Mag.* **2**, 669.

Stokes, A. R. (1948). *Proc. Phys. Soc. (London)* **61**, 382.

Stratton, R. P., and Kitchingman, W. J. (1966). *Brit. J. Appl. Phys.* **17**, 1039.

Suhrmann, R., Wedler, G., Wilke, H. G., and Reusmann, G. (1960). *Z. Phys. Chem.* **26**, 85,

Taylor, A. (1961). "X-Ray Metallography." Wiley, New York.

Umrath, W. (1967). *Z. Angew. Phys.* **22**, 406.

Vermaak, J. S., and Kuhlmann-Wilsdorf, D. (1968). *J. Phys. Chem.* **72**, 4150.

Vook, R. W. (1968). *In* "Hybrid Microelectronics Symposium" (S. F. Ribich, ed.), pp. 257–268. Western Periodicals, North Hollywood, California.

Vook, R. W., and Otooni, M. A. (1968a). *J. Appl. Phys.* **39**, 2471.

Vook, R. W., and Otooni, M. A. (1968b). Unpublished research.

Vook, R. W., and Schoening, F. R. L. (1963). *Rev. Sci. Instrum.* **34**, 792.

Vook, R. W., and Witt, F. (1965a). *J. Vac. Sci. Technol.* **2**, 49, 243.

Vook, R. W., and Witt, F. (1965b). *J. Appl. Phys.* **36**, 2169.

Vook, R. W., and Witt, F. (1966). *In* "Basic Problems of Thin Film Physics" (R. Niedermayer and H. Mayer, eds.), pp. 188–193. Vandenhoeck and Rupprecht, Göttingen.

Vook, R. W., Schoening, F. R. L., and Witt, F. (1964). *In* "Single Crystal Films" (M. Francombe and H. Sato, eds.), pp. 69–78. Pergamon, Oxford.

Vook, R. W., Parker, T., and Wright, D. (1967). *In* "Surfaces and Interfaces" (J. Burke, N. Reed, and V. Weiss, eds.), Vol. I, Chemical and Physical Characteristics, pp. 347–358. Syracuse Univ. Press, Syracuse, New York.

Vook, R. W., Ouyang, S., and Otooni, M. A. (1972). *Surface Sci.* **29**, 277.

Wagendristel, A., Weidlich, R., Ebel, H., and Biber, R. (1969). *Z. Angew Phys.* **26**, 309.

Wagner, C. N. J. (1957). *Acta Met.* **5**, 427, 477.

Wagner, C. N. J. (1966a). Tech. Rep., NONR–609 (43). DDC AD 800548.

Wagner, C. N. J. (1966b). *In* "Local Atomic Arrangements Studied by X-Ray Diffraction" (J. B. Cohen and J. E. Hilliard, eds.), pp. 219–269. Gordon and Breach, New York.

Wagner, C. N. J. (1968). *Advan. X-Ray Anal.* **12**, 50.

Wagner, C. N. J. (1969). Private communication.

Wagner, C. N. J., and Aqua, E. N. (1964). *Advan. X-Ray Anal.* **7**, 47.

Wagner, C. N. J., Tetelman, A. S., and Otte, H. M. (1962). *J. Appl. Phys.* **33**, 3080.

Wagner, C. N. J., Light, T. B., Halder, N. C., and Lukens, W. E. (1968). *J. Appl. Phys.* **39**, 3690.

Wainfan, N., and Parratt, L. G. (1960). *J. Appl. Phys.* **31**, 1331.

Walker, G. A. (1970). *J. Vac. Sci. Technol.* **7**, 465.

Walker, G. A., and Goldsmith, C. C. (1970). *J. Vac. Sci. Technol.* **7**, 569.

Warren, B. E. (1960). *J. Appl. Phys.* **31**, 2237.

Warren, B. E. (1963). *J. Appl. Phys.* **34**, 1973.

Warren, B. E. (1969). "X-Ray Diffraction." Addison-Wesley, Reading, Massachusetts.

Warren, B. E., and Averbach, B. L. (1950). *J. Appl. Phys.* **21**, 595.

Warren, B. E., and Averbach, B. L. (1952). *J. Appl. Phys.* **23**, 497.

Wedler, G., and Wissmann, P. (1968a). *Z. Naturforsch.* **23a**, 1537.

Wedler, G., and Wissmann, P. (1968b). *Z. Naturforsch.* **23a**, 1544.

Wilkinson, P. G., and Birks, L. S. (1949). *J. Appl. Phys.* **20**, 1168.

Wilson, A. J. C. (1942). *Proc. Roy. Soc.* **A180**, 277.

Witt, F., and Vook, R. W. (1968). *J. Appl. Phys.* **39**, 2773.

Witt, F., and Vook, R. W. (1969). *J. Appl. Phys.* **40**, 709.

Witt, F., Vook, R. W., and Schwartz, M. (1965). *J. Appl. Phys.* **36**, 3686.

Witt, F., Vook, R. W., and Luszcz, S. (1968). *Trans. Met. Soc. AIME* **242**, 1111.

3.6 Measurement of Microstrains in Thin Epitaxial Films

D. L. Allinson[†]

Department of Physics
University of the Witwatersrand
Johannesburg, South Africa

I. Introduction

A rigorous definition of the term "microstrain" does not exist; for the purposes of this chapter it will be taken to mean the strain in regions with linear dimensions several orders of magnitude smaller than those of the sample. When applied to epitaxial thin films this definition will only hold if these microstrains are considered as being uniform through the thickness of the film. This has been implicitly assumed by most workers in the field of epitaxy who have measured strains in very small particles or regions. In order to measure strains in regions with typical diameters $\lesssim 0.1$ μm most workers have used transmission electron microscopy techniques to obtain their results.

[†] Present address: Division of Inorganic and Metallic Structure, National Physical Laboratory, Teddington, England.

An implication of the term "microstrain" is that these strains are hetero-geneous, since, if they were not, the size of the region in which they are deter-mined could be increased to the total specimen size. However, heterogeneity does not necessarily imply that the distribution of these strains is random, and in Sections II,A,C and III,B mention will be made of microstrain measure-ments where the strains vary with the thickness of the overgrowth. As with all differential techniques the value of the heterogeneous microstrain in any region in which measurements are made is assumed to be constant over this region; the size of this region will depend on either the physical extent of the region or on the final accuracy required from the specific method employed.

II. Microstrain Determination Techniques Not Employing Moiré Fringes

In this section a brief survey will be presented of some of the work done to determine microstrains in evaporated thin films using X-ray and electron-diffraction techniques as well as transmission electron microscopy methods other than those involving the use of moiré fringes.

A. X-Ray and Electron-Diffraction Techniques

X-ray and electron-diffraction studies of evaporated thin films have been performed by many workers for the determination of the average strain (and hence stress) existing in the plane of the films. Some workers Fourier-analyzed the diffraction peaks obtained from polycrystalline thin evaporated films and found that the line broadening was largely due to heterogeneous microstrains (Halteman, 1952; Rymer, 1956; Rymer *et al.*, 1956). If the method of Warren and Averbach (1952) is used, the rms strain determined from line profile analysis can be relied on, but if only one diffraction line is obtained, micro-strain determination using the method proposed by Smith (1960) has been shown to be unreliable (Segmuller, 1962). In the field of epitaxial thin films, Freedman (1962), using X-ray techniques to determine the mean strain in Ni–NaCl films, reported marked line broadening due to heterogeneous strains, but no attempt was made to measure their magnitude or account for their origin.

An X-ray investigation of the strain in films of Cu_2O grown epitaxially on a Cu single crystal showed that the mean strain decreased as the film thickness increased (Borie *et al.*, 1962). In addition, analysis of the peak profiles showed that, for a given oxide thickness, the strain in the film decreased linearly with distance away from the oxide–metal interface. Thus, although the strain in the overgrowth was uniform over any plane parallel to the interface, the results can still be classified as microstrain measurements due to the strain variation along the film normal.

B. MICROSTRAINS FROM FILM CURVATURES

A technique for determining the three-dimensional structure of electron microscope specimens using dark-field identification of bend contours has been described by Rang (1953). Although a strain analysis was not carried out, the technique was applied to obtain the shape of lenticular voids in mica. Since these voids were typically 1 μm in diameter, the technique can be described as one from which microstrains and stresses can be calculated. Similar curved regions were found in an epitaxial Au film and the local radius of curvature was determined by the above technique (Pashley, 1959). In this instance the curvature was ascribed to the presence of a contamination layer which caused buckling of the specimen, but the stress pattern was not determined.

Later work (Pashley and Presland, 1962) showed that these regions of microstrain were indeed produced by contamination and that the stress in the film was proportional to the thickness of the contaminant layer. In partially alloyed epitaxial bicrystal films of Au–Pd, many regions showing spherical curvature were found due to the climb of misfit dislocations (Matthews and Crawford, 1965). The strain in these small regions was determined from the curvature of these zones.

C. MICROSTRAINS DETERMINED FROM DISLOCATION DENSITIES

It has been established that pseudomorphism exists in the very early stages of growth of metal–metal epitaxial films (see, for example, Jesser and Matthews, 1967, 1968a–d; Matthews, 1970; Fedorenko and Vincent, 1971). For overgrowth deposition at room temperature, completely continuous films are formed and, when energetically favored, misfit dislocations can only be introduced into the interface by slip of preexisting dislocations that penetrate the overgrowth and substrate. Due to the inefficiency of this process, these overgrowths have been found to be in a state of strain greater than that theoretically predicted by van der Merwe (1963b).

In epitaxial films where the overgrowth is continuous and has a thickness gradient, the misfit dislocation density varies along the gradient. If the linear density and Burgers vector of the misfit dislocations is known in any small area of the film, the residual strain in the area can be determined from the difference between the inherent misfit and the accommodation due to the dislocations. Clearly this is a microstrain technique, the accuracy of the strain determination depending on the local misfit dislocation density.

This technique has been used by Matthews (1970), Matthews and Crawford (1970), and Fedorenko and Vincent (1971) in the study of Co–Cu films. The residual strain determined by these workers has been shown to be markedly different from that expected for equilibrium spacing of the misfit dislocations.

III. Moiré Fringe Techniques for Microstrain Determinations

Moiré fringes are frequently seen in electron micrographs of bicrystal films. The first to describe these fringes and to account for their origin were Mitsuishi *et al.* (1951), while Pashley *et al.* (1957) were among the first to observe them in epitaxial films. Two types of moiré fringes can be differentiated: rotation moiré fringes, formed during electron transmission through a bicrystal film comprising two superimposed identical films which are slightly rotated with respect to one another (Bassett *et al.*, 1958) and parallel moiré fringes, formed similarly when the components of the bicrystal film have slightly different lattice constants but are oriented parallel to one another (Kamiya and Uyeda, 1961; Matthews, 1960; Bassett, 1960).

Since rotation moiré fringes seen in bicrystals of epitaxial films require that both the components must be liberated from their substrates prior to examination in the electron microscope, fringes of this type are not useful for in situ microstrain determinations. Tricrystal films can also exhibit moiré fringes (Hashimoto *et al.*, 1961), but their interpretation is considerably more complex: Nonetheless an epitaxial film attached to its substrate can be used with a third film superimposed to form rotation moiré fringe patterns, thereby permitting in situ strain investigations. However, provided that the misfit does not exceed $\sim 15\%$ and that the substrate can be thinned (or grown) to a suitable thickness for easy electron penetration in most electron microscopes, parallel-type moiré fringes can usually be resolved. Consequently all the investigations into microstrain in epitaxial films using moiré fringes have utilized parallel-type fringes.

The geometrical relationship between moiré fringes and the reciprocal lattice vectors of the bicrystal components is given by $\Lambda = 1/|\Delta \mathbf{g}|$, where $\Delta \mathbf{g} = \mathbf{g}_o - \mathbf{g}_s$ and the fringes are directed perpendicularly to $\Delta \mathbf{g}$. The reciprocal lattice vectors $\mathbf{g}_o$ and $\mathbf{g}_s$ that operate to create the fringes are associated with the overgrowth and substrate, respectively, and Λ is the spacing of the fringe pattern (Rang, 1960; Gevers, 1962). This result ignores phase effects introduced either by film curvature (Cowley, 1959; Hashimoto *et al.*, 1961; Allinson, 1968) or from thickness variations (Hashimoto *et al.*, 1961; Jacobs *et al.*, 1966).

For the case of parallel moiré fringes, the fringes are parallel to the lattice planes giving rise to the pattern. In terms of the lattice plane spacings d_o and d_s of the overgrowth and substrate, respectively, the moiré fringe spacing is given by

$$\Lambda = d_o d_s/(d_o - d_s)$$

When strains are present, altering the magnitude but not the direction of d_o (i.e., isotropic strain in the plane of the film), d_o becomes $d_o' = d_o(1 + \varepsilon)$.

Here d_s is assumed to be unaltered by the overgrowth strain. Substituting into the above equation, the new moiré fringe spacing Λ' is given by

$$\Lambda' = \frac{d_o(1+\varepsilon)d_s}{d_o(1+\varepsilon) - d_s} \simeq \frac{d_o d_s(1+\varepsilon)}{d_o - d_s} = \Lambda(1+\varepsilon)$$

Thus $\varepsilon = (\Lambda' - \Lambda)/\Lambda$.

The above result is the fundamental relationship for determining strains from spacing measurements made on a single set of parallel-type moiré fringes.

A. STRAINS IN DISCRETE NUCLEI

Various workers (Jesser *et al.*, 1966; Jesser and Kuhlmann-Wilsdorf, 1967a–c; Vincent, 1969) have used moiré fringes to determine the variation of strain in discrete nuclei with the mean nucleus diameter during the early stages of epitaxial growth. Although the strain in any particular nucleus has been regarded as being homogeneous, the measurements typify microstrain measurement techniques due to the very small nuclear sizes. From these measurements confirmation of the growth theories of Cabrera (1964, 1965) for epitaxial hemispherical nuclei was obtained, based on the work of van der Merwe (1963a, b).

The results obtained by Vincent (1969) for the system Sn–SnTe are of particular interest and are shown in Fig. 1. The variation of mean strain with island radius R was shown to have an R^{-1} dependence, as proposed by Cabrera (1965) and Jesser *et al.* (1966). In addition, a saw-tooth perturbation was found. The points at which the strain decreased suddenly were found to coincide with the formation of misfit dislocations at the nuclei edges. The clarity of the result is probably due to the fact that for this system only one misfitting direction exists and the misfit of the system is small.

Recent work by Matthews (1971) has shown that this effect is most probably far more common than is usually thought to be the case. Assuming discrete nucleation of misfit dislocations in the interface of a misfitting nucleus, Matthews has extended the work of Vincent (1969) to the case of nuclei which have misfit in two perpendicular directions. Under these conditions as the nucleus grows, misfit dislocations are introduced alternately along the two directions of misfit. This alternate change in the direction of maximum strain causes the reciprocal lattice point associated with certain diffracting planes in the overgrowth to oscillate about the strain-free position. These planes are normal to the film and at an angle to the misfit dislocation line directions. This oscillation manifests itself as an alteration in the direction of these moiré fringes as well as their spacing. The fringe rotations brought about by this process are, however, far larger (and therefore more easy to measure) than the changes in spacing, always excepting the case when a nucleus transforms from

NUMBER OF DISLOCATIONS

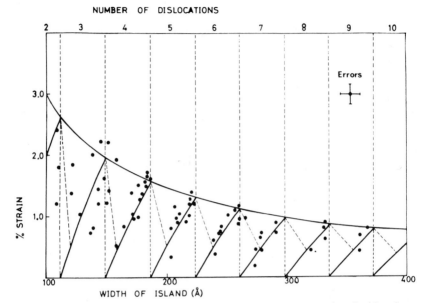

FIG. 1. Results of measurements of strain in tin islands grown on a tin telluride substrate as a function of the widths of the islands. The deposition temperatures ranged between 140 and 200°C. A proposed saw-tooth variation of the strain in the tin nuclei as misfit dislocations are introduced into the interface is superimposed on the experimental results (from Vincent, 1969).

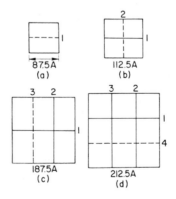

FIG. 2. Generation of misfit dislocations during the growth of nuclei. The outer continuous lines are the borders of the nucleus seen from above. Inner continuous lines are the images of already existing misfit dislocations and the dashed lines represent misfit dislocations that have either just been formed or are about to be made ($v = 0.33$, $b = 2.5$ Å, misfit 2.5%, thickness-to-width ratio 0.25). The sides are considered to be parallel to the 100 directions (from Matthews, 1971).

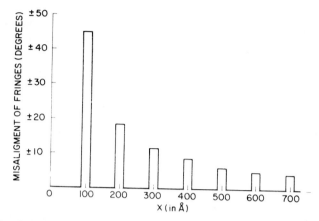

FIG. 3. Graph showing the change in orientation of the 110 moiré fringes during the growth of the nucleus shown in Fig. 2. Each discontinuous change in orientation coincides with the creation of a misfit dislocation (from Matthews, 1971).

being pseudomorphic to having misfit present. A simple model of this process is shown in Fig. 2, where the nucleus is constrained to remain square and parallel to [100] and [010]. The sizes at which the misfit dislocations are introduced depend on the misfit of the model system. The calculated change in orientation of moiré fringes of 110 type as the misfit dislocations are introduced into the interface according to the scheme shown in Fig. 2 is shown in Fig. 3. The 110 moiré fringes in the smallest nuclei show the largest misalignment on the introduction of misfit dislocations in the interface; as the number of misfit dislocations increases, the excursions away from the strain-free direction decrease.

The important aspect of these calculations is that they could yield another method for determining the state of strain in very small nuclei, besides offering an explanation for the apparent rotations of nuclei into and away from epitaxial alignment deduced from observing moiré fringes. On this mechanism the nucleus is never actually badly misaligned; the moiré fringes merely rotate as the direction of maximum strain varies. Besides reassessing previously published data on nucleus misorientations (Matthews, to be published), evidence for the mechanism has been obtained in thin films of nickel on copper (Matthews, 1971).

Although it is possible to measure strains in regions with smaller dimensions than those set by the inherent moiré fringe spacings associated with a specific overgrowth–substrate combination if direct lattice resolution is obtained, the theoretical difficulty in interpreting these images due, inter alia, to Fresnel effects, would make the measurement of microstrain very difficult. As it is, the moiré fringe spacings in small nuclei so far reported have not been corrected for displacements due to the variation in thickness of the hemispheres near their edges (Hashimoto *et al.*, 1961; Gevers, 1962; Jacobs *et al.*, 1966). However, the errors of measurement are usually considerably greater than the error introduced by neglecting these effects.

B. Homogeneous Strains in Continuous Films

A technique for measuring residual strains in epitaxial films with a continuous overgrowth and a thickness gradient was described in Section II, C. These same strains can be measured from spacing measurements of moiré fringes seen in these films. Chronologically the moiré fringe technique was the first one used to measure these strains (Matthews, 1961, 1966). The strain in any region is determined from the basic strain equation in Section III.

Fig. 4. Electron micrograph of a deposit of platinum on gold. The lines parallel to the [1$\bar{1}$0] and [$\bar{1}$10] directions are the traces left by misfit dislocations that moved by glide into the interface between the two metals. The lines roughly parallel to $\langle 100 \rangle$ are type 200 moiré fringes. The moiré fringe spacing is smallest in the vicinity of the slip traces, showing that the dislocation reduced the misfit accommodated by elastic strain (from Matthews, 1966).

The measurement of the mean spacing of moiré fringes in a continuous film of Au–Pd with a known thickness ratio (Matthews, 1966) resulted in the first indication that epitaxial films of a continuous nature were strained in excess of the value predicted by van der Merwe (1963b). Moiré fringes have also been used to demonstrate qualitatively the reduction of misfit strains in the vicinity of misfit dislocations when the latter are not uniformly spaced. This is shown in Fig. 4, where a Pt–Au film, with traces of misfit dislocations visible, is seen in moiré fringe contrast. The 200-type moiré fringes have the smallest spacing (approaching the value for an unstrained bicrystal) in the vicinity of the $\frac{1}{2}a \langle 1\bar{1}0 \rangle$ misfit dislocations and the mean spacing increases away from these zones. The angular variation shown by the fringes can also be accounted for qualitatively on the basis of Fig. 3.

For microstrain determination purposes the size of the area over which measurements are made is generally smaller than that required for an equivalent determination on the basis of linear misfit dislocation density measurements. However, the inhomogeneity of the fringe spacings when many misfit dislocations are present increases the error of the technique for these cases. It is for this reason that later work (Matthews, 1970; Fedorenko and Vincent, 1971) employs the dislocation density technique rather than moiré fringe spacings to determine the microstrains in these films.

C. Heterogeneous Strains in Semicontinuous and Continuous Films

For many epitaxial systems the homogeneous strains in discrete nuclei, brought about by the minimization of the total nucleus energy, are extremely small when radii exceed ~200 Å. This radius, at which homogeneous strains tend to be below the detection limit of moiré fringes, will depend on the specific epitaxial system being examined. As the misfit of an epitaxial system decreases, this limiting radius will increase. However, for many combinations of a metallic overgrowth on a nonmetallic substrate, the above value of ~200 Å will roughly correspond to the limiting radius at which homogeneous strains are no longer detectable by moiré fringe techniques.

This apparent strain-free state of the overgrowth does not continue indefinitely, however, and in some epitaxial systems once the overgrowth reaches the stage of film growth at which "snakelike" nuclei are common, moiré fringes once again reveal strains in the film, but now of a heterogeneous nature. These heterogeneous strains persist into the "network" and "hole and channel" stages of film growth (Pashley et al., 1964; Jacobs et al., 1966), only disappearing at the continuous film stage when the strain pattern is that of a material with a high density of penetrating dislocations in most instances. Examples of a "snakelike" nucleus and a bridge in the "network" stage of film growth are shown in Figs. 5a and b, respectively, where large heterogeneous strains are shown by the moiré fringes. For this system, Ag–MoS_2, the misfit dislocation spacing is ~ 33 Å and so, for the nuclei shown in Figs. 5a and b, changes in the direction and spacing of the moiré fringes can only have been brought about by the presence of heterogeneous strains.

In an attempt to determine these heterogeneous strains, and from the results to obtain the state of stress in various regions of the films, Allinson (1974) has developed a technique utilizing two sets of moiré fringes.

It has recently been shown (Allinson, 1974) that an epitaxial film can be regarded as being in a state of plane stress to a good approximation, provided that the film thickness t is somewhat greater than p, the misfit dislocation spacing of the overgrowth–substrate combination. With reference to Fig. 6, this implies that, apart from a layer close to the interface, the only stresses in an epitaxial film with thickness $t \gg p$ are σ_{xx}, σ_{yy}, and σ_{xy}. Under these conditions only three components of the strain tensor are required to solve for the complete state of stress in films with overgrowth material of high symmetry.

In order to analyze heterogeneous strains in epitaxial films with "snake-like" nuclei or of a semicontinuous nature, the overgrowth is considered as

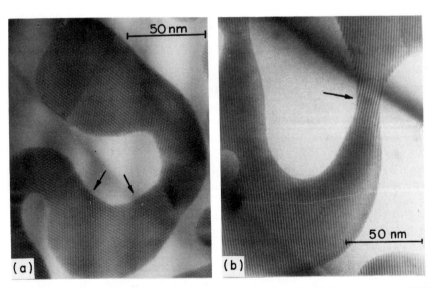

FIG. 5. Heterogeneous elastic strains in epitaxial overgrowths of silver grown at 300°C on a molybdenite substrate. Regions of severe strain, indicated by the curvature of the $2\bar{2}0$-type moiré fringes, are marked by arrows: (a) "snakelike" nucleus in a film of mean deposit thickness ~ 50 Å; (b) strains in a narrow bridge in a film at the "network" stage of film growth, with mean deposit thickness ~ 100 Å.

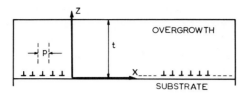

FIG. 6. Schematic representation of a section perpendicular to the interface of an epitaxial overgrowth–substrate combination. Misfit dislocations, with a spacing p, are shown in the interface, together with two of the axes used in the development of the equations relating strains in the overgrowth to moiré fringe displacements (from Allinson, 1974).

consisting of a large number of small zones wherein the local strains are considered to be constant and small in comparison with unity. By considering the influence of the strains in these small zones on the vectors defining the planes in the overgrowth which participate in the formation of moiré fringe patterns, the following expressions were obtained for the components of $\Delta\mathbf{g}$ along two perpendicular axes lying in the film plane (see Fig. 6):

$$\Delta g_x = (g_{sx}-g_x) = (g_s-g_o)\cos\theta - g_o[\varepsilon_{xx}\cos\theta + (\varepsilon_{xy}-\omega_{xy})\sin\theta] \quad (1)$$

$$\Delta g_y = (g_{sy}-g_y) = (g_s-g_o)\sin\theta - g_o[(\varepsilon_{xy}+\omega_{xy})\cos\theta + \varepsilon_{yy}\sin\theta] \quad (2)$$

Here g_s and g_o are the moduli of the unstrained reciprocal lattice vectors giving rise to the moiré fringe pattern, defined by the reciprocal lattice vector $\Delta\mathbf{g}$; ε_{xx}, ε_{yy}, and ε_{xy} are three components of the strain tensor; and ω_{xy}, a component of the antisymmetric rotation tensor, defines the body rotation in the film plane of the small zone selected for analysis. The angle between the axis $0X$ and g_o is represented by θ; the change in θ due to the presence of strains is very small and can be neglected. Equations (1) and (2) are only valid for normal incidence of the electron beam.

Since there are four unknowns in Eqs. (1) and (2), two sets of moiré fringes are required in the same area in order to yield four spacing values along the axes $0X$ and $0Y$. These spacings are inversely related to the components of the reciprocal lattice vector defining the moiré fringe pattern. All transformations between real and reciprocal space used here only apply to cartesian axes, where real and reciprocal lattices transform identically into one another.

In deriving Eqs. (1) and (2), it was assumed that the changes in the vectors defining the overgrowth planes that participate in moiré fringe formation lie in the film plane. It has been shown, however, that the correction to be made to $|\Delta\mathbf{g}|$ when this condition is relaxed is only of second-order small quantities, and these quantities have been ignored throughout (Allinson, 1974). It has also been assumed that $\mathbf{g}_s$, the substrate reciprocal lattice vector associated with the moiré fringes, is unaltered by the presence of strains in the overgrowth.

This is only true for a very thick substrate and the assumption is merely used for simplicity; however, the relative strain between overgrowth and substrate is unaffected by the validity of this assumption. When the strains measured by this technique are to be used to determine the stresses existing in the films, however, the applicability of this assumption must be carefully assessed.

As has been mentioned, Eqs. (1) and (2) have been derived on the basis of constant small local strains in a small zone of the overgrowth. In practice the physical size of these zones will depend on the number of fringes required to give a certain predetermined accuracy to the moiré fringe spacing measurements. If the state of stress in the overgrowth changes continuously with position, as is the case for heterogeneous stresses, the application of this strain- (and hence stress-) measuring technique implies that, within a zone where measurements are made, the stresses have a constant value that changes discontinuously across the zone "boundary" to another constant value for the next zone.

The above equations have also been generalized to take into account a nonnormal direction of incidence of the electron beam, but the resulting equations are complicated and three sets of moiré fringes are required to obtain a solution. In addition, the orientation of the beam relative to the axes must be known very accurately since an electron micrograph is a two-dimensional projection of a three-dimensional structure. For most epitaxial systems this extension is not necessary since moiré fringes are frequently observed when the film normal is closely parallel to the incident beam direction.

An example of the application of this technique for the determination of heterogeneous stresses will be given for a region of a film of silver grown on molybdenite at a substrate temperature of 300°C. The silver was deposited at a moderately high evaporation rate of ~ 25 Å sec^{-1} (Allinson et $al.$, to be published). The stress and rotation map shown in Fig. 7b was determined from measurements of the spacings of the two sets of moiré fringes in small zones centered on the labeled areas shown in Fig. 7a. In the hole H_1 there are four $\frac{1}{2}a[1\bar{1}0]$ incipient dislocations, but because no set of fringes remained in contrast round holes H_2 and H_3, no information about the incipient dislocations in these latter holes could be obtained.

At points A, B, and D the normal stresses measured are compatible with the incipient dislocations in H_1, but at C the change in the normal stresses can only have arisen from localized slip on the substrate. The increased shear stress at C tends to confirm this hypothesis. The constituent nucleus BC presumably divided the hole H_1 from H_2 with the junction plane C–D being the site of the last coalescence. The inhomogeneous stresses found in this region probably occurred as a consequence of torques operating on the coalescing faces of the nucleus BC.

Studies were made on many regions such as that shown in Fig. 7a taken

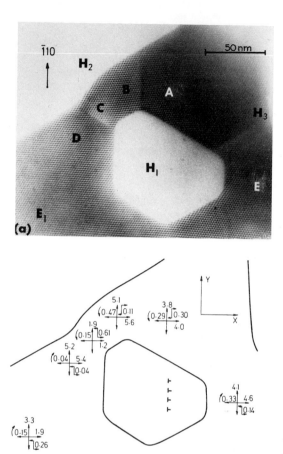

(a)

(b)

Fɪɢ. 7. (a) Electron micrograph of a portion of a silver deposit grown epitaxially on molybdenite at 300°C. The mean deposit thickness was ∼100 Å and the substrate was cleaved to permit electron penetration prior to the silver evaporation. Holes in the over-growth are labeled H_1, H_2, and H_3; four $\frac{1}{2}a[1\bar{1}0]$ incipient dislocations are in H_1. (b) Stress and rotation map of the region shown in Fig. 7a. Normal stresses are indicated by arrows parallel to the axes, shear stresses are indicated by torque symbols, and body rotations are indicated by the curved arrows; stresses ×10^8 N·m⁻²; rotations in degrees.

from films in the "snakelike" nucleus, "network," and "continuous with holes" stages of film growth. It was found that the heterogeneous shear stress gradients were larger on the average in films during the "network" stage of film growth than at the other stages mentioned. There was no significant change with film thickness in the mean stresses σ_{xx} and σ_{yy} for the films investigated. After correcting for stresses due to the differential thermal contractions, the mean stresses mentioned above had values of ∼1×10^8 N·m⁻². The accuracy

377

of the technique for the measurement of these stresses is very poor in comparison with other methods for determining the mean stress in an evaporated film (Blackburn and Campbell, 1963), but the shear stresses (which can only be measured by techniques such as that described above) could be determined to an accuracy of $\pm 1 \times 10^7$ N·m^{-2}.

One result of the investigation was that, over very small distances, large stress gradients could be tolerated by the films. Occasionally values of the resolved shear stress on {111} slip planes approaching G/100 were found, but no evidence of dislocation formation to relieve these stresses was obtained. In addition, an implied slip of the overgrowth on the substrate, deduced from stress patterns, such as that shown in Fig. 7b, was found to be rare. The heterogeneous shear stresses, in general, were consistent with the operation of torques on the coalescing faces of slightly misaligned large nuclei, but no definite evidence was found for a specific mechanism responsible for the other stresses in the film plane.

IV. Conclusion

Microstrains have been measured in epitaxial films by various techniques, possibly the most useful of which are those involving the analysis of moiré fringes. The main applications of these microstrain determinations have been in the field of experimental verification of theoretical models for epitaxial growth. In some instances, however, some additional information about epitaxial films has been found.

To account for saw-tooth variation of microstrains such as those found by Vincent (1969), the shear stress at the edge of a nucleus has been calculated. This shear stress is sufficient to nucleate a dislocation when the strain in the nucleus is at a maximum, i.e., when the nucleus is on the point of introducing another misfit dislocation into the interface. The good agreement between the theoretical and experimental strain variation shown in Fig. 1 showed that there was no effective energy barrier to the nucleation of misfit dislocations at the edges of the growing nuclei. This is quite different from the generation of misfit dislocations in the interface of an epitaxial overgrowth which grows from a pseudomorphic layer of large lateral dimensions (see Sections II, C and III, B).

Finally, the results of the investigations outlined in Section III, C have been used to show that inclined microtwin generation to relieve internuclear stresses in epitaxial fcc metal films [a mechanism proposed by Burbank and Heidenreich (1960)] is not a common process. This is based on the observations that, first, the microtwin density and resolved shear stress maxima do not occur at the same stages of film growth (Matthews and Allinson, 1963; Allinson et al., to be published) and second, the majority of the inclined microtwins, when at their maximum density, are found at sites where internuclear stresses do not exist.

References

Allinson, D. L. (1968). *Phil. Mag.* **17**, 339.
Allinson, D. L. (1974). In press.
Allinson, D. L., Matthews, J. W., and Nabarro, F. R. N. (1974). In press.
Bassett, G. A. (1960). *Proc. Eur. Reg. E.M. Conf., Delft.* **1**, 270.
Bassett, G. A., Menter, J. W., and Pashley, D. W. (1958). *Proc. Roy. Soc.* **A246**, 345.
Blackburn, H., and Campbell, D. S. (1963). *Phil. Mag.* **8**, 823.
Borie, B., Sparks, C. J., and Cathcart, J. V. (1962). *Acta Met.* **10**, 691.
Burbank, R. D., and Heidenreich, R. D. (1960). *Phil. Mag.* **5**, 373.
Cabrera, N. (1964). *Surface Sci.* **2**, 320.
Cabrera, N. (1965). *Rev. Met. Mem. Sci.* **62**, 205.
Cowley, J. M. (1959). *Acta Crystallogr.* **12**, 367.
Fedorenko, A. I., and Vincent, R. (1971). *Phil. Mag.* **24**, 55.
Freedman, J. F. (1962). *IBM J. Res. Develop.* **6**, 449.
Gevers, R. (1962). *Phil. Mag.* **7**, 1681.
Hashimoto, H., Mannami, M., and Naiki, T. (1961). *Phil. Trans. Roy. Soc.* **A253**, 490.
Halteman, E. K. (1952). *J. Appl. Phys.* **23**, 150.
Jacobs, M. H., Pashley, D. W., and Stowell, M. J. (1966). *Phil. Mag.* **13**, 129.
Jesser, W. A., and Kuhlmann-Wilsdorf, D. (1967a). *Phys. Status Solidi* **19**, 95.
Jesser, W. A., and Kuhlmann-Wilsdorf, D. (1967b). *Phys. Status Solidi* **21**, 533.
Jesser, W. A., and Kuhlmann-Wilsdorf, D. (1967c). *J. Appl. Phys.* **38**, 5128.
Jesser, W. A., and Matthews, J. W. (1967). *Phil. Mag.* **15**, 1097.
Jesser, W. A., and Matthews, J. W. (1968a). *Phil. Mag.* **17**, 461.
Jesser, W. A., and Matthews, J. W. (1968b). *Phil. Mag.* **17**, 475.
Jesser, W. A., and Matthews, J. W. (1968c). *Phil. Mag.* **17**, 595.
Jesser, W. A., and Matthews, J. W. (1968d). *Acta Met.* **16**, 1307.
Jesser, W. A., Matthews, J. W., and Kuhlmann-Wilsdorf, D. (1966). *Appl. Phys. Lett.* **9**, 176.
Kamiya, Y., and Uyeda, R. (1961). *Acta Crystallogr.* **14**, 70.
Matthews, J. W. (1960). *Proc. Eur. Reg. E.M. Conf., Delft.* **1**, 276.
Matthews, J. W. (1961). *Phil. Mag.* **6**, 1347.
Matthews, J. W. (1966). *Phil. Mag.* **13**, 1207.
Matthews, J. W. (1970). *Thin Solid Films* **5**, 369.
Matthews, J. W. (1971). *Surface Sci.* **31**, 241.
Matthews, J. W., and Allinson, D. L. (1963). *Phil. Mag.* **8**, 1283.
Matthews, J. W., and Crawford, J. L. (1965). *Phil. Mag.* **11**, 977.
Matthews, J. W., and Crawford, J. L. (1970). *Thin Solid Films* **5**, 187.
Mitsuishi, T., Nagasaki, H., and Uyeda, R. (1951). *Proc. Jap. Acad.* **27**, 86.
Pashley, D. W. (1959). *Phil. Mag.* **4**, 324.
Pashley, D. W., and Presland, A. E. B. (1962). *Phil. Mag.* **7**, 1407.
Pashley, D. W., Menter, J. W., and Bassett, G. A. (1957). *Nature (London)* **179**, 752.
Pashley, D. W., Stowell, M. J., Jacobs, M. H., and Law, T. J. (1964). *Phil. Mag.* **10**, 127.
Rang, O. (1953). *Optik* **10**, 90.
Rang, O. (1960). *Z. Kristallogr.* **114**, 98.
Rymer, T. B. (1956). *Proc. Roy. Soc.* **A 235**, 274.
Rymer, T. B., Fayers, F. J., and Hewitt, S. J. (1956). *Proc. Phys. Soc. (London)* **B69**, 1059.
Segmuller, A. (1962). *IBM J. Res. Develop.* **6**, 464.
van der Merwe, J. H. (1963a). *J. Appl. Phys.* **34**, 117.
van der Merwe, J. H. (1963b). *J. Appl. Phys.* **34**, 123.
Vincent, R. (1969). *Phil. Mag.* **19**, 1127.
Warren, B. E., and Averbach, B. L. (1952). *J. Appl. Phys.* **23**, 497.

INDEX

1

A
B 5
C 6
D 7
E 8
F 9
G 0
H 1
I 2
J 3